Internet of Things and Secure Smart Environments

Chapman & Hall/CRC Big Data Series

Series Editor:
Sanjay Ranka

A Tour of Data Science
Learn R and Python in Parallel
Nailong Zhang

Big Data Computing
A Guide for Business and Technology Managers
Vivek Kale

Big Data Management and Processing
Kuan-Ching Li, Hai Jiang, Albert Y. Zomaya

Frontiers in Data Science
Matthias Dehmer, Frank Emmert-Streib

High Performance Computing for Big Data
Methodologies and Applications
Chao Wang

Big Data Analytics
Tools and Technology for Effective Planning
Arun K. Somani, Ganesh Chandra Deka

Smart Data
State-of-the-Art Perspectives in Computing and Applications
Kuan-Ching Li, Qingchen Zhang, Laurence T. Yang, Beniamino Di Martino

Internet of Things and Secure Smart Environments
Successes and Pitfalls
Uttam Ghosh, Danda B. Rawat, Raja Datta, Al-Sakib Khan Pathan

For more information on this series please visit: https://www.crcpress.com/Chapman–HallCRC-Big-Data-Series/book-series/CRCBIGDATSER

Internet of Things and Secure Smart Environments

Successes and Pitfalls

Edited by
Uttam Ghosh
Vanderbilt University, USA

Danda B. Rawat
Howard University, USA

Raja Datta
Indian Institute of Technology (IIT) Kharagpur, India

Al-Sakib Khan Pathan
Independent University, Bangladesh

CRC Press is an imprint of the
Taylor & Francis Group, an **informa** business
A CHAPMAN & HALL BOOK

First edition published 2021
by CRC Press
6000 Broken Sound Parkway NW, Suite 300, Boca Raton, FL 33487-2742

and by CRC Press
2 Park Square, Milton Park, Abingdon, Oxon, OX14 4RN

Library of Congress Cataloging-in-Publication Data
Names: Ghosh, Uttam, editor.
Title: Internet of things and secure smart environments : successes and
pitfalls / edited by Uttam Ghosh, Vanderbilt University, USA, [and three others].
Description: First edition. | Boca Raton : CRC Press, 2021. |
Series: Chapman & Hall/CRC big data series | Includes bibliographical
references and index.
Identifiers: LCCN 2020037222 | ISBN 9780367266394 (hardback) |
ISBN 9780367276706 (ebook)
Subjects: LCSH: Internet of things—Security measures.
Classification: LCC TK5105.8857 I63 2021 | DDC 005.8—dc23
LC record available at https://lccn.loc.gov/2020037222

ISBN: 978-0-367-26639-4 (hbk)
ISBN: 978-0-367-27670-6 (ebk)

Typeset in Minion
by codeMantra

“Dedicated to the memory of my father, late Gopal Chandra Ghosh, who always believed in my ability to be successful in the academic arena. Although he is not here, I always feel his presence that used to urge me to strive to achieve my goals in life.” – Uttam Ghosh

“Dedicated to my family.” – Danda Rawat

“Dedicated to my family.” – Raja Datta

“To my loving wife, Labiba Mahmud, whose cooperation always helps me do various projects with ease.” – Al-Sakib Khan Pathan

Contents

List of Figures

List of Tables

Preface

INTRODUCTION

The Internet of Things (IoT) refers to the billions of intelligent physical devices with some processing power and storage capacity, which have unique identifiers which connect them to the Internet for collecting and sharing data. The main goal of IoT is to make secure, reliable, and fully automated smart environments, e.g., buildings, smart homes, smart cities, smart healthcare, smart grids, smart agriculture, and smart vehicles. However, there are many technological challenges in deploying IoT. This includes connectivity and networking, timeliness, power and energy consumption dependability, security and privacy, compatibility and longevity, and network/protocol standards, etc., with respect to resource-constrained embedded sensors and devices.

AIM & SCOPE OF THE BOOK

The major aim of this book is to provide a comprehensive base to the recent research and open problems in the areas of IoT research. We have felt that it is the much-needed ingredient to the researchers/scholars who are about to start their research in this emerging area. State-of-the-art problems, present solutions, and open research directions will always be positively welcomed by all the research groups, postgraduate programs, and doctoral programs around the globe in the field of computer science and information technology.

TARGET GROUP

This book will provide a comprehensive aid to prospective researchers and scholars who are considering working in this area from both industry and academia. The major aim of this book is to provide a comprehensive base to the recent research and open problems in the areas of IoT research as well as fundamental IoT architecture. Being an emerging area,

IoT and related aspects have become a matter of great interest of budding researchers in the area. Collecting and streamlining all these works in a single book require large collaborative efforts from the community across the globe. It has also been envisaged that this book will continue to become a source of great interest for the time to come and would require adaptations from time to time.

ABOUT THE TOPICS

The book aims to systematically collect and present quality research findings in the areas of recent advances in IoT to give wide benefit to a huge community of researchers, educators, practitioners, and industry professionals. We provide a typical list of chapters/areas to cover both fundamental and advanced areas of IoT for a secure and smart environment. Submissions were solicited on, but not limited to, the following topics:

1. IoT architectures for smart environments
2. Networking technology in IoT (SDN, 5G, LTE, Narrowband, WBAN/BAN/BSN, WSN, RFID, NFC, etc.)
3. IPv6 Addressing, Naming, and Discovery in IoT
4. Infrastructure identity management and access control systems in IoT for smart environments
5. Distributed computing aspect of IoT for smart environments
6. Data handling and management (accumulation, abstraction, storage, processing, encryption, fast retrieval, security, and privacy) in IoT for smart environments
7. Availability and resilience aspects in IoT for smart environments
8. Cloud computing and fog computing in IoT for smart environments
9. Energy management and green IoT for smart environments
10. Quality assurance in IoT for smart environments
11. Trust management in IoT for smart environments
12. Incentive mechanisms of IoT applications for smart environments

13. Threat and attack models of IoT applications for smart environments

After careful selections, we have been able to put together a total of 12 chapters in this book. Here, we discuss the chapters in brief.

Chapter 1: *Wireless Localization for Smart Indoor Environments* analyzes the fundamental techniques of wireless localization on the notable applications with different underlying wireless standards. A qualitative assessment of the localization methods for their compatibility with the indoor localization has also been provided here.

Chapter 2: *An Approach towards GIS Application in Smart City Urban Planning* provides an overall implication of Geographical Information System (GIS) technology in urban management and development and information about evolution and redefines satisfactory areas for improving urban safety. Some recent advancements along with the current impact of GIS in Smart City development have also been explained.

Chapter 3: *A Review of Checkpointing and Rollback Recovery Protocols for Mobile Distributed Computing Systems* elaborates the impact of checkpointing in Mobile Distributed Computing Systems (MDCS) for achieving smartness of the system. Along with some solutions based on a traditional distributed environment, this chapter elaborates the present classifications of innovative approaches specifically for MDCS. It also includes the scope of improvements, which were previously absent and have now been suitably handled by MDCS.

Chapter 4: *Softwarized Network Function Virtualization for 5G: Challenges and Opportunities* gives a detailed and fundamental idea regarding the collaborative impact of Software Defined Networking and Network Function Virtualization in next-generation 5G telecommunication networks. It provides a detailed analysis regarding the crucial challenges inside present-day network and how those can be eliminated in future networks using Softwarized Network Function Virtualization (SNFV) technology. It ends with some interesting open research areas and future scope for the readers in this field.

Chapter 5: *An Effective Deployment of SDN Controller in Smart City Renovation* deals with energy-efficient controller placement in Smart City environment using Software Defined Networking (SDN) technology. How the controller placement supports traffic management and weather monitoring has been explained nicely. Here, the proposed suitable network framework has come up with satisfactory accuracy when compared with the TCP/IP and existing GPS system.

Chapter 6: *Flying Ad Hoc Networks: Security, Authentication Protocols, and Future Directions* introduces Unmanned Aerial Vehicle (UAV) technology and its revolutionary impacts on various industry and academic sectors. Its versatile roles including military and disaster management have been explained. This chapter also gives an insight into the latest trends in the field of authentication and the possibilities for developing lightweight and novel mechanisms for Flying Ad Hoc Network applications.

Chapter 7: *Investigating Traffic of Smart Speakers and IoT Devices: Security Issues and Privacy Threats* demonstrates the risks involved in machine-learning-capable techniques to develop black-box models to automatically classify traffic and implement privacy leaking attacks. It also provides the suitable advantages of smart speakers in controlling the IoT devices. Here, some practical and experimental results collected from different realistic scenarios along with the possible countermeasures are discussed.

Chapter 8: *Hardware Security in the Context of Internet of Things: Challenges and Opportunities* showcases the unique challenges of secure IoT implementations, especially from the perspective of hardware security and their solutions. A survey based on different kinds of security attacks and vulnerabilities on IoT devices and some proposed methods for solution has been done here. The Physically Unclonable Function (PUF)-based lightweight IoT security solution concludes the chapter.

Chapter 9: *Security Challenges in Hardware Used for Smart Environments* addresses the hardware vulnerabilities in IoT devices, which cause major threats while transferring information. An overview of various hardware security issues in IoT-based devices used in smart environments is given along with discussions on the essential countermeasures that can be implemented in order to secure the future IoT concepts and architectural models.

Chapter 10: *Blockchain for Internet of Battlefield Things: A Performance and Feasibility Study* delivers the effective use of evolving Blockchain technology to improve operational efficiency of mission-oriented tasks in the battlefield. Here, a permissioned blockchain, namely Hyperledger Sawtooth, has been introduced, and its performance analysis has been done in order to determine its ability in the context of Internet of Battlefield Things (IoBT). Different advantages and disadvantages of different types of blockchains along with their characteristics, feasibility, and integration challenges in IoBT environment are provided here.

The chapter concludes with some plans for future research directions in the blockchain technology.

Chapter 11: *Internet of Things in 5G Cellular Networks: Radio Resource Perspective* aims to deliver the advanced modulation schemes supported with different codec standards to find their impacts on cell capacity in order to improve the performance of IoT-enabled 5G networks. How the system capacity has been improved with lower codec modulation schemes has also been elaborated here. Finally, the chapter draws the conclusion with future implications of the modulation schemes and scope of implementing the concepts in future IoT-enabled cellular networks.

Chapter 12: *An SDN–IoT–Based Framework for Future Smart Cities: Addressing Perspective* discusses a software-defined network (SDN)-based framework for future smart cities. It also comprises a distributed addressing scheme to facilitate the allocation of addresses to devices in the smart city dynamically. The proposed addressing scheme, a new IoT device, receives its IP address from one of their existing neighboring peer devices. This allows devices in the city to act as a proxy and generate a set of unique IP addresses from their own IP addresses, which can then be assigned to new (joining) devices. This scheme significantly reduces addressing overhead and latency.

We hope that the chapters will help both the experts and general readers of the field gain some important insight and current knowledge relevant to their interests.

Acknowledgements

The impact of our parents on each of our lives is inexpressible. We are very much indebted to their continuous encouragement for us to grow in the field that we love. Of course, without the lifetime given by the Almighty, none of our works would have been possible to be completed. It has been another incredible book editing experience, and our sincere gratitude is to the publisher for facilitating the process. This book editing journey enhanced our patience, communication, and tenacity. We are thankful to all the contributors, critics, and the publishing team. Last but not the least, our very best wishes are for our family members whose support and encouragement contributed significantly to complete this project.

It should be specifically noted that for this project, Danda B. Rawat has been partly supported by the U.S. NSF under grants CNS 1650831 and HRD 1828811, and by the U.S. Department of Homeland Security under grant DHS 2017-ST-062-000003 and DoE's National Nuclear Security Administration (NNSA) Award DE-NA0003946.

Uttam Ghosh, PhD
Vanderbilt University, USA

Danda B. Rawat, PhD
Howard University, USA

Raja Datta, PhD
Indian Institute of Technology, Kharagpur, India

Al-Sakib Khan Pathan, PhD
Independent University, Bangladesh

Editors

Uttam Ghosh is working as an Assistant Professor of the Practice in the Department of Electrical Engineering and Computer Science, Vanderbilt University, Nashville, TN, USA. Dr Ghosh obtained his PhD in Electronics and Electrical Engineering from the Indian Institute of Technology Kharagpur, India, in 2013, and has post-doctoral experience at the University of Illinois in Urbana-Champaign, Fordham University, and Tennessee State University. He has been awarded the 2018–2019 Junior Faculty Teaching Fellow (JFTF) and has been promoted to a Graduate Faculty position at Vanderbilt University. Dr Ghosh has published 50 papers at reputed international journals including IEEE Transaction, Elsevier, Springer, IET, Wiley, InderScience and IETE, and also in top international conferences sponsored by IEEE, ACM, and Springer. Dr Ghosh has conducted several sessions and workshops related to Cyber-Physical Systems (CPS), SDN, IoT and smart cities as co-chair at top international conferences including IEEE Globecom, IEEE MASS, SECON, CPSCOM, IEMCON, ICDCS, and so on. He has served as a Technical Program Committee (TPC) member at renowned international conferences including ACM SIGCSE, IEEE LCN, IEMCON, STPSA, SCS SpringSim, and IEEE Compsac. He is serving as an Associate Editor of the *International Journal of Computers and Applications* (Taylor & Francis), and also as a reviewer for international journals including IEEE Transactions (Elsevier, Springer, and Wiley). Dr Ghosh is contributing as a guest editor for special issues with *IEEE Journal of IoT, IEEE Transaction on Network Science and Engineering* (TNSE), *ACM Transactions on Internet Technology* (TOIT), *Springer MTAP, MDPI Future Internet*, and *Wiley Internet Technology Letters* (ITL). He is a Senior Member of the IEEE and a member of AAAS, ASEE, ACM, and Sigma Xi. He is actively working on seven edited volumes on Emerging CPS, Security, and Machine/Machine Learning with CRC Press, Chapman Hall Big Data Series, and Springer. His main research interests include Cybersecurity, Computer Networks, Wireless Networks, Information Centric Networking, and Software-Defined Networking.

Danda B. Rawat is a Full Professor in the Department of Electrical Engineering and Computer Science (EECS), founding director of Howard University Data Science and Cybersecurity Center (DSC2) and Cybersecurity and Wireless Networking Innovations (CWiNs) Research Lab, graduate program director of Howard-CS Graduate Programs, and director of Graduate Cybersecurity Certificate Program at Howard University, Washington, DC, USA. Dr Rawat is engaged in research and teaching in the areas of cybersecurity, machine learning, and wireless networking for emerging networked systems including CPS, IoT, smart cities, software defined systems, and vehicular networks. His professional career comprises more than 15 years in academia, government, and industry. He has secured over $4 million in research funding from the US National Science Foundation, US Department of Homeland Security, Department of Energy, National Nuclear Security Administration (NNSA), DoD Research Labs, Industry (Microsoft, Intel, etc.), and private Foundations. Dr Rawat is the recipient of the NSF CAREER Award in 2016, the US Air Force Research Laboratory (AFRL) Summer Faculty Visiting Fellowship in 2017, Outstanding Research Faculty Award (Award for Excellence in Scholarly Activity) at GSU in 2015, the Best Paper Awards, and Outstanding PhD Researcher Award in 2009. He has delivered over 15 keynotes and invited speeches at international conferences and workshops. Dr Rawat has published over 200 scientific/technical articles and 9 books. He has been serving as an editor/guest editor for over 30 international journals. He has been in the Organizing Committees for several IEEE flagship conferences such as IEEE INFOCOM, IEEE CNS, IEEE ICC, IEEE CCNC, and so on. Dr Rawat received his PhD degree from Old Dominion University (ODU), Norfolk, Virginia. Dr Rawat is a Senior Member of IEEE and ACM, a member of ASEE, and a Fellow of the Institution of Engineering and Technology (IET).

Raja Datta completed his MTech and PhD from Indian Institute of Technology (IIT) Kharagpur, India. He is a professor in the Department of Electronics and Electrical Communication Engineering (E & ECE) at IIT Kharagpur and is currently the head of G. S. Sanyal School of Telecommunication. Earlier, he worked with North Eastern Regional Institute of Science and Technology, Itanagar, India, where he was also the head of the Department of Computer Science and Engineering (CSE) for several years. Prof. Datta is a Senior Member of IEEE and has to his credit a number of publications in high impact factor journals (that includes *IEEE Transactions*, IET, Elsevier, Springers, etc.) and international conferences. Apart from being a consultant to several organizations in India, he has been associated with many institutes and universities as examiner, member of Academic Boards, etc. Prof. Datta was the Chairman of IEEE Kharagpur Section in 2014. He was also the secretary and vice chair of IEEE Kharagpur Section in 2012 and 2013, respectively. Apart from organizing a lot of activities, the section also received the best small section award in Region 10 during his tenure as an office bearer. From February 2011, he has been the Prof-in-charge of the Technology Telecom Centre (TTC) of IIT Kharagpur. His main research interests include Computer Communication Networks, Network Function Virtualization (NFV), 5G Edge Computing, Vehicular Networks, Mobile Ad-hoc and Sensor Networks, Optical Elastic Networks, Inter Planetary Networks, Computer Architecture, Distributed Operating Systems, and Distributed Processing.

Al-Sakib Khan Pathan is a professor of CSE. Currently, he is with the Independent University, Bangladesh, as an adjunct professor. He received his PhD degree in computer engineering in 2009 from Kyung Hee University, South Korea, and BSc degree in computer science and information technology from Islamic University of Technology (IUT), Bangladesh, in 2003. In his academic career so far, he worked as a faculty member in the CSE Department of Southeast University, Bangladesh, during 2015–2020, Computer Science Department, International Islamic University Malaysia (IIUM), Malaysia during 2010–2015; at BRAC University, Bangladesh from 2009–2010, and at New South University (NSU), Bangladesh, during 2004–2005. He was a guest lecturer for the STEP project at the Department of Technical and Vocational Education, IUT, Bangladesh, in 2018. He also worked as a researcher at Networking Lab, Kyung Hee University, South Korea, from September 2005 to August 2009, where he completed his MS leading to PhD. His research interests include Wireless Sensor Networks, Network Security, Cloud Computing, and e-Services Technologies. Currently, he is also working on some multidisciplinary issues. He is a recipient of several awards/best paper awards and has several notable publications in these areas. So far, he has delivered over 20 keynotes and invited speeches at various international conferences and events. He has served as a General Chair, Organizing Committee Member, and TPC member in numerous top-ranked international conferences/workshops like INFOCOM, GLOBECOM, ICC, LCN, GreenCom, AINA, WCNC, HPCS, ICA3PP, IWCMC, VTC, HPCC, SGIoT, etc. He was awarded the IEEE Outstanding Leadership Award for his role in IEEE GreenCom'13 conference. He is currently serving as the Editor-in-Chief of *International Journal of Computers and Applications*, Taylor & Francis, UK; Associate Technical Editor of *IEEE Communications Magazine*; Editor of *Ad Hoc and Sensor Wireless Networks*, Old City Publishing; *International Journal of Sensor Networks*, Inderscience Publishers; and *Malaysian Journal of Computer Science*; Associate Editor of *Connection Science*, Taylor & Francis, UK; *International Journal of Computational Science and Engineering*, Inderscience; Area Editor of *International Journal of Communication Networks and Information Security*; Guest Editor of many special issues of top-ranked journals, and Editor/Author of 21 books. One of his books has been included twice in Intel Corporation's Recommended Reading List for Developers, second half 2013 and first half of 2014; three books were included in IEEE Communications Society's (IEEE ComSoc) Best Readings in Communications and Information Systems Security, 2013; two other books were indexed with all the titles (chapters) in Elsevier's acclaimed abstract and citation database, Scopus, in February 2015 and a seventh book has been translated to simplified Chinese language from English version. Also, two of his journal papers and one conference paper were included under different categories in IEEE ComSoc's Best Readings Topics on Communications and Information Systems Security, 2013. He also serves as a referee of many prestigious journals. He received some awards for his reviewing activities, such as one of the most active reviewers of IAJIT several times and Elsevier Outstanding Reviewer for Computer Networks, Ad Hoc Networks, FGCS, and JNCA in multiple years. He is a Senior Member of the IEEE, USA.

Contributors

Jaime C. Acosta earned his PhD from the University of Texas at El Paso (UTEP) in 2009. He has worked for the army as a government civilian for over 15 years and as a visiting researcher and adjunct professor at UTEP for over 8 years. He is the ARL South site lead at UTEP, where he spearheads the Cybersecurity Rapid Innovation Group (CyberRIG) consisting of faculty, members of the regional FBI's Cyber and Forensics groups, the Infragard, the Department of Homeland Security, and the El Paso Water and Electric Utilities, among others. Dr Acosta's primary focus is to collectively and collaboratively solve critical cybersecurity problems through research, tool development, and rapid innovation activities.

Venkataramana Badarla conducts his research broadly in Computer Networks, specifically in Wireless Networks, Cloud Computing, IoT, and Named Data Networking. He is also interested in the application of ICT to solve the problems in the areas of Precision Agriculture and Smart Infrastructure. He has published over 40 research papers in various peer-reviewed journals and conferences and won the best of the conference award for his work on low-cost infrastructure for rural communication. He has successfully guided 4 PhD theses and 12 master's thesis and over 70 BTech projects. He has been a reviewer and also served as a member of the TPC for various top-tier journals and conferences. He is a senior member of the professional bodies ACM and IEEE. He obtained his PhD from IIT Madras, India, in 2007 for his work on reliable data transport solutions for multi-hop wireless networks. After his PhD, he worked as a postdoctoral researcher at the Hamilton Institute, Ireland, for 4 years. Then he worked in the Department of CSE at IIT Jodhpur for 6 years. Since 2017, he has been working as an associate professor in the Department of CSE at IIT Tirupati, India, where he is also chairing the department.

Sourav Banerjee holds a PhD degree from the University of Kalyani, India, in 2017. He is currently an assistant professor at the Department of CSE of Kalyani Government Engineering College at Kalyani, West Bengal, India. He has authored numerous reputed journal articles, book chapters, and international conferences. His research interests include Big Data, Cloud Computing, Cloud Robotics, Distributed Computing and Mobile Communications, IoT. He is a member of IEEE, ACM, IAE, and MIR Labs as well. He is a SIG member of MIR Labs, USA. He is an editorial board member of Wireless Communication Technology. His profile can be found here: https://www.kgec.edu.in/department/CSE/.

Deborsi Basu is working as a junior researcher at the Communication Networks Lab in the Department of E & ECE, IIT Kharagpur with the Joint collaboration of G.S. Sanyal School of Telecommunication, IIT, Kharagpur & Department of EECS, University of Vanderbilt, Nashville, Tennessee, USA. He completed his MTech from Kalyani Government Engineering College, Kalyani, West Bengal, India in Department of Electronics and Communication Engineering in 2018 and BTech from Heritage Institute of Technology, Kolkata, West Bengal, India in Department of ECE in 2016. He is a Graduate Student Member of IEEE and a student member of IET and ACM. The broad area of his research is Wireless Communication and Network Performance Optimization. Currently, he is working on SDN, OpenFlow Protocol Architecture Design, NFV, Network Scalability in Distributed System, Techno-Economic Architecture Development for 5G, and beyond.

Padmalochan Bera is working as an assistant professor in the Department of CSE in IIT, Bhubaneswar, India. His research interests include Network Security, Cryptography, Access Control, SDNs, Cloud Computing, Formal Verification, and Optimization. He completed his PhD from IIT, Kharagpur, India and he was a postdoctoral fellow in CyberDNA Research Center, University of North Carolina Charlotte, USA from 2010 to 2011.

Davide Caputo is a second-year PhD student in computer science at the University of Genova, Genova, Italy. He obtained both his BSc and MSc in computer engineering at the University of Genova, and is now working under the supervision of Prof. Alessio Merlo. His research topic focuses on Mobile Security and IoT Security.

Luca Caviglione received his PhD in electronics and computer engineering from the University of Genoa, Genoa, Italy. He is a research scientist with the Institute for Applied Mathematics and Information Technologies of the National Research Council of Italy, Genoa. His research interests include Optimization of Large-Scale Computing Frameworks, Wireless and Heterogeneous Communication Architectures, and Network Security. He is an author or co-author of more than 140 academic publications and several patents in the field of p2p and energy-aware computing. He has been involved in many research projects funded by the European Space Agency, the European Union, and the Italian Ministry of Research. He is a Work Group Leader of the Italian IPv6 Task Force, a contract professor in the field of networking/security and a professional engineer. He is a member of the TPC of many international conferences. From 2016, he has been an associate editor of *International Journal of Computers and Applications*, Taylor & Francis.

Chinmay Chakraborty is an assistant professor in the Department of Electronics and Communication Engineering, Birla Institute of Technology (BIT), Mesra, India. His primary areas of research include Wireless Body Area Network, Internet of Medical Things, Energy-Efficient Wireless Communications and Networking, and Point-of-Care Diagnosis. He has authored more than 50 articles in reputed journals, books, book chapters, and international conferences. He received a Young Research Excellence Award, Global Peer Review Award, Young Faculty Award, and Outstanding Researcher Award. He was the speaker for AICTE, DST sponsored FDP, and CEP Short Term Course. His profile can be found here: https://sites.google.com/view/dr-chinmay-chakraborty.

Rajat Subhra Chakraborty received his PhD in computer engineering from Case Western Reserve University. He is currently an associate professor in the CSE Department, IIT Kharagpur, India. He has held positions with National Semiconductor, Bangalore, India, and Advanced Micro Devices, Inc., Santa Clara, California, USA. His research interests include Hardware Security, VLSI Design, Digital Content Protection, and Digital Forensics. He holds two US patents, is the co-author of three books, and has close to 100 publications in international journals and conferences of repute, which have received about 3,200 citations till date. He is a recipient of IBM Faculty Award (2012), Royal Academy of Engineering (UK) "RECI Fellowship" (2014), IBM SUR Award (2015), IEI Young Engineers Award (2016), and Outstanding Faculty Award from IIT Kharagpur (2018). He is currently an associate editor of *IEEE TCAD* and *IEEE TMSCS* journals, and has previously served as a guest editor of *ACM TECS*. He also regularly serves in the program committee of top international conferences. Dr Chakraborty is a Senior Member of both IEEE and ACM.

Pushpita Chatterjee is a research consultant at Old Dominion University (ODU), Norfolk, Virginia, USA. Pushpita received her PhD from IIT Kharagpur, India in 2012 under the supervision of Prof. Indranil Sengupta and Prof. Soumya K Ghosh. She received her MTech in computer engineering and MS in computer and information science from the University of Calcutta, India, in 2004 and 2002, respectively. Prior to joining ODU, she was a senior research lead in SRM Institute of Science and Technology (a Unit of SRM University, Chennai), Bangalore, India. She was responsible for leading a group of 20+ researchers who were working for OpenFlow and SDN and Deep Learning related application research with NEC, Japan, and NTT, Japan. She has a good number of publications to her credit in international journals, conferences, and books. Her research interests include Mobile Computing, Distributed and Trust Computing, Wireless Ad Hoc and Sensor Networks, Information-Centric Networking and SDN. She is a Member of the IEEE.

Debashis Das is currently working as a university research scholar at Kalyani University, India. He completed the M Tech in CSE from Kalyani Government Engineering College, India. He also completed B Tech in CSE from the Government College of Engineering and Leather Technology. His fascinating research areas are Cloud Computing, IoT, and Applications of Blockchain. He is currently engaged in broad research on blockchain applications.

Sajal K. Das is a professor of computer science and the Daniel St. Clair Endowed Chair at the Missouri University of Science and Technology, Rolla, Missouri, USA, where he was the Chair of Computer Science Department from 2013–2017. He served the NSF as a Program Director in the Computer Networks and Systems division during 2008–2011. Prior to 2013, he was a University Distinguished Scholar Professor of CSE and founding director of the Center for Research in Wireless Mobility and Networking at the University of Texas at Arlington. His research interests include Wireless and Sensor Networks, Mobile and Pervasive Computing, Mobile Crowdsensing, CPS and IoT, Smart Environments (e.g., smart city, smart grid, smart transportation, and smart health care), Distributed and Cloud Computing, Cybersecurity, Biological and Social Networks, and Applied Graph Theory and Game Theory. He has published extensively in these areas with over 700 research articles in high quality journals and refereed conference proceedings. Dr Das holds 5 US patents and co-authored four books – *Smart Environments: Technology, Protocols, and Applications* (John Wiley, 2005), *Handbook on Securing Cyber-Physical Critical Infrastructure: Foundations and Challenges* (Morgan Kauffman, 2012), *Mobile Agents in Distributed Computing and Networking* (Wiley, 2012), and *Principles of Cyber-Physical Systems: An Interdisciplinary Approach* (Cambridge University Press, 2020). His h-index is 85 with more than 31,500 citations according to Google Scholar. He is a recipient of 10 Best Paper Awards at prestigious conferences like ACM MobiCom and IEEE PerCom, and numerous awards for teaching, mentoring, and research, including the IEEE Computer Society's Technical Achievement Award for pioneering contributions to sensor networks and mobile computing, and University of Missouri System President's Award for Sustained Career Excellence. Dr Das serves as the founding Editor-in-Chief of Elsevier's *Pervasive and Mobile Computing Journal*, and as associate editor of several journals including *IEEE Transactions on Dependable and Secure Computing*, *IEEE Transactions on Mobile Computing*, and *ACM Transactions on Sensor Networks*. He is an IEEE Fellow.

Abel O. Gomez Rivera is a research assistant in computer science at The University of Texas at El Paso (UTEP), Texas, USA, and is working toward the PhD degree in computer science at UTEP, where his dissertation focus is toward developing a blockchain empowered platform for identity management and process integrity for sensors in fossil power plants. Prior to that he was a Service Information Developer II at DXC Technology. He received his MSc degree in software engineering from UTEP in 2018. His research interests include Identity Management Mechanisms, Physical Unclonable Functions, Blockchain, Secure Communication Protocols, Internet of Battlefield Things, and CPS.

Houssem Mansouri received his engineering degree in computer science from the University of Farhat Abbes, Sétif, Algeria, in 2004, and his masters and PhD, both in computer science, from the University of Abderrahmane Mira, Bejaia, Algeria, in 2007 and 2015, respectively. He is working as an associate professor at the Computer Science Department,

Faculty of Sciences, Ferhat Abbas Sétif University 1, Algeria. Sétif. His research interests are Fault Tolerance Techniques for Distributed Systems in Mobile and Ad Hoc Networks.

Jimson Mathew received his master's degree in computer engineering from Nanyang Technological University, Singapore, and PhD in computer engineering from the University of Bristol, Bristol, UK., Currently, he is the head of the CSE Department, IIT Patna, India. He is also an Honorary Visiting Fellow with the Department of CSE, University of Bristol. He has held positions with the Centre for Wireless Communications, National University of Singapore; Bell Laboratories Research Lucent Technologies North Ryde, Australia; Royal Institute of Technology KTH, Stockholm, Sweden; and Department of Computer Science, University of Bristol, United Kingdom. His research interests include Fault-Tolerant Computing, Computer Arithmetic, Hardware Security, VLSI Design and Automation, and Design of Nano-Scale Circuits and Systems.

Alessio Merlo received his PhD in computer science in 2010 and he is currently serving as a senior (tenured) assistant professor at the University of Genoa. His main research interests focus on Mobile and IoT Security, and he leads the Mobile Security research group at the University of Genoa. He has published more than 100 scientific papers in international conferences and journals.

Aiswarya S. Nair is a doctoral research scholar at the Centre for Research and Innovation in Cyber Threat Resilience (CRICTR), Indian Institute of Information Technology and Management-Kerala (IIITM-K), India. She received her MTech degree in Embedded Systems from Amrita Vishwa Vidyapeetham University, Coimbatore, India, in 2012 and BTech in Electronics and Communication from Mahatma Gandhi University, Kottayam, India, in 2010. Her research interests include IoT, Cybersecurity, and Embedded System Design.

Ajay Pratap is an assistant professor with the Department of CSE, National Institute of Technology Karnataka (NITK), Surathkal, India. Before joining the NITK, he was a postdoctoral researcher at the Department of Computer Science, Missouri University of Science and Technology, Rolla, Missouri, USA, from August 2018 to December 2019. He completed his PhD in computer science and engineering from IIT, Patna,

India, in July 2018; MTech in CSE from IIIT Bhubaneswar, India, in 2014; and BTech in CSE from Uttar Pradesh Technical University, Lucknow, India, in 2011. His research interests include QoS level issues at the MAC layer, resource allocation, and algorithm design for next-generation advanced wireless networks. His current work is related to HetNet, Small Cells, Fog Computing, IoT, and D2D communication underlaying cellular 5G and beyond.

Madhukrishna Priyadarsini is currently pursuing her PhD in the Department of CSE in IIT, Bhubaneswar, India. Her current work includes Computer Network Management, SDN, and Security Issues in SDN. She is also interested in Image Processing and Game Theoretical Approaches. She is an active IEEE student member and has organized certain workshops in real-time implications of SDN.

Sree Ranjani Rajendran is a postdoctoral researcher in RISE Lab, Department of CSE, IIT Madras, India. She received her PhD from Amrita Vishwa Vidyapeetham, Coimbatore, India. She is a passionately curious researcher, in pursuit of knowledge and expertise in the broader domain of Hardware Security, with specific interest in "Design for a secured hardware." She completed her internship in the research project titled "Design and Implementation of Trojan checker circuit for Hardware Security," sponsored by Microsoft Research India, undertaken in the Department of CSE of IIT KGP. She also completed "Hardware Security" (Grade: 91.5%) from the University of Maryland in Coursera. She has more than 9 years of professional experience in research and university teaching. She received a best paper award for her paper titled "A Novel Logical Locking Technique against Key-Guessing Attacks" and third prize in VLSID 2019 design contest for the project titled "IoT based Smart Vehicle Automation and Control with Enhanced Safety, Security and Tracking System using Wireless Sensors Networks." Her broader area of interest includes VLSI Testing; Hardware Verification; Hardware Security and Trust, with emphasis on anti-counterfeiting, anti-Trojan, authentication, and anti-piracy solutions. She has authored about 20+ research articles published in refereed conference proceedings and renowned journal publications. Her research impact is 57+ citations and h-index is 3. She has been a member of IEEE since 2016.

Pranesh Santikellur is currently a PhD student in the Department of CSE, IIT Kharagpur, India, under the guidance of Dr Rajat Subhra

Chakraborthy. He received his BE degree in CSE from SDM College of Engineering and Technology, Dharwad (India) in 2010. He worked as "Firmware Engineer" with a total of 6 years of industry experience at Horner Engineering India, Bangalore and Processor Systems, Bangalore. His research interests include Deep Learning, Hardware Security, Network Security, and PLC Automation.

Ravi Sharma is currently a research scholar with the Department of CSE at IIT Jodhpur, India. His work is oriented towards designing multi-objective optimization strategies for Indoor Beacon Placement Problems. From January to March 2020, he has worked as a senior project engineer in IIT Kanpur, India. From July 2013 to May 2014, he was also a visiting assistant professor with the Department of CSE, The LNMIIT Jaipur, Rajasthan, India. He received his master's degree (MTech) in 2013 from IIT Kanpur and graduate degree (BTech) in 2010 from KIIT University, India. He is a recipient of travel awards by Microsoft Research India, IEEE COMSOC, ACM SIGBED, and COMSNETS India for presenting his research at international and national venues.

Sachin Shetty is an associate professor in the Virginia Modeling, Analysis and Simulation Center at Old Dominion University (ODU), Norfolk, Virginia, USA. He holds a joint appointment with the Department of Modeling, Simulation, and Visualization Engineering and the Center for Cybersecurity Education and Research. He received his PhD in Modeling and Simulation from ODU in 2007 under the supervision of Prof. Min Song. Prior to joining ODU, he was an associate professor in the Electrical and Computer Engineering Department at Tennessee State University, Nashville, USA. He was also the associate director of the Tennessee Interdisciplinary Graduate Engineering Research Institute and directed the CyberSecurity laboratory at Tennessee State University. He also holds a dual appointment as an engineer at the Naval Surface Warfare Center, Crane, Indiana. His research interests lie at the intersection of Computer Networking, Network Security, and Machine Learning. His laboratory conducts cloud and mobile security research and has received over $10 million in funding from the National Science Foundation, Air Office of Scientific Research, Air Force Research Lab, Office of Naval Research, Department of Homeland Security, and Boeing. He is the site lead on the DoD CyberSecurity Center of Excellence, the Department of Homeland Security National Center of Excellence, the Critical Infrastructure Resilience Institute (CIRI), and Department of

Energy, Cyber Resilient Energy Delivery Consortium (CREDC). He has authored and co-authored over 140 research articles in journals and conference proceedings and two books. He is the recipient of DHS Scientific Leadership Award and has been inducted in Tennessee State University's million-dollar club. He has served on the TPC for ACM CCS and IEEE INFOCOM.

Sabu M. Thampi is a professor at IIITM, Kerala, India. His current research interests include Cognitive Computing, IoT, Biometrics, and Video Surveillance. He is currently serving as the editor for *Journal of Network and Computer Applications* (JNCA) and associate editor for *IEEE Access*. He is a Senior Member of the IEEE and ACM.

Deepak Tosh is an assistant professor in computer science at The University of Texas at El Paso (UTEP), Texas, USA. Prior to that, he was a cybersecurity researcher at the DoD Sponsored Center of Excellence in Cybersecurity, Norfolk State University (NSU), Norfolk, Virginia. His research is focused on establishing data provenance mechanisms in cloud computing in addition to addressing research challenges in the area of distributed system security, blockchain, cyber-threat information sharing, cyber-insurance, and Internet of Battlefield Things (IoBT). Specifically, he is interested in distributed consensus models in Blockchain technology, cyber-resiliency in battlefield environments, and various practical issues in cloud computing security. He received his PhD in CSE from the University of Nevada, Reno, USA, where his dissertation was focused on designing market-based models to enable cybersecurity information sharing among organizations, which received the Outstanding Thesis Award from the CSE Department.

Luca Verderame obtained his PhD in electronic, information, robotics, and telecommunication engineering at the University of Genoa, Italy, in 2016, where he worked on mobile security. He is currently working as a post-doc research fellow at the Computer Security Laboratory (CSEC Lab), and he is also the CEO and co-founder of Talos, a cybersecurity startup and university spin-off. His research interests mainly cover information security applied, in particular, to mobile and IoT environments.

CHAPTER 1

Wireless Localization for Smart Indoor Environments

Ravi Sharma

Indian Institute of Technology Jodhpur

Venkataramana Badarla

Indian Institute of Technology Tirupati

CONTENTS

AMONG THE POSSIBLE TECHNOLOGIES to provide the foundation for a smart solution, Indoor Localization carries a vast application potential in designing automation and control systems for smart cities, buildings, and homes. Also, this decade has witnessed a manifold growth in the research and development over localization for the use cases such as health care, appliance control, surveillance, navigation, and tracking. Due to the vast end-user availability of handheld and wearable devices having wireless communication capabilities, technologies such as WiFi, Bluetooth (basic rate (BR), enhanced data rate (EDR), low energy (LE)), Radio Frequency Identification (RFID), and Zigbee have been a significant role player in delivering smartness to indoor spaces. With this context, this chapter begins by detailing the reader about the fundamental techniques of localization. It expands on the notable applications of indoor localization with underlying different wireless standards. A qualitative assessment of localization techniques and wireless standards for their suitability in indoor localization is also provided. This assessment is intended to serve as a guideline for readers in rationalizing the needs of their research domain. The chapter concludes with a prospect of upcoming and ongoing development in this domain to gyrate innovative ideas.

1.1 INTRODUCTION

The development of communication technology over the past few decades has spread its roots in every aspect of human life. Facilitation of automated services is no longer a concern restricted to industries due to the increasing deployment of smart electronic units and advanced machinery in commercial and residential indoor environments. Several attempts have been made in the past to identify structural requirements and formulate solution frameworks for automating indoor building environments. In the hype cycle[1] of emerging technologies, it is visible that the trend of connecting *things* and inducing smartness in indoor environments has shown sufficient potential for innovation and upholds expectations for service provisioning. Among the possible technologies to provide a foundation for an Indoor automation system, localization carries a vast application potential. For the residential environments, the following are few of the promising use cases:

- *Child Care*: Monitoring the health and activity of infants and kids in an indoor environment is a challenging task spanning the entire day and night cycle. A wearable device or distant sensing unit to monitor and inform the vitals and activities of a child towards a hazardous situation can be of help to the family [1].

- *Elderly Monitoring*: Assistance in navigation and operation is a primary concern arising with the old in a domestic environment. Presence of location-aware devices with intelligent functionalities can prove to be of substance [2–8].
- *Appliance Control*: Maintenance of temperature conditioning in the house, controlling appliances such as washing machines and refrigerators, and adjusting illumination based on the human presence requires location-related information to automate intelligently. This requirement gives an open floor for localization to come into play [9–14].
- *Security*: Accessibility control for admission to restricted places and locating security breaches require precise positioning and tracking methods. Safeguarding against theft and planning for emergencies can be assisted with location-based services [15–18].

Similarly, non-residential infrastructures, where horizontal and vertical constructions are frequent, present additional spatial requirements for automation [19,20]. In such environments, resource and service provisioning [21–25] is of great concern due to the underlying cost [26–29]. This requires support of optimization techniques to achieve efficient operation goals such as energy conservation. In the following, we present a few use cases that demonstrate the applicability of indoor localization for commercial and public spaces:

- *Occupancy Analysis*: Conserving electric energy is of great concern for commercial environments. Activities of employees and assisting appliances trigger continuous electricity consumption at times, raising the need for optimal energy usage frameworks. Such intelligent techniques require knowledge about occupants in the space of consideration. Localization can be useful in identifying and locating the inhabitants for objectives such as automated lighting control and heating, ventilation, and air conditioning (HVAC) [30–33].
- *Patient Monitoring*: Recording and monitoring the vitals and activities of a patient in a medical facility are of high importance for doctors and nursing staff. This requires the dissemination of accurate information in real time. Hence, detection of

critical conditions and informing them to the relevant available nursing stations or doctors can be of great help with location-based information [34–40].

- *Assisting Handicapped*: For a person with physical disability, especially with vision impairment, the assist of a localization system for identification, navigation, and interaction can turn out to be of great help [41–43].

- *Underground Locations*: Establishments such as mineral mines require a lot of human and machine mobility in challenging environmental conditions. The location information in the form of applications, delivering digitized maps and navigation assistance, can be of great help for the workers and machinery deployed in such depths [44–53].

- *Surveillance*: Defence establishments and government organizations of high importance require a robust service framework for monitoring every private entity without fail. Hence, positioning and tracking based applications can play a vital role in such scenarios [54–61].

The above description explains the motivation behind the need for indoor localization solutions. The concept of an indoor localization system is bound to be formed upon the implementation of positioning methods. Past researches in this regard have experimented localization with technologies such as Radio communication standards, Visible Light, Magnetic Field, Vibration, Sound, and Inertial sensors deployed with suitable algorithms for position estimation. Each implementation of a localization solution can be segregated into two parts, namely method and technology. Methods of positioning are mostly algorithms, built to work on the technologies as mentioned above. They can be categorized into the following two types depending on the metric of observation undertaken:

- *Transmission*: The usual parameters associated with this type of method are time of transmission, angle of arrival (AoA), and phase of arrival (PoA).

- *Characteristic*: This division utilizes the signal strength and channel information of the underlying communication method to localize.

The earlier efforts in summarizing the positioning technologies have adopted different perspectives. The literature by Mao et al. [62] presents localization techniques for Wireless Sensor Networks (WSNs). They classified the localization algorithms based on the number of hops, i.e., single or multi-hop, and further investigated the underlying approaches from the connectivity and distance-based categorization. Liu et al. [63] categorized the positioning methods by underlying estimation techniques based on triangulation, scene analysis, and proximity. They presented the performance evaluation metrics for a localization approach to be accuracy, precision, complexity, scalability, robustness, and cost. Their analysis concluded with the juxtaposition of the then existing wireless localization solutions in reference to aforementioned metrics.

Gu et al. [64] have presented a systematic comparison of 17 commercial and research based solution for indoor positioning. They have analyzed the methods and technologies used for position measurements and proposed evaluation criteria to be security and privacy, cost, performance, robustness, complexity, user preference, commercial availability, and limitations. Similar to Liu et al. [63], Farid et al. [65] have presented three main categories of positioning measurements along with further dividing of the triangulation-based methods to be based on direction and distance. They presented the performance evaluation metrics to be accuracy, responsiveness, coverage, adaptiveness, scalability, cost and complexity of the positioning system. Their analysis considered the positioning technologies to be of four types, namely Global Positioning, Infra-red, Radio frequency, and Ultrasound. A comparative analysis of methods and systems based on multiple metrics is also presented.

Mainetti et al. [66] have reviewed the enabling technologies for positioning from the objective of tracking animal in an indoor environment. They have presented a comparison among the wireless and vision based techniques and proposed a hybrid approach. Iliev et al. [67] have presented a detailed analysis of localization algorithms and implementation with an intention to foster low-power WSNs. They divided the algorithms into two categories, namely centralized and distributed, which utilize machine learning techniques for learning and classification. Also, their literature presents a combined review of methods and communication technologies falling under two major categories of interferometric and non-interferometric. It also presents a layered architecture for localization from the algorithmic point of view comprising three layers, namely application, location discovery, and physical layer. This approach lacks a system outlook of communication technology.

Dwiyasa et al. [68] have presented three categories of positioning methods based on different parameters calculated: power, time, and angle. Further they demonstrated parametric and non-parametric models for utilizing these parameter values to result in location estimates.

Over the past few decades, based on the technologies used in different application domains for positioning and navigation such as global positioning system (GPS), Laser Scanning, and Radio Frequencies, as referenced in Figure 1.1, it is quite a clear notion by now that each approach has its limitations of accuracy and coverage. Thus, rationalizing a hybrid solution seems justifiable in every sense.

With this brief overview of the domain, the rest of the chapter is structured as follows: A categorical explanation of the positioning methods which form the core of any localization technique is presented in Section 1.2. Section 1.3 surveys the technologies used for localization and describes the underlying methodologies. At the end of Sections 1.2 and 1.3 each, we critically analyze the methods and technologies respectively with their limitations and opportunities. A suitability analysis of localization methods and technologies for different indoor environments is presented in Section 1.4. The chapter concludes in Section 1.5 with the summary of its contributions with future directions.

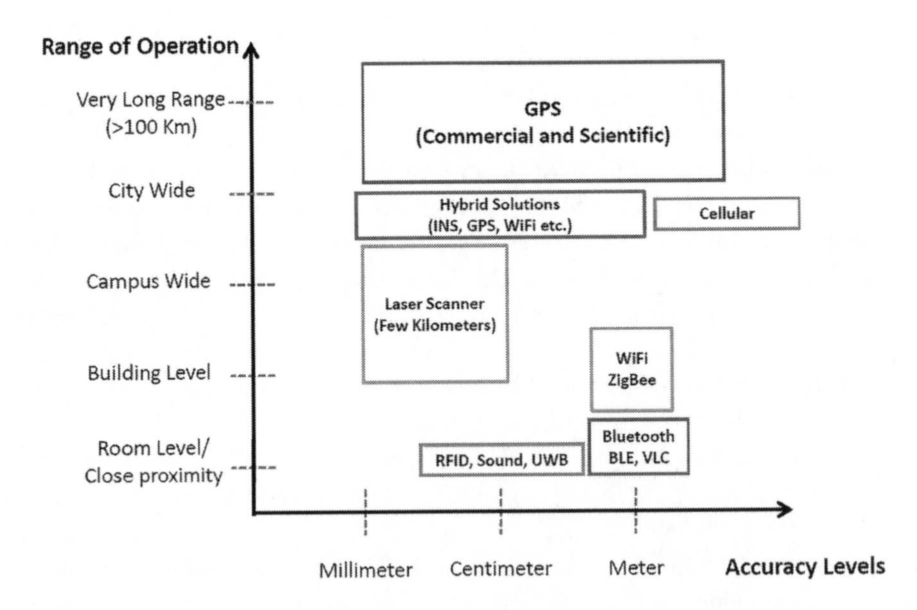

Figure 1.1 Limitations with classical positioning technologies.

1.2 POSITIONING METHODS

This section is intended to classify the mathematical foundation of positioning. As mentioned earlier, all the forms of communication techniques utilized for positioning require a medium which is essentially a wave signal. A transmitter and receiver can communicate via a signal by observing its physical attributes or modifying its characteristic properties. In other words, for positioning purposes, a communication signal can be analyzed to provide transmission related observations such as time, direction, and phase or its strength and channel characteristics. To process above gathered data two types of approaches are generally followed, namely absolute and relative. Following is the description of the two:

- *Absolute*: These methods approach to calculate the coordinates of the location under observation with reference to the system's primary coordinate system, which is then mapped to the relevant location information.
- *Relative*: In case of a hybrid environment or due to the limitation of infrastructure, these approaches calculate the location of the position by classifying the signal strength measurements to result in relative location information. These methods typically deploy machine learning and probabilistic approaches over the raw observations.

In the following subsection, a categorized description of various positioning methods used for localization have been explained with their fundamental methodology. Figure 1.2 presents the categorical classification of positioning methods.

1.2.1 Transmission

This category concerns with calculating the position estimate of requested location based on the to and fro communication between service points and users. Following are the major categories for transmission-based approaches:

1.2.1.1 Time

Methods that calculate time as a parameter of observation for positioning are typically researched with the titles of Time of Flight (ToF) [69], Time of Arrival (ToA) [70], Round Trip Time (RTT) [71], and

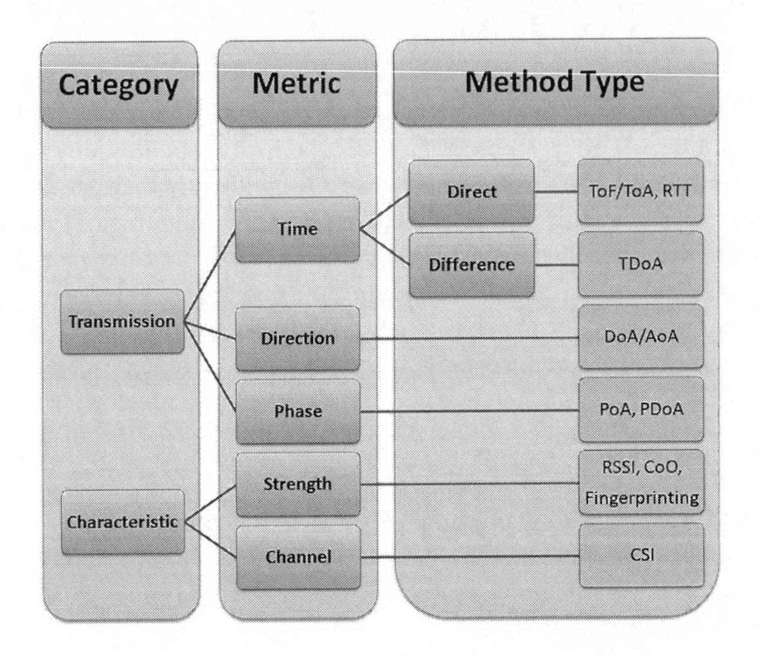

Figure 1.2 Categories of positioning methods with examples.

Time Difference of Arrival (TDoA) [72,73]. Position computation by these methods utilizes the fundamental equation of $distance = time \times velocity$ for ranging in combination with multilateration. Further, we categorize these methods based on the type of measurements undertaken, namely Direct and Difference.

- *Direct*: ToA/ToF method typically involves the two ends of communication to be time synchronized as represented in Figure 1.3. In these methods, the time of transmission start is embedded in the signal, which upon reception at the receiver side is used to compute the duration of flight. This information is used to calculate the distance between the two nodes. On the other hand, as shown in Figure 1.4, RTT approach doesn't require time synchronization. The signal is intended to be transmitted to target receiver and returned back, which gives the total raw RTT information at the sending reference station. Also, the packet processing at the target and other protocol overhead delays must be subtracted from the raw RTT to calculate the effective transmission time which can further be converted to distance measurements. The two types of methods explained above have their intrinsic pros and cons as listed in the following:

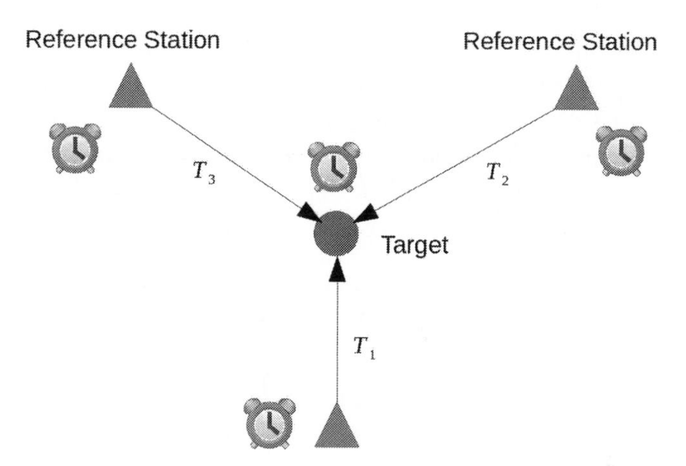

Figure 1.3 An example representation of time synchronization between stations in ToA/ToF methods. The times of signal transmission, i.e., T_1, T_2, T_3, from reference stations to target along with the speed of signal c is used to formulate the system of range equations by calculating Euclidean distances. This system is solved for target coordinates by methods such as multilateration.

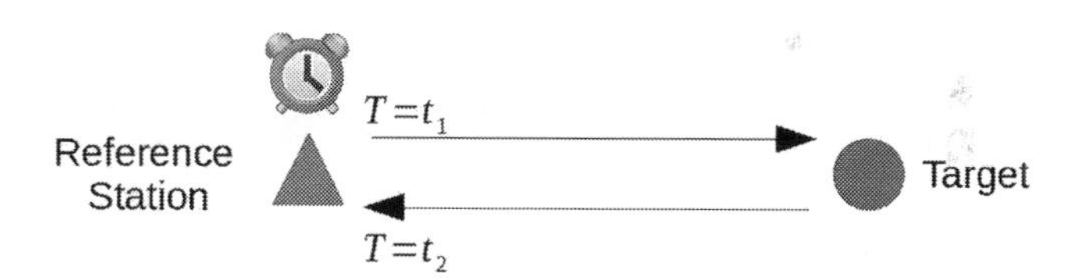

Figure 1.4 An example representation of methods involving RTT calculation of packet transmission. A single clock at reference station minimizes the effort in synchronization, though, the amount of packet processing at target must be calculated/calibrated.

- *Synchronization*: ToF or ToA methods are efficient in terms of avoiding consecutive propagation of a signal as it happens in RTT, though the overhead of synchronization is there. Sensor networks of high density and population are prone to latency and demand high precision in synchronization to avoid error propagation. Moreover, the overhead of keeping two clocks at both the receiving and sending sides increases with the sensor count, which is absent in the RTT approach.

  - *Dynamic Nodes*: Computing RTT at multiple devices in parallel can induce unwanted delays, which gets even severe in case of moving sensor nodes. Though the effect of multiple simultaneous calculation gets mitigated while choosing ToA calculation over RTT due to one way communication at the cost of time synchronization among sensor nodes.

- *Difference*: TDoA method doesn't require the time synchronization between the sender and the receiver. Rather, a signal originating from an unknown clocked target T gets delivered to mutually synchronized reference stations R_i, $\forall i = 1..n$. From this, the time difference of signal arrival can be formulated as a difference in distance measurements of each (T, R_i) pair. This removes the requirement of users to be synchronized with the communication infrastructure. From here, as shown in Figure 1.5, the system of parabolic equations $d_{i,j}^2 = ||T - R_i|| - ||T - R_j||$, $\forall i, j = 1..n$, can be solved using methods such as linearization and regression for the unknown target location. The issue of synchronization among the reference stations can be mitigated by managing the choice of precision of synchronization among the hierarchy of the installed infrastructure.

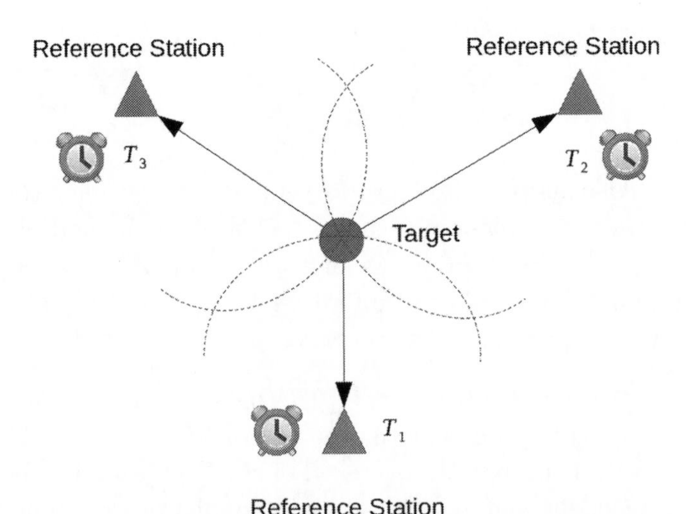

Figure 1.5 An example representation of a TDoA method where difference in arrival times of signals on reference stations from the target is used to formulate range equations solved for target coordinates.

1.2.1.2 Direction

Directional methods involve calculation of angular position of an unknown transmitting node by relative geometric angular measurements of known receiving antennas. These approaches usually implement additional array of antennas to improve positional accuracy. Direction of Arrival (DoA) [74] and AoA [75–77] are frequently used names for the methods falling in this category. These methods are often implemented along with the TDoA method between the antenna array to improve the location accuracy. As a variation of this approach, angular measurements can also be carried out using a microphone array to localize by sound waves. The practical implementation of this methodology gets limited by its design requirement of multiple antennas. Also, the spatial separation among the antennas makes it an unworthy candidate when space and cost requirement for additional hardware are limiting factors.

1.2.1.3 Phase

Phase-based localization has been attempted with either PoA [78,79] or Phase Difference of Arrival (PDoA) [80] techniques. In both cases a sinusoidal model relating the phase and the distance is assumed. As shown in Figure 1.6, in PoA approach, an unknown node transmits a signal to multiple receivers with known locations available in its proximity. Upon reception of the signal, based on the signal model, the location is resolved via multiple observations collected by different receivers. On the other hand, in PDoA, a source transmits two signals to the receiver having slight offset in time. On the arrival of the signal at the receiver, the PDoA in conjunction with the signal model concludes the location. Though effective, these methods involve the overhead of computation and additional hardware.

1.2.2 Characteristics

1.2.2.1 Strength

The methods of this domain have been most researched of all the localization approaches. Received Signal Strength (RSS) of a transmission from an unknown sender to a known receiver is the fundamental of all the calculations herein. One of the two approaches of RSS measurements assumes a path loss model for the signal transmission and later utilizes triangulation, lateration, or neighborhood algorithms to

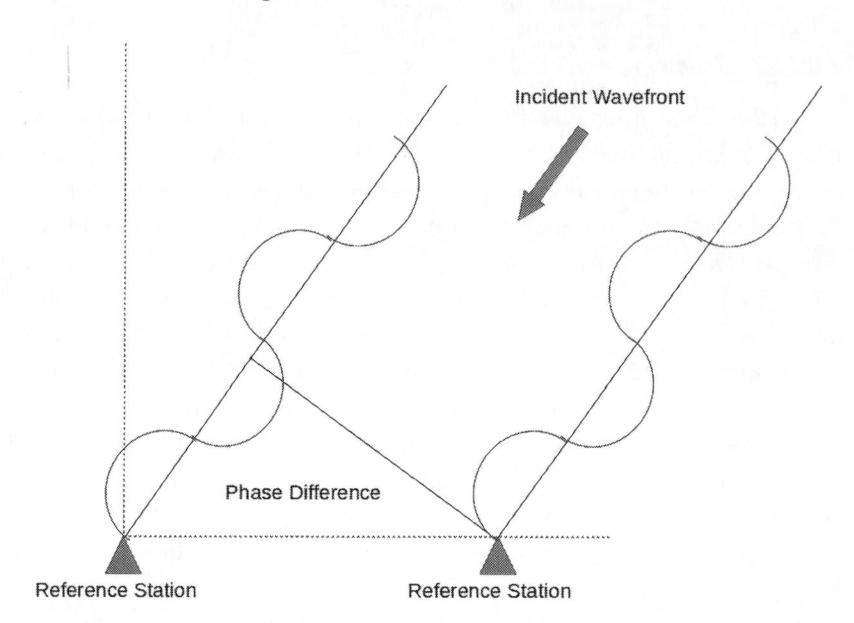

Figure 1.6 An example representation of PoA/PDoA methods where the difference in phase difference of the incident wavefronts on the antenna array of reference stations is used to measure the location information.

calculate the location. This method is typically referred to as the RSS Indication (RSSI) [81,82] in literature. The other method utilizes the RSS in two phases, namely offline and online. In the offline phase, the signal strength map of the area under consideration is constructed based on the received strength from the transmitter to different grid points, typically other transmitting nodes of known location. Later, in the online phase, the unknown node's strength is localized based on the grid map available from the offline phase. This method is often termed as Fingerprinting [83–86] in literature. Another method similar to the fingerprinting utilizes the concept of Cell of Origin (CoO) [87] of the signal. A polygonal periphery for each transmitter distinguishes itself from others and hence provides location for the unknown node based on the RSS value. RSSI has the advantage of the absence of preprocessing as required with the fingerprinting. However fingerprinting remains less computationally intense and thus quick in response. CoO has the drawback of uncertainty of cell association of a requesting node in the presence of multiple access points. All the aforementioned methods also suffer from the environment generated issues such as multipath.

1.2.2.2 Channel

Channel State Information (CSI) [88,89] is the other category utilizing the phase and amplitude information in addition to strength from a signal transmission. Due to the availability of additional parameters, its effectiveness gets into a trade-off with the overheads of computational and hardware requirements.

1.2.3 Critique

As referenced in Table 1.1, all the methods discussed in previous subsections have their limitations of implementation with one or more of the following parameters:

- *Cost*: The cost incurred in localization can be attributed to the underlying software or hardware requirements. More precisely, inherent algorithmic complexities or additional infrastructure requirements are two major sources that we explain in the following:
  - *Computation*: Computational costs can be associated to the demand of processing speed and storage capacity embedded with a method. For example, techniques such as fingerprinting fall into the category of quick responding procedures due to inherent simple algorithm but requires high storage for offline spatial learning. As we move into the techniques of localization by the signal transmission category, the additional accuracy comes at a cost of complex algorithm requiring coordinated effort from additional infrastructure at the receiver. This parameter also contributes to the timeliness of the information required.
  - *Deployment*: Need for additional deployments in a localization system can be generated by the method's inherent observational requirements or infrastructure scaling. For example, based on the earlier description of localization methods, CSI-based approach falls into the former category while AoA and PDoA belong to the latter.
- *Accuracy*: The most important concern with any localization method remains the resultant positional accuracy. Credibility of derived system depends on the precision and correctness of

TABLE 1.1 Comparative Analysis of Limitations in Localization Methods

Type	**Computation**	**Deployment**	**Accuracy**	**Remarks**
Time	Moderate	Moderate	High	Tend to deliver meter and sub-meter level accuracies but frequently suffers from fading effects.
Direction	High	Moderate	Moderate	Compatible to time-based methods in accuracy but involve additional antenna infrastructure and susceptible to noise.
Phase	High	High	High	Accuracy compatible to time and direction based approaches but require additional electronic circuitry for observation.
Strength	Low	Low	Low	Known for its ease of applicability but suffer from propagation losses. Overheads of offline processing (fingerprinting) and channel modeling. Results in meter level accuracy or higher.
Channel	Moderate	Moderate	Moderate	More accurate than strength type approaches but stands weaker to transmission methods. Observing multiple channel parameters demand additional hardwares.

the positioning method. Methods falling into the category of signal transmission usually tend to deliver high accuracy while characteristic-based approaches fail to reach that level.

The following section will now present the review of past research attempts focusing on the communication technologies used to implement the aforementioned localization methods.

1.3 POSITIONING TECHNOLOGIES

It can be intuitively inferred by now that methods of localization typically involve a signal for implementation. Indoor location estimation has been implemented most by the radio wave communication technologies such as WiFi, Bluetooth, Zigbee, and RFID. Additionally, the use of Visible light, Magnetic Field, Sound, Vibration and inertial sensors have been experimented to provide stand-alone and hybrid solutions. The following subsections will attempt to review various communication technologies utilized for localization over this decade.

1.3.1 Radio Waves

Radio waves which have evolved in the form of multiple communication technologies will be covered in the following illustration.

- *WiFi*: Modern systems build to communicate over Wireless Local Area Network (WLAN) follow the IEEE 802.11 standard popularly known as WiFi. Undoubtedly, the major research in the field of localization has happened over this standard due to its ease and availability. Today's computer systems, smartphones, and tablets are readily found compatible with WiFi and access points. The breadth and depth of WiFi's expansion have made it a de facto technology for indoor implementation.

  Though time-based localization approaches are popular in terms of their achievable accuracy, the focus of research has recently gained the direction of analyzing and removing sources of underlying error. Schauer et al. [90] has analyzed the ToF approach for implementation over WiFi. It presented an improvised timing approach and stated the clock's precision to be the main limitation. Probabilistic approaches have been extensively used with the location estimation of range calculations. Yang et al. [91] has presented a grid scan based ToA calculation method to improve maximum likelihood estimation by avoiding local optimum stagnation. Ciurana et al. [92] has experimented the issue of time synchronization by modifying the operating system configuration for a WLAN client. It concluded the increase in stability of ranging with the suggested enhancements. A Prior Non-Line-of-Sight (PNMC) method with particle filtering has been implemented in [93] to improve the performance of indoor tracking by RTT positioning. As mentioned before, TDoA method alleviates the need for time synchronization between sender and receiver. While solving the relative measurements between receivers, TDoA tends to stuck with the local minima in gradient descent method or diverges in using Newton's method. To address this issue an algorithm of Iterative cone alignment is presented in [94], which solves the TDoA positioning iteratively as a non-linear optimization. Another attempt presented in [95] improves the TDoA method by utilizing the Channel Impulse Responses (CIRs) from multipath components as a transmission from virtual source. A Kalman filter (KF) incorporating maximum likelihood estimates is implemented for identifying multipath components.

AoA-based positioning has been experimented in [96] with ray tracing to mitigate the effect of multipath. It has also demonstrated the effect of spatial smoothing and its requirement as a must for other AoA estimation algorithms such as Beamscan, MVDR, Multiple signal classification (MUSIC) and ESPRIT in the presence of multipath environments. The concept of Distance Vector (DV) containing range, AoA, and the direction of earth's gravity is presented in [97]. A method of calculating complete DV uses linear antenna array and ranging in a distributed manner.

A combination of Signal Strength based approaches with WLAN has been researched heavily in the past. Due to the plethora of such attempts we will present in brief the diverse attempts made. Zaki et al. [98] described an algorithm that calculates the weighted average of the signal strengths of known anchor nodes received at a node to calculate its location. In [18], a hybrid approach for combining data from inertial sensors such as gyroscope and accelerometer for Pedestrian Dead Reckoning (PDR) along with the WiFi and magnetic fingerprinting are presented. With the intention of mitigating the overhead of preprocessing of the sight to generate signal strength map, a method of path surveys is presented in [100], which works with discrete fingerprints collected by a moving node. In [101], a machine learning based fingerprinting approach is presented to optimize the performance in multi-building and multi-floor environments. In [102], the measure of condition entropy is claimed to be a better estimator of localization error that usually gets calculated by the mean error distance which tends to follow a uni-modal distribution. Different definitions of fingerprinting involving the measures of RSSI, access point visibility, and hybrid have been presented and experimented in [103]. It has also analyzed the effect of choosing different frequency bands and virtual access points over vertical and horizontal infrastructure span. In [104], a computational optimization of a Rank-Based Fingerprinting (RBF) algorithm is presented. This algorithm ranks the access points in a dynamic environment based on the strength of signal received. A fusion of security cameras and crowd sourced fingerprint data is performed in [105] for simultaneous tracking of multiple persons in an indoor environment with the algorithm title as MOLTIE (Multi Object Location Trajectory and Identity Estimator). [106] and [107] have presented an approach for tracking a person using WiFi-based fingerprinting in addition to data from inertial and environment sensors. Radio maps utilized for fingerprinting are susceptible to the surrounding and tend to vary over time. In such directions, a method of maintaining radio maps for fingerprinting without the need of human intervention is presented in [108]. For the benchmarking of evaluating fingerprinting algorithms, an approach dividing the process into multiple stages has been presented with experimentation in [109]. A database UJIIndoorLoc for WLAN fingerprinting based indoor localization methods has been presented in [110]. A method implementing variant of k-Nearest Neighbor (kNN) is presented in [111] to form a robust fingerprinting solution. Fingerprinting is highly sensitive to the obstacles in a signal travel path. On the similar track, the effect of human presence resulting in an increase in the error of positioning is analyzed by [112]. In an organizational environment, it is common to receive signals from multiple access points at a location. The access points, having weak signal strength at a location, provide unnecessary computation overhead with no or less improvement in positional accuracy. A method for filtering useless fingerprinting observations and respective access points is presented in [113], describing the potential for overall performance improvement. Though fingerprinting has been tested quite largely, the response of heterogeneous devices varies while communicating from the same source as suggested in [114]. Variation in recorded signal strengths and their spatio-temporal correlation were studied with reference to multiple chipsets. It also concluded that the selection of 5 GHz band signals has the potential to improve the positioning and stability over the 2.4 GHz channel.

RSSI-based techniques have been quite famous for delivering good accuracy in comparison to fingerprinting. Usually, a logarithmic path loss model similar to Equation (1.1) is utilized to transform the RSS measurements to distance traveled.

$$S_{Rec} = S_{Ref} - f(log(distance)) + noise \tag{1.1}$$

Here, S_{Rec} is the signal strength received at an unknown node. S_{Ref} is the signal power at the origin which is usually provided with the hardware. $f(log(distance))$ represents a logarithmic function of distance traveled by signal and $noise$ variable reflects a random error value.

As mentioned earlier, the methods of localization based on signal strength measurements have been researched sufficiently in the past, and our focus remains with the attempts displaying enhancements over the fundamental method. Hybrid approaches have gained much attention of the research community in recent years. A method combining the RSSI and RTT approaches is presented in [115] that computes the distance measurements from both techniques. Trilateration is performed over all the measurements and refined for the outliers using a median filter. A combination of RSSI and fingerprinting is presented in [116] in the form of a two stage process to enhance the accuracy.

To improve the accuracy of positioning, selection of reliable reference nodes is important. A utility function based approach augmented with RSSI localization is presented in [117]. It involved simulation to assist accuracy enhancement for algorithms using iterative multilateration. Simultaneous localization and mapping (SLAM) is a challenging topic in terms of dynamic accuracy requirements. A method to combine RSS measurements with foot mounter inertial sensor data by Bayesian filtering is experimented in [118] as an improvisation over SLAM with the title WiSLAM. In [119], RSSI has been adopted as a measure of increasing accuracy for Self-Organizing Maps (SOMs) with a reported decrease in the cost of computation and communication. Algorithms of

RSS suffer from varying strength measurements over time due to multipaths. A method to mitigate these temporal variations is presented in [120] using linear antenna array with reported improvement by simulations.

A hybrid approach using RSSI and PDR with the available sensors on smartphones alone is devised in [121]. It concluded the resulting increase in accuracy with the increase in cooperative participants. A likelihood model for RSS values can be generated by Gaussian processing (GP). In [122] a signal propagation model for GP is presented and compared with other models establishing a higher correlation and accuracy by experiments. A different approach utilizing both the fingerprinting and RSSI by an Extended KF (EKF) is presented in [123] to form a robust localization system. It reported the improvement of tracking accuracy in comparison to the application of EKF over fingerprinting alone. Buildings having higher degrees of automation usually contain a mesh of communication access points. In case of a real-time environment, the amount of data arriving on a node can affect the performance of WiFi-based RSSI measurements. With the same context, a distributed algorithm with Support Vector Machine (SVM) implementation is presented in [124] with a potential of deployment for network gateways in a smart infrastructure. It reported a performance improvement with stable positioning accuracy. RSSI in combination with smartphone's Inertial Measurement Unit (IMU) Sensors have been experimented in [125] to provide a real-time tracking solution. A particle filter is implemented to fuse the information from both the data sources, and resulting accuracy was reported to outperform the regular PDR approach. Shopping areas such as multi-storied malls usually tend to have a network of wide and densely placed WiFi access points. In such an environment, a method of positioning and pedestrian flow estimation using probe request signals was implemented by [126]. Estimating positions in a multi-story commercial or non-commercial buildings opens the need for easily accessible and configurable solutions. Smartphone-based positioning in combination with pressure sensor for floor detection is attempted in [127]. Position estimates were calculated by offline processing without any need for prior calibration or mapping stage. A similar problem has been attempted in [128] using RSS information alone obtained from multiple smartphones. The process first calculates the position by Bayesian filtering, and then a relation between RSS values and positioning is established by GP. Finally, a particle filter combines the output of the above two steps for indoor localization without needing any true map. Readers can find additional references for the research and development on WiFi-based localization in [129–140].

- *Bluetooth and BLE*: Under our consideration, both classical and low energy Bluetooth variants have been mainly experimented for RSS-based approaches. In [141], an algorithm implementing fingerprinting with Bluetooth is presented. This algorithm modifies the classical fingerprinting by adding the speed detection to filter the positional information. An RSSI-based localization using BLE is demonstrated in [142]. In addition to that, a distance weighted filter to reduce RSSI ambiguity and a collaborative localization method using Taylor series expansion is also incorporated. As a supportive enhancement, BLE beacon based RSSI positioning has been used in [143] to provide a robust indoor solution. It presented a novel method to rasterize the easily available floor plan and utilized the IMU data from smartphones for positioning and navigation. A more practical implementation is presented in [144] to identify the location of a customer in the form of grocery section of its vicinity. A network of BLE beacons with standard mobile phone hardware is utilized for collecting RSSI measurements. It verified the three methods of nearest-beacon, exponentially averaged weighted range and particle filtering for localization and concluded with the particle filter outperforming others. In case of an RSSI estimation the characteristics of receiving devices can cause degradation in the positioning. A method for online calibration of the propagation model that rectifies the path loss parameters with real-time range estimates is presented in [145]. In an attempt to create a smart lightning system for indoor environment, RSSI localization with BLE beacons is experimented in [146]. Each room is designated with an anchor node holding its settings, and a new arriving node can locate and adjust itself with reference to an anchor. A room level localization using RSSI with BLE is presented in [147]. A pair of threshold signal strength values is assigned to each BLE beacon. These values are used to decide the presence of a user inside a room by RSS readings. A hybrid approach that combines BLE and Ultra-Wide Band (UWB) in combination with IMU and map matching techniques is presented in [148] for the objective of indoor pedestrian navigation. A particle filter is implemented to combine the measurements from different input sources. Additional resources describing the development on localization by Bluetooth and BLE technology can be obtained from [149–161].

- *Zigbee and IEEE 802.15.4*: Zigbee is a communication suite developed over IEEE 802.15.4 standard, designated for creating low-power indoor automation solutions. An attempt to perform a low-power localization (LPL) by RSSI over IEEE 802.15.4 is presented in [162]. It presents two protocols, namely LPL and Optimized Listening Period (OLP) to optimize the power consumption among mobile and anchor nodes by modifying the underlying communication and synchronization. Another method using Zigbee for estimating RSSI values for localization has been presented in [163]. The algorithm calibrates the RSSI model among the reference nodes for calculating the position every time a request arrives from an unknown node. A hybrid approach involving user mounted IMU and GPS units for PDR within a Zigbee-enabled WSN is presented in [164]. KF is implemented in the same to fuse the RSSI measurements with the available sensor data. To reduce the power consumption involved in configuration and synchronization among the end nodes of a Zigbee wireless network, a method using RSSI estimations is attempted in [165]. This method works in two phases. First, it estimates the location of an unknown node based on its proximity with the nearest reference node. Second, it utilizes RSSI measurements from other nodes to calculate its position by relatively complex algorithms of triangulation and fingerprinting. Another tracking and positioning solution based on IEEE 802.15.4 radio network is demonstrated in [166]. This works with a user-mounted tag to locate via RSSI values in combination with inertial sensor and pressure sensor for activity and floor detection, respectively. A network of IEEE 802.15.4 nodes is utilized in [167] to improve the positioning accuracy in non-line-of-sight (NLOS) situations which frequently arrive while locating passengers within transportation vehicles. The approach uses PoA technique in combination with semi-definite programming (SDP) to optimize the solution. Pelka et al. [168] has also utilized ranging between nodes by phase measurements with multiple frequencies. A mathematical model for location estimation was created, and different sources of errors were analyzed in the same. An IEEE 802.15.4 compliant

WSN is used in [169] that implements ranging and positioning approaches by combining both RSSI and ToF measurements. The solution complements the two positioning methods and concluded the effectiveness of RSSI and ToF for short and long ranges, respectively. Another IEEE 802.15.4 based WSN is utilized in [170] to approach a hybrid solution using TDoA methods in combination with user mounted accelerometer and gyroscope sensors. An EKF was implemented to fuse the data from different input sources. Another implementation of RSSI-based localization is experimented with IEEE 802.15.4 compliant network in [171]. It used singular value decomposition (SVD) for positioning by multilateration along with a KF for accuracy improvement. Further implementations of localization by Zigbee standard are available in [172–180].

- *Radio Frequency Identification (RFID)*: RFID is a well-established principle in the field of identification and asset management [181–185]. Recent years have drawn attention towards RFID for stitching small range localization solutions. Gunawan et al. [186] has implemented RFID in assisting the job of creating a WiFi fingerprinting database. Locations were identified based on their cell id as RFID tag interacts with the scanner. A method for locating continuously moving target with RFID is presented in [187]. An algorithm titled Continuous Moving Communication Range Recognition (CM-CRR) is designed to enhance the earlier attempt of Swift-CRR (S-CRR) [188]. S-CRR's stop and wait approach for identification is relieved by performing multiple short and long range measurements of neighboring RFID tags. Positioning and orientation have jointly been attempted by least square modeling in [189]. Machine learning approaches such as Weighted Path Loss (WPL) and Artificial Neural Network (ANN) have also been experimented for localization with RFID in [190] and [191], respectively. Hybrid and cooperative localization approaches involving other technologies in combination with RFID such as camera, UWB, WLAN, and IMU have also been attempted in [192–195]. As an indirect application RFID has also been utilized for obstacle avoidance in robot navigation [196].
- *Ultra-Wideband (UWB)*: Characteristics of low energy levels and high bandwidth have enabled UWB for short range positioning applications. ToA-based positioning methods usually require the measurements from three receivers in line of sight to the

unknown location. A method with the title Reflection-Aided Maximum Likelihood (RAML) over UWB communication is attempted in [197] that utilizes multipath signals from walls and ceilings. Multipath assisted approaches have also been attempted with UWB via cooperative localization in [198,199]. Utilizing UWB for fingerprinting in a typical office environment is experimented in [200]. The high bandwidth of UWB transmission assists the fingerprints to overcome multipath propagation by providing a high temporal resolution. Following are the few applied domains for UWB localization:

– Human motion tracking using ToA [201]
– Robotics using TDoA [202–204]
– Radar imaging [205]
– Railways using TDoA [206]
– Internet of Things (IoT) using ToF [207]

A ToA-based experimental validation of UWB ranging technique that uses energy detector for received signal pulse is presented in [208]. Applications of transmission methods based UWB localization using inertial sensors have also been attempted in [209–212], which demonstrates its potential for hybrid PDR scenarios. An interesting enhancement to UWB positioning is implemented in [213] that uses frequency modulated continuous wave (FMCW) radar principle experimented with 7.5 GHz center frequency with 1 GHz bandwidth.

- *Cellular*: Cellular networks have become an integral part of today's human life due to availability and capability. Standards such as Global System for Mobile Communications (GSM), Universal Mobile Telecommunications System (UMTS), and Long-Term Evolution (LTE) have gone through stages of evolution and application domains which leave localization as no exception. On the same track, a 3G Wide-Code Division Multiple Access (W-CDMA) is experimented in [214] for detecting sub-cell movement. This method utilizes the time difference of arriving signals from the cell under consideration to calculate positional variations. Similarly, LTE femtocells were experimented in [215] for estimation of location by a relative time difference of arrival (RTDoA) method. With the availability of sufficient cells, this computes the relative time difference of signals arriving from cells having known and unknown locations for themselves. Ease of implementing fingerprinting has also gained attention in cellular applications. In the CERN's Large Hadron Collider (LHC), the tunnels implement a GSM network by radiating wires. This infrastructure was utilized to implement and analyze the applicability of fingerprinting with GSM signals underneath in [216,217]. LTE signals have also been utilized with ToA for positioning and tracking which combines the multiple measurements with an EKF. In [218], a TDoA-based positioning approach over Third generation partnership project(3GPP)-LTE is presented that compares the location estimates with static and dynamic recursive Bayesian Cramér-Rao lower bounds. LTE has also been experimented in [219] for positioning and scheduling using CSI over RSS values. Cellular networks have also shown an integrable potential with other positioning techniques such as WiFi and Bluetooth in [220,221].

1.3.2 Sound

Sound-based systems have already been in use for underwater positioning [222]. However, more recently, the research community has gained interest in the application of sound-based systems with time-based localization approaches. Kohler et al. [223] have presented an architecture to support localization in Ambient assisted living (AAL) with the combination of RF and ultrasound system. Position calculation is performed by TDoA measurements with multilateration. An acoustic pose estimation involving position and orientation calculation has been presented in [224]. This system utilizes the ToF measurements for estimation within the infrastructure of fixed loudspeakers and moving microphones. Another approach that utilizes TDoA estimations with sound is presented in [225]. It has used the iterative cone alignment method for optimizing the hyperbolic multilateration positioning. An analytical view on the capabilities of mobile phones' speakers to be potentially utilized for indoor localization is presented in [226]. A comparison of ultrasound positioning with other indoor technologies such as WiFi, Bluetooth, and Inertial sensors is also presented therein. MUSIC algorithm is a known method estimating DoA. To alleviate its inherent complexity, another approach Time Space additional Temporal (TSaT)-MUSIC, with the capability of being faster in estimating DoA's of multi-carrier ultrasonics is presented in [227]. This three-dimensional localization was verified with simulations and experimentations. Schweinzer et al. [228] presented a method of localizing and tracking static and mobile devices under the title LOSNUS (Localization of Sensor Nodes by Ultrasound). Its positioning server utilized TDoA for positioning. A fingerprinting approach with ultrasound is experimented in [229], which uses transmitter arrays in a room level arrangement to locate a user receiver inside. A comparative analysis of ToF- and RSSI-based sound positioning has been performed in [230], exemplifying its suitability for robotic and indoor applications, respectively.

As mentioned earlier, time-based approaches have been of major interest with acoustic positioning system. More precisely, location and tracking application have been experimented with ToA/ToF [231–238] and TDoA [239–248] for indoor localization, which demonstrates its potential in parallel to other radio-based approaches such as WiFi and Bluetooth.

1.3.3 Magnetic Field

Magnetic localization is mainly focused around its strength values. Every surrounding ferromagnetic substance has its unique magnetic properties [249]. Whether the method utilizes artificial or earth's magnetic field, in case of targeting magnetic waves to estimate ranges, the ferromagnetic elements in the environment can create anomalies contributing to positioning errors. Though artificial magnetic fields generated by coils tend to be affected lesser by most obstacles in the surrounding than the geomagnetism, the magnetic disturbances still remain a challenge along with the infrastructure requirement for such deployment [250]. Similarly, an improved version of geo-magnetic fingerprinting is implemented in [251] using particle filter. Song et al. [252] has proposed a method of fingerprint data generation via geo-magnetic field. Dependence of magnetic field in a building environment on the type of construction material is taken into account in the same. Magnetic fingerprinting in combination with PDR is used for self-localization in [253]. The positioning considers stable anomalies for magnetic matching and detectable landmarks for PDR.

In addition, magnetic localization, due to its performance uncertainty has mostly been used as a component of hybrid solutions such as with WiFi, smartphone sensors, and RFID in [18,254–260].

1.3.4 Visible Light

Availability of light-based equipments in an indoor environment is a common practice providing sufficient potential to be used as a measure for positioning. In the same context, a localization approach by Visible Light Communication (VLC) using Infra-red light emitting diodes (Ir-LED) is presented in [261]. This method uses the received power for fingerprinting and timestamp of impulse responses for positioning. Fingerprinting has also been used in [262] which combines unmodified fluorescent lights (FLs) and smartphones as light sensors for positioning. Another optical positioning for tracking vehicles in a logistic environment is experimented in [263]. This method implements a camera in combination with an IR-LED and a photo detector. Camera images compute the AoA measurements, while photo detector records the coordinates of a global beacon which are later utilized for positioning via triangulation with the AoA values. Another relatively simpler approach in [264] uses VLC via LED in combination with smartphone camera to record and process the modulated messages to locate. A similar approach is adopted in [265] that uses fisheye camera for positioning. A similar approach was utilized in [266] to perform VLC-based positioning for LED tracking and ID estimation. VLC has also been utilized as an assisting measure in [267]. An acoustic ToA localization scheme is implemented therein which uses a video camera to capture modulated LED signals for time synchronization. The extension of this work is presented later with the title SyncSync in [234]. Another approach of positioning via modulated light signals is attempted in [268] that uses LED transmitted coordinates to be recorded by a phone's camera. Recorded images are processed by Neural network to extract distance between camera and LED that gets utilized with decoded modulated coordinates to achieve camera position. A method implementing RSSI over VLC is experimented in [269], which used distance illumination model and trilateration for localization. Further approaches using visible light for localization can be found in [270–273].

1.3.5 Vibration

Sensing vibrational waves have taken up a recent spot in the field of localization. In the same streak, a method implementing TDoA over vibrations is experimented in [274]. A smart building infrastructure equipped with sufficient accelerometer count to access the velocity and timestamp was utilized. As a test bed, the hammer strikes were localized using the

TDoA. Another approach that claims to overcome the difficulty in calculation of vibration velocity is presented in [275]. It implemented a piezo sensor based approach utilizing piezo component to be installed on the floor. The method is similar to fingerprinting in terms of recognizing the vibration type in real time with the analysis from a vibration type database calculated by means of feature analysis.

1.3.6 Evaluation Metrics

It is understandable now that the spread of a technology in an environment controls its degree of implementation. Communication technologies such as WiFi and Bluetooth are likely to be implemented and experimented more than other contemporaries in upcoming years due to their presence in near human devices. However, a rising potential has been seen with the technologies such as BLE, Zigbee, RFID, and VLC due to their cost-effective possible usage in human proximity. We summarize a comprehensive outlook of aforementioned positioning technologies with the following metrics.

- *Coverage*: Technology's coverage is a deciding parameter for establishing any localization system. Standards such as WiFi, Zigbee, and Bluetooth tend to cover greater distances in LoS conditions while RFID and BLE prove to be efficient in short ranges. Other approaches such as sound, magnetic field, and visible Light are suitable for indoor applications, but their coverage gets affected by surroundings obstacles as mentioned in an earlier explanation.
- *Cost*: Consumption of electric power and hardware requirements is inherent to communication standards. We explain the meaning of these metrics in the following.
  - *Power*: One of the primary concern for today's smart automation system is to control energy consumption. Standards such as RFID and BLE stand promising for short-range near-user interaction to optimize power usages.
  - *Deployment*: Standards such as WiFi, Bluetooth, and Zigbee have been in sufficient exposure in the past and require regular deployment infrastructure. On the other hand VLC and magnetic field based localization demands high establishment costs, as mentioned while explaining the technologies earlier in the chapter.

- *Accuracy*: This concern is inherent to the localization method used by the technology. Fading effects and noise are the major concerns with the deliverable accuracy of a communication technology. All the technologies with short range positioning, mostly utilizing LoS situation, stand capable of delivering sub-meter level accuracy. However, as the spatial coverage increases, approaches with WiFi, Zigbee, and Bluetooth become suitable.

We now present our scrutiny in Table 1.2 with reference to the above metrics for positioning technologies mentioned in the previous subsection. Each metric carries three degrees, namely High, Moderate, and Low that are derived with the understanding of reviewed literatures as defined above.

1.3.7 Critique

In addition to the technical evaluation, technologies need to be critically analyzed for the environment under consideration. Based on the same context, we summarize our analysis in the following points:

- *Reachability*: Whether it is the attenuation of a traveling signal or obstacles in its path, any positioning technology has this inevitable limitation. Modeling the signal propagation for such scenarios typically approximate the transmission and adopt inaccuracies inherently. These accuracies are liable to expand in practical unconstrained deployments in contrast to controlled or synthetic experimentations. Indoor localization demands a context aware solution that is embedded with the need of an omnipresent service.

- *Ease*: This category contains ease of deployment, access, and maintenance as the potential question marks on positioning technologies. Although the requirement of additional hardware depends mostly on the underlying algorithm, it is unavoidable. This issue demands the minimal requirements to establish a solution. Thus, the technologies such as WiFi and Bluetooth which readily announce their presence within the human life, present a better floor for experimentation. On the other hand, technologies such as RFID, BLE, and VLC show a considerable potential to alleviate the radius of integration requirement with a trade-off between better accuracy provisioning and subtle infrastructure additions.

TABLE 1.2 Suitability of Positioning Technologies

Type	Coverage	Power	Deployment	Accuracy	Remarks
WiFi	High	Moderate	Moderate	Moderate	Available in most of today's smart devices makes it a primary candidate for such implementations.
Bluetooth	Moderate	Moderate	Moderate	Moderate	Available in most of today's smart devices but due to low range an auxiliary technique.
BLE	Low	Low	Moderate	Moderate	Promising technology, potential for short range deployments, need for hybrid and stand-alone experimentation.
Zigbee	High	Low	Moderate	Moderate	Yet to be explored standard though implemented discretely for hybrid solutions, carries potential for short and long range applications.
RFID	Low	Low	Low	High	Promising standard for short range identification. Potential candidate for meter level localization in hybrid solutions.
UWB	Low	Low	Moderate	High	Applicability to short range scenarios due to low energy and high bandwidth characteristic, preferable with cooperative positioning.

(*Continued*)

TABLE 1.2 (*Continued*) Suitability of Positioning Technologies

Type	Coverage	Power	Deployment	Accuracy	Remarks
Cellular	High	High	High	Low	Unreliable measurements over time, high infrastructure cost.
Sound	Low	Moderate	Low	High	Effective for LOS conditions, potential for combination with WiFi, Bluetooth, and Zigbee style communications.
Magnetic field	Moderate	High	High	Moderate	Varying characteristic with uncertain measurements, additional hardware cost.
Visible light	Moderate	High	Moderate	Moderate	Upcoming constructions with energy-saving establishments will have technologies such as LED as must; moreover, visible light is one the basic needs.

- *Availability*: The sense of availability in contrast to the reachability is the way of service provisioning. The expansion of localization solutions to different elements of an indoor infrastructure with different environment considerations require rigorous sensing and processing. For example, a residential place is likely to have hall rooms, bedrooms, toilet rooms, and kitchen, while a commercial environment contains conference rooms, employee cabins, recreation arenas, and canteens in general. The characteristic of an environment demands the solution to be versatile giving a hint for hybrid solutions. Also, the paradigm of cloud computing with capabilities of big data warehousing, efficient computing, and seamless communication foreseeably awaits for its role to be formulated.

1.4 A VIEW FOR RESEARCH PROSPECTS

The main parameter deciding the success of a technology is its demand. This is a value that reaches beyond the experimenting laboratories and faces the cost-effectiveness in modern era. Localization, an undoubtedly promising technology to assist future smart systems, is approaching its peak of expectation as demonstrated in the aforementioned review. In the following, we will try to project the specifics of research directions that can be taken upon to deliver the next generation of solutions.

- *Resource Optimization*: Accuracy and cost are the two dominant concerns when it comes to evaluating the feasibility of a localization solution for indoor environments. In this direction, for an optimal localization sensor network, maximizing coverage and accuracy while minimizing deployment cost and energy should be analyzed beforehand. This domain has been typically posed as an optimization problem but carries an open front of efforts for realistic environment designs and user constraints.

- *Standardization*: The vast variety of localization methods and technologies, implemented in so far, makes it a challenging task to visualize a general architecture for the localization process. In the age when we are talking of connecting things seamlessly, a general framework that can identify and structure the elements of an indoor positioning solution is much needed. For an example, the advent of global positioning solutions such as GPS and Global Navigation Satellite System (GLONASS) have seen decades of improvement in their service delivery architecture while research on the technique still continues. All the approaches of localization more or less contain sensors, middleware, and computing servers. This heterogeneity makes indoor localization weak in terms of adopting variety of context-effective solutions.

In the era of availability of smart handheld equipment with most, it is becoming easier to deliver information via applications. Services can be delivered to users via web or application interfaces. Both the use cases require active network connectivity, making radio specific localization solution to be more relevant. An open need, after the analysis of past researches, remains in the area of a formal web-service portal/architecture along with governing protocol standards. Classical communication standards such as hypertext transfer protocol (HTTP) need to be refined for real-time and reliable service objectives. Management of databases with characterized intelligent data is the utmost requirement to handle information from multiple sources. To address aforementioned issues, we compile following three potential domains to foster research and development for delivering the next generation of localization systems:

- *Application Interface*: It refers to a general interface and data flow for delivering localization as a service to users in indoor environment. This process requires spatial information to travel between client and server as a web service. Web standards such as HTTP-Representational State Transfer (REST) and Constrained Application Protocol (CoAP) have shown their capacity in sensor network applications. Research and development on such standards with optimal messaging techniques to work between smart devices and sensing units for indoor environments is intuitive.
- *Methods and Technology*: As mentioned earlier localization solutions are restricted by the surrounding's possibilities, making them quite difficult for researchers to standardize. In this context, for a localization system to be successful, its deployment should survive in a formal environment in terms of its architecture. In other words, residential places are highly varying with their design plans and infrastructure than commercial environments such as institutions and shopping malls. Hence, the localization systems should begin with their intended assist to non-residential buildings. This vision is an

undeniable objective of smart infrastructure. A comparative analysis of requirement for residential and commercial/public environments is presented in Table 1.3.

Though non-residential environments possess greater challenges in requirement provisioning, they provide sufficient infrastructure heterogeneity to test and implement standardized localization service solutions. On the other hand residential environments have less complex surrounding elements in addition to peculiar control and monitoring requirements.

Also, as mentioned in Section 1.2, accuracy and cost of a method always encounter while suitability of methods for a localization system is assessed. The cost of underlying infrastructure and its maintenance should be traded off with desired accuracy. Residential applications are less demanding of coordinate accuracy than public and commercial infrastructures. On the other hand, system's cost is an essential metric for home establishments but bothers less to non-residential applications due to their scale of application and available commercial support.

Hence, approaches based on characteristics such as fingerprinting, RSSI, and CSI suit better for the applications where accuracy requirements can be relaxed to few meters. However, applications requiring sub-meter level accuracies must utilize transmission-type localization methods.

- *Security*: Identity and location information of an entity itself is a concern of privacy. Any localization system transmitting spatial location and identity of an individual must be secured in a distributed flow among network nodes such as routers and Session Border Controllers (SBCs). Moreover, in the era of automation where we intend to connect multiple things over the internetwork, WSN should be researched for such connected units devising new protocols and security standards to handle heterogeneity.

- *Reliability of Service (RoS)*: As pointed out earlier, any designed automation system for an environment can claim its worth based on its suitability and accuracy of operation. This metric represents the applicability of a combination of methods and technology for a particular environment and line of sight condition. Table 1.3

TABLE 1.3 Mapping Combinations of Methods and Technologies for Applicability in Different Conditions

	Time	Direction	Phase	Strength	Channel
WiFi	All indoor, LoS	All indoor, LoS	All indoor, LoS	All indoor, both	All indoor, both
Bluetooth	All indoor, LoS	All indoor, LoS	All indoor, LoS	All indoor, both	All indoor, both
BLE	All indoor, LoS	All indoor, LoS	All indoor, LoS	LoS, both	All indoor, LoS
Zigbee	All indoor, LoS	All indoor, LoS	All indoor, LoS	All indoor, both	All indoor, both
RFID	Non-residential, LoS	Non-residential, LoS	Non-residential, LoS	Non-residential, LoS	Non-residential, both
UWB	Non-residential, both	Non-residential, both	Non-residential, both	Non-residential, both	Non-residential, both
Cellular	Non-residential, NLoS	Non-residential, NLoS	Non-residential, NLoS	Non-residential, NLoS	Non-residential, NLoS
Sound	All indoor, LoS	All indoor, LoS	All indoor, LoS	All indoor, both	All indoor, both
Magnetic	Non-residential, NLoS	Non-residential, NLoS	Non-residential, NLoS	Non-residential, NLoS	Non-residential, NLoS
Visible light	All indoor, LoS	All indoor, LoS	All indoor, LoS	All indoor, LoS	All indoor, LoS

presents such recommendation based on the analysis in the previous sections.

- *Robustness*: Localization system is likely to be parenting a lot of services in the upcoming smart era. The system's immunity to malfunctioning and security threats is undeniable. A secure communication and data management with auxiliary measures is challenging. Network security standards and middleware are going to play an important role for the same.

1.5 CONCLUSION

This chapter presents the application, research, and development over the two pillars, i.e., methods and technology used with indoor localization. It is by far a clear understanding that no technology alone carries the potential to provide a complete standardized, optimum to environment, easily accessible, and robust solution for localization. This need gives rise to research and architect a framework for localization that should have procedures for hybrid infrastructure deployment and policies for service provisioning. This chapter has presented a view to reconsider the localization as a systematic approach. A localization solution must consider environment's restrictions and requirements to validate the worth of related hardware and software deployments. We have attempted in this chapter to present localization as a subject matter fostering systematic approaches for it in the future.

Note

[1] https://www.gartner.com/doc/3383817.

Bibliography

[1] Geoffrey Poon, Kin Chung Kwan, Wai Man Pang, and Kup Sze Choi. Towards using tiny multi-sensors unit for child care reminders. In *Proceedings - 2016 IEEE 2nd International Conference on Multimedia Big Data, BigMM 2016*, pages 372–376, 2016.

[2] A. Marco, R. Casas, J. Falco, H. Gracia, J. I. Artigas, and A. Roy. Location-based services for elderly and disabled people. *Computer Communications*, 31(6):1055–1066, 2008.

[3] Andreas Lorenz and Reinhard Oppermann. Mobile health monitoring for the elderly: Designing for diversity. *Pervasive and Mobile Computing*, 5(5):478–495, 2009.

[4] Chung Chih Lin, Ming Jang Chiu, Chun Chieh Hsiao, Ren Guey Lee, and Yuh Show Tsai. Wireless health care service system for elderly with dementia. *IEEE Transactions on Information Technology in Biomedicine*, 10(4):696–704, 2006.

[5] Bart Jansen, Ankita Sethi, and Pieter Van Den Bergh. Position Based Behaviour Monitoring of Elderly Patients With A 3D Camera. In *Europe BMI 2009*, 2009.

[6] M. Karunanithi. Monitoring technology for the elderly patient. *Expert Review of Medical Devices*, 4(2):267–277, 2007.

[7] Andreas Lorenz, Dorit Mielke, Reinhard Oppermann, and Lars Zahl. Personalized mobile health monitoring for elderly. *International conference on Human computer interaction with mobile devices and services (MobileHCI)*, pages 297–304, 2007.

[8] Nagender Kumar Suryadevara and Subhas Chandra Mukhopadhyay. Wireless sensor network based home monitoring system for wellness determination of elderly. *IEEE Sensors Journal*, 12(6): 1965–1972, 2012.

[9] Hiroshi Kanma, Noboru Wakabayashi, Ritsuko Kanazawa, and Hiromichi Ito. Home appliance control system over Bluetooth with a cellular phone. *IEEE Transactions on Consumer Electronics*, 49(4):1049–1053, 2003.

[10] Yosuke Tajika, Takeshi Saito, Keiichi Teramoto, Naohisa Oosaka, and Masao Isshiki. Networked home appliance system using Bluetooth technology integrating appliance control/monitoring with internet service. *IEEE Transactions on Consumer Electronics*, 49(4): 1043–1048, 2003.

[11] Takeshi S. Aitoh, Tomoyuki O. Saki, Ryosuke K. Onishi, and Kazunori S. Ugahara. Current Sensor Based Home Appliance and State of Appliance Recognition. *SICE Journal of Control, Measurement, and System Integration*, 3(2):086–093, 2010.

[12] Ming Wang, Guiqing Zhang, Chenghui Zhang, Jianbin Zhang, and Chengdong Li. An IoT-based appliance control system for smart

homes. In *Proceedings of the 2013 International Conference on Intelligent Control and Information Processing, ICICIP 2013*, pages 744–747, 2013.

[13] Malik Sikandar, Hayat Khiyal, Aihab Khan, and Erum Shehzadi. SMS Based Wireless Home Appliance Control System (HACS) for Automating Appliances and Security Preliminaries Home Appliance Control System (HACS). *Issue s In Science and Information Technology*, 6:887–894, 2009.

[14] Jianwen Shao, Clifford Ortmeyer, and David Finch. Smart home appliance control. In *Conference Record - IAS Annual Meeting (IEEE Industry Applications Society)*, 2008.

[15] Jeong-ah Kim, Min-kyu Choi, Rosslin John Robles, Eun-suk Cho, and Tai-hoon Kim. A review on security in smart home development. *Security*, 15:13–22, 2010.

[16] Nikos Komninos, Eleni Philippou, and Andreas Pitsillides. Survey in smart grid and smart home security: Issues, challenges and countermeasures, *IEEE Communications Surveys Tutorials*, 16(4):1933–1954, 2014.

[17] Changmin Lee, Luca Zappaterra, Kwanghee Choi, and Hyeong Ah Choi. Securing smart home: Technologies, security challenges, and security requirements. In *2014 IEEE Conference on Communications and Network Security, CNS 2014*, pages 67–72, 2014.

[18] Michael Schiefer. Smart home definition and security threats. In *Proceedings - 9th International Conference on IT Security Incident Management and IT Forensics, IMF 2015*, pages 114–118, 2015.

[19] Kehua Su, Jie Li, and Hongbo Fu. Smart city and the applications. In *2011 International Conference on Electronics, Communications and Control (ICECC)*, pages 1028–1031. IEEE, sep 2011.

[20] a.H. Buckman, M. Mayfield, and Stephen B.M. Beck. What is a Smart Building? *Smart and Sustainable Built Environment*, 3(2): 92–109, sep 2014.

[21] Abhishek Roy, Sajal K. Das, and Kalyan Basu. A predictive framework for location-aware resource management in smart homes. *IEEE Transactions on Mobile Computing*, 6(11):1270–1283, 2007.

[22] A. Roy, S. K. Das Bhaumik, A. Bhattacharya, K. Basu, D. J. Cook, and S. K. Das. Location aware resource management in smart homes. *Proceedings of the First IEEE International Conference on Pervasive Computing and Communications, 2003. (PerCom 2003)*, pages 481–488, 2003.

[23] Silviu Nistor, Jianzhong Wu, Mahesh Sooriyabandara, and Janaka Ekanayake. Cost optimization of smart appliances. In *IEEE PES Innovative Smart Grid Technologies Conference Europe*, 2011.

[24] Marko Jurmu, Mikko Perttunen, and Jukka Riekki. Lease-Based Resource Management in Smart Spaces. *Fifth Annual IEEE International Conference on Pervasive Computing and Communications Workshops (PerComW'07)*, pages 622–626, 2007.

[25] Ji Yeon Son, Jun Hee Park, Kyeong Deok Moon, and Young Hee Lee. Resource-aware smart home management system by constructing resource relation graph. *IEEE Transactions on Consumer Electronics*, 57(3):1112–1119, 2011.

[26] Suha Alawadhi, Armando Aldama-Nalda, Hafedh Chourabi, J. Ramon Gil-Garcia, Sofia Leung, Sehl Mellouli, Taewoo Nam, Theresa A. Pardo, Hans J. Scholl, and Shawn Walker. Building understanding of smart city initiatives. In *Lecture Notes in Computer Science (including subseries Lecture Notes in Artificial Intelligence and Lecture Notes in Bioinformatics)*, volume 7443 LNCS, pages 40–53, 2012.

[27] Rodger Lea and Michael Blackstock. Smart Cities. *Proceedings of the 2014 International Workshop on Web Intelligence and Smart Sensing - IWWISS '14*, pages 1–2, 2014.

[28] G. Kortuem, F. Kawsar, D. Fitton, and V. Sundramoorthy. Smart objects as building blocks for the Internet of Things. *Internet Computing, IEEE*, 14(1):44–51, 2010.

[29] Terence K.L. Hui, R. Simon Sherratt, and Daniel Díaz Sánchez. Major requirements for building smart homes in smart cities based on Internet of Things technologies. *Future Generation Computer Systems*, 2016.

[30] Robert H. Dodier, Gregor P. Henze, Dale K. Tiller, and Xin Guo. Building occupancy detection through sensor belief networks. *Energy and Buildings*, 38(9):1033–1043, 2006.

[31] Tobore Ekwevugbe, Neil Brown, and Denis Fan. A design model for building occupancy detection using sensor fusion. *2012 6th IEEE International Conference on Digital Ecosystems and Technologies (DEST)*, pages 1–6, June 2012.

[32] Timilehin Labeodan, Wim Zeiler, Gert Boxem, and Yang Zhao. Occupancy measurement in commercial office buildings for demand-driven control applications - A survey and detection system evaluation. *Energy and Buildings*, 93, pages 303–314, 2015.

[33] Zhenyu Han, Robert X. Gao, and Zhaoyan Fan. Occupancy and indoor environment quality sensing for smart buildings. In *2012 IEEE I2MTC - International Instrumentation and Measurement Technology Conference, Proceedings*, pages 882–887, 2012.

[34] Pravin Pawar, Val Jones, Bert Jan F. van Beijnum, and Hermie Hermens. A framework for the comparison of mobile patient monitoring systems. *Journal of Biomedical Informatics*, 45(3):544–556, 2012.

[35] Sweta Sneha and Upkar Varshney. Enabling ubiquitous patient monitoring: Model, decision protocols, opportunities and challenges. *Decision Support Systems*, 46(3):606–619, 2009.

[36] Alexsis Bell, Paul Rogers, Chris Farnell, Brett Sparkman, and Scott C. Smith. Wireless patient monitoring system. In *2014 IEEE Healthcare Innovation Conference, HIC 2014*, pages 149–152, 2014.

[37] K. Wac, R. Bults, B. Van Beijnum, I. Widya, V. M. Jones, D. Konstantas, M. Vollenbroek-Hutten, and H. Hermens. Mobile patient monitoring: The MobiHealth system. In *Proceedings of the 31st Annual International Conference of the IEEE Engineering in Medicine and Biology Society: Engineering the Future of Biomedicine, EMBC 2009*, pages 1238–1241, 2009.

[38] Yuan Hsiang Lin, I. Chien Jan, Patrick Chow In Ko, Yen Yu Chen, Jau Min Wong, and Gwo Jen Jan. A wireless PDA-based physiological monitoring system for patient transport.

IEEE Transactions on Information Technology in Biomedicine, 8(4):439–447, 2004.

[39] Upkar Varshney. A framework for supporting emergency messages in wireless patient monitoring. *Decision Support Systems*, 45(4):981–996, 2008.

[40] Upkar Varshney. Managing comprehensive wireless patient monitoring. In *2006 Pervasive Health Conference and Workshops, PervasiveHealth*, 2007.

[41] Tony Gentry. Smart homes for people with neurological disability: State of the art. *NeuroRehabilitation*, 25(3):209–17, 2009.

[42] Ali Hussein, Mehdi Adda, Mirna Atieh, and Walid Fahs. Smart home design for disabled people based on neural networks. In *Procedia Computer Science*, 37:117–126, 2014.

[43] Rachid Kadouche, Mounir Mokhtari, Sylvain Giroux, and Bessam Abdulrazak. Personalization in smart homes for disabled people. In *Proceedings of the 2008 2nd International Conference on Future Generation Communication and Networking, FGCN 2008*, volume 2, pages 411–415, 2008.

[44] Abdellah Chehri, Paul Fortier, and Pierre Martin Tardif. Application of Ad-hoc sensor networks for localization in underground mines. In *IEEE Wireless and Microwave Technology Conference, WAMICON 2006*, 2006.

[45] Shehadi Dayekh, Sofiène Affes, Nahi Kandil, and Chahé Nerguizian. Cooperative localization in mines using fingerprinting and neural networks. In *IEEE Wireless Communications and Networking Conference, WCNC*, 2010.

[46] Shehadi Dayekh, Sofiène Affes, Nahi Kandil, and Chahé Nerguizian. Cooperative geo-location in underground mines: A novel fingerprint positioning technique exploiting spatio-temporal diversity. In *IEEE International Symposium on Personal, Indoor and Mobile Radio Communications, PIMRC*, pages 1319–1324, 2011.

[47] Shehadi Dayekh, Sofiène Affes, Nahi Kandil, and Chahe Nerguizian. Radio-localization in underground narrow-vein mines using neural networks with in-built tracking and time diversity.

In *2011 IEEE Wireless Communications and Networking Conference, WCNC 2011*, pages 1788–1793, 2011.

[48] Zhu Daixian and Yi Kechu. Particle filter localization in underground mines using UWB ranging. In *Proceedings - 4th International Conference on Intelligent Computation Technology and Automation, ICICTA 2011*, volume 2, pages 645–648, 2011.

[49] Zhongmin Pei, Zhidong Deng, Shuo Xu, and Xiao Xu. Anchor-Free Localization Method for Mobile Targets in Coal Mine Wireless Sensor Networks. *Sensors (Basel, Switzerland)*, 9(4), pages 2836–2850, 2009.

[50] Daixian Zhu and Kechu Yi. EKF localization based on TDOA/RSS in underground mines using UWB ranging. In *2011 IEEE International Conference on Signal Processing, Communications and Computing, ICSPCC 2011*, 2011.

[51] V. Hon, M. Dostál, and Z. Slanina. The use of RFID technology in mines for identification and localization of persons. In *IFAC Proceedings Volumes (IFAC-PapersOnline)*, volume 10, pages 236–241, 2010.

[52] Wang Xiaodong, Zhao Xiaoguang, Liang Zize, and Tan Min. Deploying a Wireless Sensor Network on the Coal Mines. *2007 IEEE International Conference on Networking, Sensing and Control*, pages 324–328, Apr 2007.

[53] Linus Thrybom, Jonas Neander, Ewa Hansen, and Krister Landerns. Future challenges of positioning in underground mines. In *IFAC-PapersOnLine*, volume 28, pages 222–226, 2015.

[54] Roman Pflugfelder and Horst Bischof. Localization and trajectory reconstruction in surveillance cameras with nonoverlapping views. *IEEE Transactions on Pattern Analysis and Machine Intelligence*, 32(4):709–721, 2010.

[55] Morgan Quigley, Michael A. Goodrich, Stephen Griffiths, Andrew Eldredge, and Randal W. Beard. Target acquisition, localization, and surveillance using a fixed-wing mini-UAV and gimbaled camera. In *Proceedings - IEEE International Conference on Robotics and Automation*, volume 2005, pages 2600–2606, 2005.

[56] Wee Kheng Leow. Localization and mapping of surveillance cameras. *Proceedings of the 16th ACM international conference on Multimedia (MM '08)*, pages 369–378, 2008.

[57] Dongsoo Han, Byeongcheol Moon, and Giwan Yoon. Address-based crowdsourcing radio map construction for Wi-Fi positioning systems. *IPIN 2014 - 2014 International Conference on Indoor Positioning and Indoor Navigation*, pages 58–67, 2014.

[58] D. P. Dogra, A. Ahmed, and H. Bhaskar. Interest area localization using trajectory analysis in surveillance scenes. *VISAPP 2015 - 10th International Conference on Computer Vision Theory and Applications; VISIGRAPP, Proceedings*, volume 2, pages 478–485, Mar 2015.

[59] Aran Hampapur, Lisa Brown, Jonathan Connell, Sharat Pankanti, Andrew Senior, and Yingli Tian. Smart surveillance: Applications, technologies and implications. In *ICICS-PCM 2003 - Proceedings of the 2003 Joint Conference of the 4th International Conference on Information, Communications and Signal Processing and 4th Pacific-Rim Conference on Multimedia*, volume 2, pages 1133–1138, 2003.

[60] Milo Spadacini, Stefano Savazzi, and Monica Nicoli. Wireless home automation networks for indoor surveillance: Technologies and experiments. *EURASIP Journal on Wireless Communications and Networking*, 1(6):1–17, 2014.

[61] Athanasios Tsitsoulis, Ryan Patrick, and Nikolaos Bourbakis. Surveillance issues in a smart home environment. In *Proceedings - International Conference on Tools with Artificial Intelligence, ICTAI*, volume 2, pages 18–25, 2012.

[62] Guoqiang Mao, Barış Fidan, and Brian D.O. Anderson. Wireless sensor network localization techniques. *Computer Networks*, 51(10):2529–2553, 2007.

[63] H. Liu, H. Darabi, P. Banerjee, and J. Liu. Survey of wireless indoor positioning techniques and systems. *IEEE Transactions on Systems, Man, and Cybernetics, Part C (Applications and Reviews)*, 37(6):1067–1080, Nov 2007.

[64] Y. Gu, A. Lo, and I. Niemegeers. A survey of indoor positioning systems for wireless personal networks. *IEEE Communications Surveys Tutorials*, 11(1):13–32, First 2009.

[65] Zahid Farid, Rosdiadee Nordin, and Mahamod Ismail. Recent advances in wireless indoor localization techniques and system. *Journal of Computer Networks and Communications*, 2013, 2013.

[66] L. Mainetti, L. Patrono, and I. Sergi. A survey on indoor positioning systems. In *2014 22nd International Conference on Software, Telecommunications and Computer Networks (SoftCOM)*, pages 111–120, Sep 2014.

[67] N. Iliev and I. Paprotny. Review and comparison of spatial localization methods for low-power wireless sensor networks. *IEEE Sensors Journal*, 15(10):5971–5987, Oct 2015.

[68] F. Dwiyasa and M. H. Lim. A survey of problems and approaches in wireless-based indoor positioning. In *2016 International Conference on Indoor Positioning and Indoor Navigation (IPIN)*, pages 1–7, Oct 2016.

[69] S. Lanzisera, D. T. Lin, and K. S. J. Pister. RF Time of Flight Ranging for Wireless Sensor Network Localization. *2006 International Workshop on Intelligent Solutions in Embedded Systems*, 2006.

[70] Stuart A. Golden and Steve S. Bateman. Sensor measurements for Wi-Fi location with emphasis on time-of-arrival ranging. *IEEE Transactions on Mobile Computing*, 6(10):1185–1198, 2007.

[71] Jenni Wennervirta and Torbjrn Wigren. RTT positioning field performance. *IEEE Transactions on Vehicular Technology*, 59(7):3656–3661, 2010.

[72] Soo Yong Jung, Swook Hann, and Chang Soo Park. TDOA-based optical wireless indoor localization using LED ceiling lamps. *IEEE Transactions on Consumer Electronics*, 57(4):1592–1597, 2011.

[73] Stefan Schwalowsky, Henning Trsek, Reinhard Exel, and Nikolaus Kerö. System integration of an IEEE 802.11 based TDoA localization system. In *ISPCS 2010 - 2010 International IEEE Symposium on Precision Clock Synchronization for Measurement, Control and Communication, Proceedings*, pages 55–60, 2010.

[74] R. Levorato and E. Pagello. DOA Acoustic Source Localization in Mobile Robot Sensor Networks. *2015 IEEE International Conference on Autonomous Robot Systems and Competitions (ICARSC)*, pages 71–76, 2015.

[75] Hui Tian, Shuang Wang, and Huaiyao Xie. Localization using cooperative AOA approach. In *2007 International Conference on Wireless Communications, Networking and Mobile Computing, WiCOM 2007*, pages 2416–2419, 2007.

[76] Jun Xu, Ode Ma, and Choi Look Law. AOA cooperative position localization. In *GLOBECOM - IEEE Global Telecommunications Conference*, pages 3751–3755, 2008.

[77] Junru Zhou, Hongjian Zhang, and Lingfei Mo. Two-dimension localization of passive RFID tags using AOA estimation. In *Conference Record - IEEE Instrumentation and Measurement Technology Conference*, pages 511–515, 2011.

[78] J. M. Engelbrecht, R. Weber, and O. Michler. Reduction of multipath propagation influences at poa positioning using uniform circular array antennas analyses based on measurements in vehicular scenarios. In *2016 13th Workshop on Positioning, Navigation and Communications (WPNC)*, pages 1–5, Oct 2016.

[79] Martin Scherhaufl, Markus Pichler, Dominik Muller, Andreas Ziroff, and Andreas Stelzer. Phase-of-arrival-based localization of passive UHF RFID tags. *2013 IEEE MTT-S International Microwave Symposium Digest (MTT)*, pages 1–3, 2013.

[80] Ales Povalac and Jiri Sebesta. Phase difference of arrival distance estimation for RFID tags in frequency domain. In *2011 IEEE International Conference on RFID-Technologies and Applications, RFID-TA 2011*, pages 188–193, 2011.

[81] Xiuyan Zhu and Yuan Feng. RSSI-based Algorithm for Indoor Localization. *Communications and Network*, 05(02):37–42, 2013.

[82] M. O. Gani, C. OBrien, S. I. Ahamed, and R. O. Smith. RSSI Based Indoor Localization for Smartphone Using Fixed and Mobile Wireless Node. *Computer Software and Applications Conference (COMPSAC), 2013 IEEE 37th Annual*, pages 110–117, 2013.

[83] Ramsey Faragher and Andrew Rice. SwiftScan: Efficient Wi-Fi scanning for background location-based services. *2015 International Conference on Indoor Positioning and Indoor Navigation, IPIN 2015*, 2015.

[84] Li Zhang, Xiao Liu, Jie Song, Cathal Gurrin, and Zhiliang Zhu. A comprehensive study of bluetooth fingerprinting-based algorithms for localization. In *Proceedings - 27th International Conference on Advanced Information Networking and Applications Workshops, WAINA 2013*, pages 300–305, 2013.

[85] Fazli Subhan, Halabi Hasbullah, Azat Rozyyev, and Sheikh Tahir Bakhsh. Indoor positioning in Bluetooth networks using fingerprinting and lateration approach. In *2011 International Conference on Information Science and Applications, ICISA 2011*, 2011.

[86] C. Héctor José Pérez Iglesias, Valentín Barral. Indoor Person Localization System Through Rssi Bluetooth Fingerprinting. *Iwssip 2012*, pages 11–13, Apr 2012.

[87] Y. B. Bai, R. Norman, Yang Zhao, Shuangxia Tang, Suqin Wu, Hongren Wu, G. Retscher, and Kefei Zhang. A new algorithm for improving the tracking and positioning of cell of origin. In *2015 International Association of Institutes of Navigation World Congress (IAIN)*, pages 1–6, Oct 2015.

[88] Kaishun Wu, Jiang Xiao, Youwen Yi, Dihu Chen, Xiaonan Luo, and Lionel M. Ni. CSI-based indoor localization. *IEEE Transactions on Parallel and Distributed Systems*, 24(7):1300–1309, 2013.

[89] Zheng Yang, Zimu Zhou, and Yunhao Liu. From RSSI to CSI: Indoor Localization via Channel Response. *ACM Computing Surveys (CSUR)*, 46(2):25, 2013.

[90] L. Schauer, F. Dorfmeister, and M. Maier. Potentials and limitations of wifi-positioning using time-of-flight. In *International Conference on Indoor Positioning and Indoor Navigation*, pages 1–9, Oct 2013.

[91] Y. Yang, Y. Zhao, and M. Kyas. A grid-scan maximum likelihood estimation with a bias function for indoor network localization. In *International Conference on Indoor Positioning and Indoor Navigation*, pages 1–9, Oct 2013.

[92] M. Ciurana, D. Giustiniano, A. Neira, F. Barcelo-Arroyo, and I. Martin-Escalona. Performance stability of software toa-based ranging in wlan. In *2010 International Conference on Indoor Positioning and Indoor Navigation*, pages 1–8, Sep 2010.

[93] J. Prieto, S. Mazuelas, A. Bahillo, P. Fernandez, R. M. Lorenzo, and E. J. Abril. On the minimization of different sources of error for an rtt-based indoor localization system without any calibration stage. In *2010 International Conference on Indoor Positioning and Indoor Navigation*, pages 1–6, Sep 2010.

[94] Johannes Wendeberg, Fabian Höflinger, Christian Schindelhauer, and Leonard Reindl. Anchor-free TDOA self-localization. *2011 International Conference on Indoor Positioning and Indoor Navigation, IPIN 2011*, 2011.

[95] Christian Gentner and Thomas Jost. Indoor positioning using time difference of arrival between multipath components. *International Conference on Indoor Positioning and Indoor Navigation*, pages 1–10, Oct 2013.

[96] Stijn Wielandt, Jean Pierre Goemaere, Bart Nauwelaers, and Lieven De Strycker. Study and Simulations of an Angle of Arrival Localization System for Indoor Multipath Environments. *Proceedings of the 11th International Symposium on Location-Based Services*, pages 203–211, Nov 2014.

[97] P. Mazurkiewicz, A. Gkelias, and K. K. Leung. Linear antenna array, ranging and accelerometer for 3d gps-less localization of wireless sensors. In *2010 International Conference on Indoor Positioning and Indoor Navigation*, pages 1–5, Sep 2010.

[98] Farhan Zaki and Rashid Rashidzadeh. An Indoor Location Positioning Algorithm for Portable Devices and Autonomous Machines. pages 4–7, Oct 2016.

[99] You Li, Peng Zhang, Haiyu Lan, Yuan Zhuang, Xiaoji Niu, and Naser El-Sheimy. A modularized real-time indoor navigation algorithm on smartphones. *2015 International Conference on Indoor Positioning and Indoor Navigation, IPIN 2015*, pages 13–16, Oct 2015.

[100] C. Gao and R. Harle. Easing the survey burden: Quantitative assessment of low-cost signal surveys for indoor positioning. In *2016 International Conference on Indoor Positioning and Indoor Navigation (IPIN)*, pages 1–8, Oct 2016.

[101] J. Torres-Sospedra, G. M. Mendoza-Silva, R. Montoliu, O. Belmonte, F. Benitez, and J. Huerta. Ensembles of indoor positioning systems based on fingerprinting: Simplifying parameter selection and obtaining robust systems. In *2016 International Conference on Indoor Positioning and Indoor Navigation (IPIN)*, pages 1–8, Oct 2016.

[102] Rafael Berkvens, Maarten Weyn, and Herbert Peremans. Localization performance quantification by conditional entropy. *2015 International Conference on Indoor Positioning and Indoor Navigation, IPIN 2015*, pages 13–16, Oct 2015.

[103] Arsham Farshad, Jiwei Li, Mahesh K. Marina, and Francisco J. Garcia. A microscopic look at WiFi fingerprinting for indoor mobile phone localization in diverse environments. *2013 International Conference on Indoor Positioning and Indoor Navigation, IPIN 2013*, pages 28–31, Oct 2013.

[104] J. Machaj and P. Brida. Optimization of rank based fingerprinting localization algorithm. In *2012 International Conference on Indoor Positioning and Indoor Navigation (IPIN)*, pages 1–7, Nov 2012.

[105] C. Nielsen, J. Nielsen, and V. Dehghanian. Fusion of security camera and rss fingerprinting for indoor multi-person tracking. In *2016 International Conference on Indoor Positioning and Indoor Navigation (IPIN)*, pages 1–7, Oct 2016.

[106] V. C. Ta, D. Vaufreydaz, T. K. Dao, and E. Castelli. Smartphone-based user location tracking in indoor environment. In *2016 International Conference on Indoor Positioning and Indoor Navigation (IPIN)*, pages 1–8, Oct 2016.

[107] A. Moreira, M. J. Nicolau, A. Costa, and F. Meneses. Indoor tracking from multidimensional sensor data. In *2016 International Conference on Indoor Positioning and Indoor Navigation (IPIN)*, pages 1–8, Oct 2016.

[108] P. Wilk, J. Karciarz, and J. Swiatek. Indoor radio map maintenance by automatic annotation of crowdsourced wi-fi fingerprints. In *2015 International Conference on Indoor Positioning and Indoor Navigation (IPIN)*, pages 1–8, Oct 2015.

[109] F. Lemic, A. Behboodi, V. Handziski, and A. Wolisz. Experimental decomposition of the performance of fingerprinting-based localization algorithms. In *2014 International Conference on Indoor Positioning and Indoor Navigation (IPIN)*, pages 355–364, Oct 2014.

[110] J. Torres-Sospedra, R. Montoliu, A. Martínez-Usó, J. P. Avariento, T. J. Arnau, M. Benedito-Bordonau, and J. Huerta. Ujiindoorloc: A new multi-building and multi-floor database for wlan fingerprint-based indoor localization problems. In *2014 International Conference on Indoor Positioning and Indoor Navigation (IPIN)*, pages 261–270, Oct 2014.

[111] C. Laoudias, M. P. Michaelides, and C. G. Panayiotou. Fault detection and mitigation in wlan rss fingerprint-based positioning. In *2011 International Conference on Indoor Positioning and Indoor Navigation*, pages 1–7, Sep 2011.

[112] S. Garcia-Villalonga and A. Perez-Navarro. Influence of human absorption of wi-fi signal in indoor positioning with wi-fi fingerprinting. In *2015 International Conference on Indoor Positioning and Indoor Navigation (IPIN)*, pages 1–10, Oct 2015.

[113] S. Eisa, J. Peixoto, F. Meneses, and A. Moreira. Removing useless aps and fingerprints from wifi indoor positioning radio maps. In *International Conference on Indoor Positioning and Indoor Navigation*, pages 1–7, Oct 2013.

[114] G. Lui, T. Gallagher, B. Li, A. G. Dempster, and C. Rizos. Differences in rssi readings made by different wi-fi chipsets: A limitation of wlan localization. In *2011 International Conference on Localization and GNSS (ICL-GNSS)*, pages 53–57, June 2011.

[115] A. Bahillo, S. Mazuelas, J. Prieto, P. Fernandez, R.M. Lorenzo, and E.J. Abril. Hybrid RSS-RTT localization scheme for wireless networks. In *2010 International Conference on Indoor Positioning and Indoor Navigation*, pages 1–7. IEEE, Sep 2010.

[116] Olga E. Segou, Stelios A. Mitilineos, and Stelios C.A. Thomopoulos. DALE: A range-free, adaptive indoor localization method enhanced by limited fingerprinting. In *2010 International Conference on Indoor Positioning and Indoor Navigation*, pages 1–8. IEEE, Sep 2010.

[117] Senka Hadzic and Jonathan Rodriguez. Utility based node selection scheme for cooperative localization. In *2011 International Conference on Indoor Positioning and Indoor Navigation*, pages 1–6. IEEE, Sep 2011.

[118] Luigi Bruno and Patrick Robertson. WiSLAM: Improving FootSLAM with WiFi. *2011 International Conference on Indoor Positioning and Indoor Navigation, IPIN 2011*, 2011.

[119] Nyein Aye Maung Maung and Makoto Kawai. Hybrid RSS-SOM localization scheme for wireless ad hoc and sensor networks. In *2012 International Conference on Indoor Positioning and Indoor Navigation (IPIN)*, pages 1–7. IEEE, Nov 2012.

[120] Gayan Attanayake and Yue Rong. RSS-based indoor positioning accuracy improvement using antenna array in WLAN environments. In *2012 International Conference on Indoor Positioning and Indoor Navigation (IPIN)*, pages 1–6. IEEE, Nov 2012.

[121] Tatsuya Iwase and Ryosuke Shibasaki. Infra-free indoor positioning using only smartphone sensors. In *International Conference on Indoor Positioning and Indoor Navigation*, pages 1–8. IEEE, Oct 2013.

[122] Alejandro Correa, Marc Barcelo, Antoni Morell, and Jose Lopez Vicario. Indoor pedestrian tracking system exploiting multiple receivers on the body. In *2014 International Conference on Indoor Positioning and Indoor Navigation (IPIN)*, pages 518–525. IEEE, Oct 2014.

[123] Peerapong Torteeka, Xiu Chundi, and Yang Dongkai. Hybrid technique for indoor positioning system based on Wi-Fi received signal strength indication. In *2014 International Conference on Indoor Positioning and Indoor Navigation (IPIN)*, pages 48–57. IEEE, Oct 2014.

[124] Yuehua Cai, Suleman Khalid Rai, and Hao Yu. Indoor positioning by distributed machine-learning based data analytics on smart gateway network. In *2015 International Conference on Indoor Positioning and Indoor Navigation (IPIN)*, pages 1–8. IEEE, Oct 2015.

[125] Jose Luis Carrera, Zhongliang Zhao, Torsten Braun, and Zan Li. A real-time indoor tracking system by fusing inertial sensor, radio signal and floor plan. In *2016 International Conference on Indoor Positioning and Indoor Navigation (IPIN)*, pages 1–8. IEEE, Oct 2016.

[126] Hiroaki Togashi, Hiroshi Furukawa, Yuki Yamaguchi, Ryuta Abe, and Junpei Shimamura. Network-based positioning and pedestrian flow measurement system utilizing densely placed wireless access points. In *2016 International Conference on Indoor Positioning and Indoor Navigation (IPIN)*, pages 1–8. IEEE, Oct 2016.

[127] Simon Burgess, Kalle Astrom, Mikael Hogstrom, and Bjorn Lindquist. Smartphone positioning in multi-floor environments without calibration or added infrastructure. In *2016 International Conference on Indoor Positioning and Indoor Navigation (IPIN)*, pages 1–8. IEEE, Oct 2016.

[128] Jaehyun Yoo, H. Jin Kim, and Karl H. Johansson. Mapless indoor localization by trajectory learning from a crowd. In *2016 International Conference on Indoor Positioning and Indoor Navigation (IPIN)*, pages 1–7. IEEE, Oct 2016.

[129] Hongbo Liu, Yu Gan, Jie Yang, Simon Sidhom, Yan Wang, Yingying Chen, and Fan Ye. Push the limit of WiFi based localization for smartphones. *Proceedings of the 18th annual international conference on Mobile computing and networking (Mobicom '12)*, page 305, 2012.

[130] Sinno Jialin Pan, Sinno Jialin Pan, Vincent Wenchen Zheng, Vincent Wenchen Zheng, Qiang Yang, Qiang Yang, Derek Hao Hu, and Derek Hao Hu. Transfer Learning for WiFi-based Indoor Localization. *Intelligence*, pages 43–48, 2008.

[131] Chin Heng Lim, Yahong Wan, Boon Poh Ng, and Chong Meng Samson See. A real-time indoor WiFi localization system

utilizing smart antennas. *IEEE Transactions on Consumer Electronics*, 53(2):618–622, 2007.

[132] Brian Roberts and Kaveh Pahlavan. Site-specific RSS signature modeling for WiFi localization. In *GLOBECOM - IEEE Global Telecommunications Conference*, 2009.

[133] Hongbo Liu, Jie Yang, Simon Sidhom, Yan Wang, Yingying Chen, and Fan Ye. Accurate WiFi based localization for smartphones using peer assistance. *IEEE Transactions on Mobile Computing*, 13(10):2199–2214, 2014.

[134] Guimar Aes. WiFi Localization as a Network Service. In *International Conference on Indoor Positioning and Indoor Navigation*, pages 21–23, Sep 2011.

[135] Timea Bagosi and Zoltan Baruch. Indoor localization by WiFi. In *Proceedings - 2011 IEEE 7th International Conference on Intelligent Computer Communication and Processing, ICCP 2011*, pages 449–452, 2011.

[136] Sujittra Boonsriwai and Anya Apavatjrut. Indoor WIFI localization on mobile devices. In *2013 10th International Conference on Electrical Engineering/Electronics, Computer, Telecommunications and Information Technology, ECTI-CON 2013*, 2013.

[137] Mohd Nizam Husen and Sukhan Lee. Indoor human localization with orientation using WiFi fingerprinting. *Proceedings of the 8th International Conference on Ubiquitous Information Management and Communication - ICUIMC '14*, pages 1–6, 2014.

[138] Chouchang Yang and Huai Rong Shao. WiFi-based indoor positioning. *IEEE Communications Magazine*, 53(3):150–157, 2015.

[139] Qiuxia Chen, Dik Lun Lee, and Wang Chien Lee. Rule-based WiFi localization methods. In *Proceedings of The 5th International Conference on Embedded and Ubiquitous Computing, EUC 2008*, volume 1, pages 252–258, 2008.

[140] Jianwei Niu, Banghui Lu, Long Cheng, Yu Gu, and Lei Shu. ZiLoc: Energy efficient WiFi fingerprint-based localization with low-power radio. In *IEEE Wireless Communications and Networking Conference, WCNC*, pages 4558–4563, 2013.

[141] Liang Chen, Heidi Kuusniemi, Yuwei Chen, Ling Pei, Tuomo Kroger, and Ruizhi Chen. Information filter with speed detection for indoor Bluetooth positioning. In *2011 International Conference on Localization and GNSS (ICL-GNSS)*, pages 47–52. IEEE, June 2011.

[142] Zhu Jianyong, Luo Haiyong, Chen Zili, and Li Zhaohui. RSSI based Bluetooth low energy indoor positioning. In *2014 International Conference on Indoor Positioning and Indoor Navigation (IPIN)*, pages 526–533. IEEE, Oct 2014.

[143] Vivek Chandel, Nasimuddin Ahmed, Shalini Arora, and Avik Ghose. InLoc: An end-to-end robust indoor localization and routing solution using mobile phones and BLE beacons. In *2016 International Conference on Indoor Positioning and Indoor Navigation (IPIN)*, pages 1–8. IEEE, Oct 2016.

[144] Patrick Dickinson, Gregorz Cielniak, Olivier Szymanezyk, and Mike Mannion. Indoor positioning of shoppers using a network of Bluetooth Low Energy beacons. In *2016 International Conference on Indoor Positioning and Indoor Navigation (IPIN)*, pages 1–8. IEEE, Oct 2016.

[145] auGrigorios G. Anagnostopoulos, AuMichel Deriaz, and AuDimitri Konstantas. Online self-calibration of the propagation model for indoor positioning ranging methods. In *2016 International Conference on Indoor Positioning and Indoor Navigation (IPIN)*, pages 1–6. IEEE, Oct 2016.

[146] Stijn Crul, Geoffrey Ottoy, and Lieven De Strycker. Location awareness enables autonomous commissioning in wireless sensor networks. In *2016 International Conference on Indoor Positioning and Indoor Navigation (IPIN)*, pages 1–6. IEEE, Oct 2016.

[147] Athanasios I. Kyritsis, Panagiotis Kostopoulos, Michel Deriaz, and Dimitri Konstantas. A BLE-based probabilistic room-level localization method. *Proceedings of 2016 International Conference on Localization and GNSS, ICL-GNSS 2016*, 2016.

[148] Pekka Peltola, Chris Hill, and Terry Moore. Particle filter for context sensitive indoor pedestrian navigation. *Proceedings of 2016 International Conference on Localization and GNSS, ICL-GNSS 2016*, 2016.

[149] Adel Thaljaoui, Thierry Val, Nejah Nasri, and Damien Brulin. BLE localization using RSSI measurements and iRingLA. In *Proceedings of the IEEE International Conference on Industrial Technology*, volume 2015, pages 2178–2183, June 2015.

[150] Meera Radhakrishnan, Archan Misra, Rajesh Krishna Balan, and Youngki Lee. Smartphones and BLE services: Empirical insights. In *Proceedings - 2015 IEEE 12th International Conference on Mobile Ad Hoc and Sensor Systems, MASS 2015*, pages 226–234, 2015.

[151] Yuan Zhuang, Jun Yang, You Li, Longning Qi, and Naser El-Sheimy. Smartphone-based indoor localization with bluetooth low energy beacons. *Sensors (Switzerland)*, 16(5), 2016.

[152] Patrick Lazik and Oliver Shih. ALPS : A Bluetooth and Ultrasound Platform for Mapping and Localization. *ACM SenSys*, pages 73–84, 2015.

[153] Myungin Ji, Jooyoung Kim, Juil Jeon, and Youngsu Cho. Analysis of positioning accuracy corresponding to the number of BLE beacons in indoor positioning system. In *International Conference on Advanced Communication Technology, ICACT*, volume 2015, pages 92–95, Aug 2015.

[154] Filippo Palumbo, Paolo Barsocchi, Stefano Chessa, and Juan Carlos Augusto. A stigmergic approach to indoor localization using Bluetooth Low Energy beacons. *2015 12th IEEE International Conference on Advanced Video and Signal Based Surveillance (AVSS)*, pages 1–6, 2015.

[155] Paul Martin, Bo-Jhang Ho, Nicholas Grupen, Samuel Muñoz, and Mani Srivastava. An iBeacon primer for indoor localization. *Proceedings of the 1st ACM Conference on Embedded Systems for Energy-Efficient Buildings - BuildSys '14*, pages 190–191, 2014.

[156] Zhenghua Chen, Qingchang Zhu, Hao Jiang, and Yeng Chai Soh. Indoor localization using smartphone sensors and iBeacons. In *2015 IEEE 10th Conference on Industrial Electronics and Applications (ICIEA)*, pages 1723–1728, 2015.

[157] Marco Altini, Davide Brunelli, Elisabetta Farella, and Luca Benini. Bluetooth indoor localization with multiple neural networks. *IEEE*

5th International Symposium on Wireless Pervasive Computing 2010, volume 2016, pages 295–300, Aug 2010.

[158] Mortaza S. Bargh and Robert de Groote. Indoor localization based on response rate of bluetooth inquiries. *Proceedings of the first ACM international workshop on Mobile entity localization and tracking in GPS-less environments*, volume 2016, pages 49–54, Nov 2008.

[159] Aswin N. Raghavan and Harini Ananthapadmanaban. Accurate Mobile Robot Localization in indoor environments using Bluetooth. *Intelligent Systems Engineering*, pages 4391–4396, 2010.

[160] S.S. Chawathe. Beacon Placement for Indoor Localization using Bluetooth. In *2008 11th International IEEE Conference on Intelligent Transportation Systems*, pages 980–985, 2008.

[161] Javier J. M. Diaz, Rodrigo De A. Maues, Rodrigo B. Soares, Eduardo F. Nakamura, and Carlos M. S. Figueiredo. Bluepass: An indoor Bluetooth-based localization system for mobile applications. In *Proceedings - IEEE Symposium on Computers and Communications*, pages 778–783, 2010.

[162] Jorge Juan Robles, Sebastián Tromer, Mónica Quiroga, and Ralf Lehnert. Enabling Low-power Localization for Mobile Sensor Nodes. In *IEEE International Conference on Indoor Positioning and Indoor Navigation (IPIN)*, pages 1–10, Sep 2010.

[163] Katrin Achutegui, Javier Rodas, Carlos J. Escudero, and Joaquin Miguez. A model-switching sequential Monte Carlo algorithm for indoor tracking with experimental RSS data. In *2010 International Conference on Indoor Positioning and Indoor Navigation*, pages 1–8. IEEE, Sep 2010.

[164] Johannes Schmid, Markus Volker, Tobias Gadeke, Pascal Weber, Wilhelm Stork, and K.D. Muller-Glaser. An approach to infrastructure-independent person localization with an IEEE 802.15.4 WSN. In *2010 International Conference on Indoor Positioning and Indoor Navigation*, pages 1–9. IEEE, Sep 2010.

[165] L. Bras, M. Oliveira, N. Borges De Carvalho, and P. Pinho. Low power location protocol based on ZigBee Wireless Sensor

Networks. In *International Conference on Indoor Positioning and Indoor Navigation IPIN 2010*, pages 15–17, Sep 2010.

[166] Gianni Giorgetti, Richard Farley, Kiran Chikkappa, Judy Ellis, and Telis Kaleas. Cortina: Collaborative indoor positioning using low-power sensor networks. In *2011 International Conference on Indoor Positioning and Indoor Navigation*, pages 1–10. IEEE, Sep 2011.

[167] Richard Weber, Uwe Gosda, Oliver Michler, and Julia Ringel. WSN-based passenger localization in severe NLOS environments using SDP. In *International Conference on Indoor Positioning and Indoor Navigation*, pages 1–7. IEEE, Oct 2013.

[168] Mathias Pelka, Christian Bollmeyer, and Horst Hellbruck. Accurate radio distance estimation by phase measurements with multiple frequencies. In *2014 International Conference on Indoor Positioning and Indoor Navigation (IPIN)*, pages 142–151. IEEE, Oct 2014.

[169] Alberto Rolando and Emanuele Amoruso. An ubiquitous positioning system based on IEEE 802.15.4 radio signals. In *International Conference on Indoor Positioning and Indoor Navigation*, pages 1–10. IEEE, Oct 2013.

[170] Julian Lategahn, Marcel Muller, and Christof Rohrig. Robust pedestrian localization in indoor environments with an IMU aided TDoA system. In *2014 International Conference on Indoor Positioning and Indoor Navigation (IPIN)*, pages 465–472. IEEE, Oct 2014.

[171] Jihoon Yang, Haeyoung Lee, and Klaus Moessner. Multilateration localization based on Singular Value Decomposition for 3D indoor positioning. In *2016 International Conference on Indoor Positioning and Indoor Navigation (IPIN)*, pages 1–8. IEEE, Oct 2016.

[172] Junpei Tsuji, Hidenori Kawamura, Keiji Suzuki, Takeshi Ikeda, Akio Sashima, and Koichi Kurumatani. ZigBee based indoor localization with particle filter estimation. In *Conference Proceedings - IEEE International Conference on Systems, Man and Cybernetics*, pages 1115–1120, 2010.

[173] Jianwei Niu, Bowei Wang, Lei Shu, Trung Q. Duong, and Yuanfang Chen. ZIL: An Energy-Efficient Indoor Localization System Using ZigBee Radio to Detect WiFi Fingerprints. *IEEE Journal on Selected Areas in Communications*, 33(7):1431–1442, 2015.

[174] Chul Young Park, Dae Heon Park, Jang Woo Park, Yang Sun Lee, and Youngeun An. Localization algorithm design and implementation to utilization RSSI and AOA of Zigbee. In *2010 5th International Conference on Future Information Technology, FutureTech 2010 - Proceedings*, 2010.

[175] Y. T. Chen and C. L. Yang. A RSSI-based algorithm for indoor localization using ZigBee in wireless sensor network. *International Journal of Digital Content Technology and its ApplicationsJournal of Digital Content Technology and its Applications*, 5(7):407–416, 2006.

[176] Xin Hu, Lianglun Cheng, and Guangchi Zhang. A Zigbee-based localization algorithm for indoor environments. In *Proceedings of 2011 International Conference on Computer Science and Network Technology, ICCSNT 2011*, volume 3, pages 1776–1781, 2011.

[177] Hyuntae Cho, Hyunsung Jang, and Yunju Baek. Practical localization system for consumer devices using Zigbee networks. *IEEE Transactions on Consumer Electronics*, 56(3):1562–1569, 2010.

[178] Wen Hsing Kuo, Yun Shen Chen, Gwei Tai Jen, and Tai Wei Lu. An intelligent positioning approach: RSSI-based indoor and outdoor localization scheme in Zigbee networks. In *2010 International Conference on Machine Learning and Cybernetics, ICMLC 2010*, volume 6, pages 2754–2759, 2010.

[179] A. Ropponen, M. Linnavuo, and R. Sepponen. Low-frequency localization and identification system with zigbee network. *International Journal on Smart Sensing and Intelligent Systems*, 4(1):75–93, 2011.

[180] Effat O. El Khashab and Hany F. Hammad. A size-reduced wearable antenna for Zigbee indoor localization. In *RWW 2012 - Proceedings: IEEE Radio and Wireless Symposium, RWS 2012*, pages 95–98, 2012.

[181] A. Schmidt, H. W. Gellersen, and C. Merz. Enabling implicit human computer interaction: A wearable rfid-tag reader. In *Digest of Papers. Fourth International Symposium on Wearable Computers*, pages 193–194, Oct 2000.

[182] M. M. Ollivier. Rfid-a new solution technology for security problems. In *European Convention on Security and Detection, 1995*, pages 234–238, May 1995.

[183] M. M. Kaleja, A. J. Herb, R. H. Rasshofer, G. Friedsam, and E. M. Biebl. Imaging rfid system at 24 ghz for object localization. In *1999 IEEE MTT-S International Microwave Symposium Digest (Cat. No.99CH36282)*, volume 4, pages 1497–1500, June 1999.

[184] N. Raza, V. Bradshaw, and M. Hague. Applications of rfid technology. In *IEE Colloquium on RFID Technology (Ref. No. 1999/123)*, pages 1/1–1/5, 1999.

[185] Vassilis Gikas, Andreas Dimitratos, Harris Perakis, Guenther Retscher, and Andreas Ettlinger. Full-scale testing and performance evaluation of an active RFID system for positioning and personal mobility. In *2016 International Conference on Indoor Positioning and Indoor Navigation (IPIN)*, pages 1–8. IEEE, Oct 2016.

[186] Michael Gunawan, Binghao Li, Thomas Gallagher, Andrew G. Dempster, and Gunther Retscher. A new method to generate and maintain a WiFi fingerprinting database automatically by using RFID. *2012 International Conference on Indoor Positioning and Indoor Navigation, IPIN 2012 - Conference Proceedings*, pages 1–6, Nov 2012.

[187] E. Nakamori, D. Tsukuda, M. Fujimoto, Y. Oda, T. Wada, H. Okada, and K. Mutsuura. A new indoor position estimation method of RFID tags for continuous moving navigation systems. In *2012 International Conference on Indoor Positioning and Indoor Navigation (IPIN)*, pages 1–8, Nov 2012.

[188] M. Fujimoto, N. Uchitomi, A. Inada, T. Wada, K. Mutsuura, and H. Okada. A novel method for position estimation of passive rfid tags; swift communication range recognition (s-crr) method. In *2010 IEEE Global Telecommunications Conference GLOBECOM 2010*, pages 1–6, Dec 2010.

[189] Fernando Seco, Antonio R. Jimnez, and Francisco Zampella. Joint estimation of indoor position and orientation from RF signal strength measurements. In *2013 International Conference on Indoor Positioning and Indoor Navigation, IPIN 2013*, pages 28–31, 2013.

[190] Mohammad Mostafa Soltani, Ali Motamedi, and Amin Hammad. Enhancing Cluster-based RFID Tag Localization using artificial neural networks and virtual reference tags. In *International Conference on Indoor Positioning and Indoor Navigation*, pages 1–10, Oct 2013.

[191] Han Zou, Lihua Xie, Qing-Shan Jia, and Hengtao Wang. An integrative Weighted Path Loss and Extreme Learning Machine approach to Rfid based Indoor Positioning. In *International Conference on Indoor Positioning and Indoor Navigation*, pages 1–5. IEEE, Oct 2013.

[192] Theresa Nick, Sebastian Cordes, Jurgen Gotze, and Werner John. Camera-assisted localization of passive RFID labels. In *2012 International Conference on Indoor Positioning and Indoor Navigation, IPIN 2012 - Conference Proceedings*, pages 13–15, Nov 2012.

[193] Francisco Zampella, Antonio R. Jimnez R., and Fernando Seco. Robust indoor positioning fusing PDR and RF technologies: The RFID and UWB case. In *2013 International Conference on Indoor Positioning and Indoor Navigation, IPIN 2013*, pages 28–31, Oct 2013.

[194] Elena Simona Lohan, Karoliina Koski, Jukka Talvitie, and Leena Ukkonen. WLAN and RFID Propagation channels for hybrid indoor positioning. In *International Conference on Localization and GNSS 2014 (ICL-GNSS 2014)*, pages 1–6. IEEE, June 2014.

[195] Fernando Seco, Antonio R. Jimenez, and Xufei Zheng. RFID-based centralized cooperative localization in indoor environments. In *2016 International Conference on Indoor Positioning and Indoor Navigation (IPIN)*, volume 1, pages 1–7. IEEE, Oct 2016.

[196] Sho Tatsukawa, Tadashi Nakanishi, Ryo Nagao, Tomotaka Wada, Manato Fujimoto, and Kouichi Mutsuura. New moving control of

mobile robot without collision with wall and obstacles by passive RFID system. In *2015 International Conference on Indoor Positioning and Indoor Navigation (IPIN)*, pages 1–7. IEEE, Oct 2015.

[197] Jan Kietlinski-Zaleski and Takaya Yamazato. UWB positioning using known indoor features - environment comparison. In *2010 International Conference on Indoor Positioning and Indoor Navigation*, pages 1–9. IEEE, Sep 2010.

[198] Josef Kulmer, Erik Leitinger, Paul Meissner, Stefan Hinteregger, and Klaus Witrisal. Cooperative localization and tracking using multipath channel information. In *2016 International Conference on Localization and GNSS (ICL-GNSS)*, pages 1–6. IEEE, July 2016.

[199] Samuel Van de Velde and Heidi Steendam. CUPID algorithm for cooperative indoor multipath-aided localization. In *2012 International Conference on Indoor Positioning and Indoor Navigation (IPIN)*, pages 1–6. IEEE, Nov 2012.

[200] Harald Kroll and Christoph Steiner. Indoor ultra-wideband location fingerprinting. In *2010 International Conference on Indoor Positioning and Indoor Navigation*, pages 1–5. IEEE, Sep 2010.

[201] Zemene W. Mekonnen, Eric Slottke, Heinrich Luecken, Christoph Steiner, and Armin Wittneben. Constrained maximum likelihood positioning for UWB based human motion tracking. In *2010 International Conference on Indoor Positioning and Indoor Navigation*, pages 1–10. IEEE, Sep 2010.

[202] Marcel Segura, Hossein Hashemi, Cristian Sisterna, and Vicente Mut. Experimental demonstration of self-localized Ultra Wideband indoor mobile robot navigation system. In *2010 International Conference on Indoor Positioning and Indoor Navigation*, pages 1–9. IEEE, Sep 2010.

[203] Amanda Prorok, Phillip Tome, and Alcherio Martinoli. Accommodation of NLOS for ultra-wideband TDOA localization in single- and multi-robot systems. In *2011 International Conference on Indoor Positioning and Indoor Navigation*, pages 1–9. IEEE, sep 2011.

[204] Benjamin Kempke, Pat Pannuto, and Prabal Dutta. Harmonium: Asymmetric, Bandstitched UWB for Fast, Accurate, and Robust Indoor Localization. In *2016 15th ACM/IEEE International Conference on Information Processing in Sensor Networks (IPSN)*, pages 1–12. IEEE, Apr 2016.

[205] Rahmi Salman and Ingolf Willms. A mobile security robot equipped with UWB-radar for super-resolution indoor positioning and localisation applications. In *2012 International Conference on Indoor Positioning and Indoor Navigation (IPIN)*, pages 1–8. IEEE, Nov 2012.

[206] B. Fall, F. Elbahhar, M. Heddebaut, and A. Rivenq. Time-Reversal UWB positioning beacon for railway application. In *2012 International Conference on Indoor Positioning and Indoor Navigation (IPIN)*, pages 1–8. IEEE, Nov 2012.

[207] Vladimir Maximov, Oleg Tabarovsky, and Dmitriy Filgus. Distributed localisation algorithm for IoT network. In *2015 International Conference on Indoor Positioning and Indoor Navigation (IPIN)*, pages 1–7. IEEE, Oct 2015.

[208] Michal M. Pietrzyk and Thomas von der Grun. Experimental validation of a TOA UWB ranging platform with the energy detection receiver. In *2010 International Conference on Indoor Positioning and Indoor Navigation*, pages 1–8. IEEE, Sep 2010.

[209] Christian Ascher, Lukasz Zwirello, Thomas Zwick, and Gert Trommer. Integrity monitoring for UWB/INS tightly coupled pedestrian indoor scenarios. In *2011 International Conference on Indoor Positioning and Indoor Navigation*, pages 1–6. IEEE, Sep 2011.

[210] Christian Ascher, Sebastian Werling, Gert F. Trommer, Lukasz Zwirello, Carina Hansmann, and Thomas Zwick. Radio-asissted inertial navigation system by tightly coupled sensor data fusion: Experimental results. In *2012 International Conference on Indoor Positioning and Indoor Navigation (IPIN)*, pages 1–7. IEEE, Nov 2012.

[211] Hanna E. Nyqvist, Martin A. Skoglund, Gustaf Hendeby, and Fredrik Gustafsson. Pose estimation using monocular vision and

inertial sensors aided with ultra wide band. In *2015 International Conference on Indoor Positioning and Indoor Navigation (IPIN)*, pages 1–10. IEEE, Oct 2015.

[212] Francisco Zampella, Alessio De Angelis, Isaac Skog, Dave Zachariah, and Antonio Jimenez. A constraint approach for UWB and PDR fusion. In *2012 International Conference on Indoor Positioning and Indoor Navigation (IPIN)*, pages 1–9. IEEE, Nov 2012.

[213] Benjamin Waldmann, Robert Weigel, Randolf Ebelt, and Martin Vossiek. An ultra-wideband local positioning system for highly complex indoor environments. In *2012 International Conference on Localization and GNSS*, pages 1–5. IEEE, June 2012.

[214] Zoltan Feher, Zalan Heszberger, and Andras Veres. Movement detection for location based network management. In *2011 International Conference on Localization and GNSS (ICL-GNSS)*, pages 81–86. IEEE, June 2011.

[215] M. Awais Amin and Sven Fischer. Position location of LTE femtocells deployed in a cluster. In *2012 International Conference on Localization and GNSS*, pages 1–6. IEEE, Jun 2012.

[216] Fernando Pereira, Adriano Moreira, and Manuel Ricardo. Evaluating location fingerprinting methods for underground GSM networks deployed over Leaky Feeder. In *2011 International Conference on Indoor Positioning and Indoor Navigation*, pages 1–6. IEEE, Sep 2011.

[217] Fernando Pereira, Christian Theis, Adriano Moreira, and Manuel Ricardo. Multi-technology RF fingerprinting with leaky-feeder in underground tunnels. *2012 International Conference on Indoor Positioning and Indoor Navigation, IPIN 2012 - Conference Proceedings*, pages 13–15, Nov 2012.

[218] C. Gentner, S. Sand, and A. Dammann. OFDM indoor positioning based on TDOAs: Performance analysis and experimental results. In *2012 International Conference on Localization and GNSS*, pages 1–7. IEEE, June 2012.

[219] Kyoungbaek Min and Jaewoo So. Scheduling and positioning for the expanded region of an indoor cell in heterogeneous networks. In *2014 International Conference on Indoor Positioning*

and Indoor Navigation (IPIN), volume 2012, pages 685–692. IEEE, Oct 2014.

[220] Karthick Nanmaran and B. Amutha. Situation assisted indoor localization using signals of opportunity. In *2014 International Conference on Indoor Positioning and Indoor Navigation (IPIN)*, pages 693–698. IEEE, Oct 2014.

[221] Marcus Dombois and Sebastian Doweling. A pipeline architecture for indoor location tracking. In *2016 International Conference on Indoor Positioning and Indoor Navigation (IPIN)*, pages 1–8. IEEE, Oct 2016.

[222] K. Vickery. Acoustic positioning systems. a practical overview of current systems. In *Proceedings of the 1998 Workshop on Autonomous Underwater Vehicles (Cat. No.98CH36290)*, pages 5–17, Aug 1998.

[223] Frank Kohler, Marcus Thoss, and Alexander Aring. An energy-aware indoor positioning system for AAL environments. In *2010 International Conference on Indoor Positioning and Indoor Navigation*, pages 1–7. IEEE, Sep 2010.

[224] Ferdinand Packi, Frederik Beutler, and Uwe D. Hanebeck. Wireless acoustic tracking for extended range telepresence. In *2010 International Conference on Indoor Positioning and Indoor Navigation*, pages 1–9. IEEE, Sep 2010.

[225] Thomas Janson, Christian Schindelhauer, and Johannes Wendeberg. Self-localization application for iPhone using only ambient sound signals. In *2010 International Conference on Indoor Positioning and Indoor Navigation*, pages 1–10. IEEE, Sep 2010.

[226] Viacheslav Filonenko, Charlie Cullen, and James Carswell. Investigating ultrasonic positioning on mobile phones. In *2010 International Conference on Indoor Positioning and Indoor Navigation*, pages 1–8. IEEE, Sep 2010.

[227] Kyohei Mizutani, Toshio Ito, Masanori Sugimoto, and Hiromichi Hashizume. Fast and accurate ultrasonic 3D localization using the TSaT-MUSIC algorithm. In *2010 International Conference*

on Indoor Positioning and Indoor Navigation, pages 1–5. IEEE, Sep 2010.

[228] Herbert Schweinzer and Mohammad Syafrudin. LOSNUS: An ultrasonic system enabling high accuracy and secure TDoA locating of numerous devices. In *2010 International Conference on Indoor Positioning and Indoor Navigation*, pages 1–8. IEEE, Sep 2010.

[229] Sverre Holm and Carl-Inge C. Nilsen. Robust ultrasonic indoor positioning using transmitter arrays. In *2010 International Conference on Indoor Positioning and Indoor Navigation*, pages 1–5. IEEE, Sep 2010.

[230] Sverre Holm. Ultrasound positioning based on time-of-flight and signal strength. In *2012 International Conference on Indoor Positioning and Indoor Navigation (IPIN)*, pages 1–6. IEEE, Nov 2012.

[231] Mohammad Syafrudin, Christian Walter, and Herbert Schweinzer. Robust locating using LPS LOSNUS under NLOS conditions. In *2014 International Conference on Indoor Positioning and Indoor Navigation (IPIN)*, pages 582–590. IEEE, Oct 2014.

[232] Mohammed Alloulah and Mike Hazas. An efficient CDMA core for indoor acoustic position sensing. In *2010 International Conference on Indoor Positioning and Indoor Navigation*, pages 1–5. IEEE, Sep 2010.

[233] Dennis Laurijssen, Steven Truijen, Wim Saeys, Walter Daems, and Jan Steckel. A flexible embedded hardware platform supporting low-cost human pose estimation. In *2016 International Conference on Indoor Positioning and Indoor Navigation (IPIN)*, pages 1–8. IEEE, Oct 2016.

[234] Takayuki Akiyama, Masanori Sugimoto, and Hiromichi Hashizume. SyncSync: Time-of-arrival based localization method using light-synchronized acoustic waves for smartphones. In *2015 International Conference on Indoor Positioning and Indoor Navigation (IPIN)*, pages 1–9. IEEE, Oct 2015.

[235] Ish Rishabh, Don Kimber, and John Adcock. Indoor localization using controlled ambient sounds. In *2012 International Conference*

on Indoor Positioning and Indoor Navigation (IPIN), pages 1–10. IEEE, Nov 2012.

[236] Yuya Itagaki, Akimasa Suzuki, and Taketoshi Iyota. Indoor positioning for moving objects using a hardware device with spread spectrum ultrasonic waves. In *2012 International Conference on Indoor Positioning and Indoor Navigation (IPIN)*, pages 1–6. IEEE, Nov 2012.

[237] Hiromichi Yoshiga, Akimasa Suzuki, and Taketoshi Iyota. An information addition technique for indoor self-localization systems using SS ultrasonic waves. In *2012 International Conference on Indoor Positioning and Indoor Navigation (IPIN)*, pages 1–6. IEEE, Nov 2012.

[238] Andrei Stancovici, Mihai V. Micea, and Vladimir Cretu. Cooperative positioning system for indoor surveillance applications. In *2016 International Conference on Indoor Positioning and Indoor Navigation (IPIN)*, volume 1, pages 1–7. IEEE, Oct 2016.

[239] Walter Erik Verreycken, Dennis Laurijssen, Walter Daems, and Jan Steckel. Firefly based distributed synchronization in Wireless Sensor Networks for passive acoustic localization. In *2016 International Conference on Indoor Positioning and Indoor Navigation (IPIN)*, pages 1–8. IEEE, Oct 2016.

[240] M. C. Perez, D Gualda, J M Villadangos, J. Urena, P Pajuelo, E. Diaz, and E. Garcia. Android application for indoor positioning of mobile devices using ultrasonic signals. In *2016 International Conference on Indoor Positioning and Indoor Navigation (IPIN)*, pages 1–7. IEEE, Oct 2016.

[241] Joan Bordoy, Patrick Hornecker, Fabian Hoflinger, Johannes Wendeberg, Rui Zhang, Christian Schindelhauer, and Leonhard Reindl. Robust tracking of a mobile receiver using unsynchronized time differences of arrival. In *International Conference on Indoor Positioning and Indoor Navigation*, pages 1–10. IEEE, Oct 2013.

[242] Ling Pei, Liang Chen, Robert Guinness, Jingbin Liu, Heidi Kuusniemi, Yuwei Chen, Ruizhi Chen, and Stefan Soderholm. Sound positioning using a small-scale linear microphone array. In *International Conference on Indoor Positioning and Indoor Navigation*, pages 1–7. IEEE, Oct 2013.

[243] Takayuki Akiyama, Masanari Nakamura, Masanori Sugimoto, and Hiromich Hashizume. Smart phone localization method using dual-carrier acoustic waves. In *International Conference on Indoor Positioning and Indoor Navigation*, pages 1–9. IEEE, Oct 2013.

[244] David Gualda, Ma Carmen Perez, Jesus Urena, Juan C. Garcia, Daniel Ruiz, Enrique Garcia, and Alejandro Lindo. Ultrasonic LPS adaptation for smartphones. In *International Conference on Indoor Positioning and Indoor Navigation*, pages 1–8. IEEE, Oct 2013.

[245] Johannes Wendeberg, Jorg Muller, Christian Schindelhauer, and Wolfram Burgard. Robust tracking of a mobile beacon using time differences of arrival with simultaneous calibration of receiver positions. In *2012 International Conference on Indoor Positioning and Indoor Navigation (IPIN)*, pages 1–10. IEEE, Nov 2012.

[246] Armin Runge, Marcel Baunach, and Reiner Kolla. Precise self-calibration of ultrasound based indoor localization systems. In *2011 International Conference on Indoor Positioning and Indoor Navigation*, pages 1–8. IEEE, Sep 2011.

[247] Fabian Hoflinger, Rui Zhang, Joachim Hoppe, Amir Bannoura, Leonhard M. Reindl, Johannes Wendeberg, Manuel Buhrer, and Christian Schindelhauer. Acoustic Self-calibrating System for Indoor Smartphone Tracking (ASSIST). In *2012 International Conference on Indoor Positioning and Indoor Navigation (IPIN)*, pages 1–9. IEEE, Nov 2012.

[248] Alexander Ens, Leonhard M. Reindl, Joan Bordoy, Johannes Wendeberg, and Christian Schindelhauer. Unsynchronized ultrasound system for TDOA localization. In *2014 International Conference on Indoor Positioning and Indoor Navigation (IPIN)*, pages 601–610. IEEE, Oct 2014.

[249] R. Indeck and E. Glavinas. Magnetic Fingerprinting. In *[1993] Digests of International Magnetics Conference*, pages CP–CP. IEEE, 1993.

[250] Binghao Li, Thomas Gallagher, Andrew G. Dempster, and Chris Rizos. How feasible is the use of magnetic field alone for indoor positioning? In *2012 International Conference on Indoor Positioning and Indoor Navigation (IPIN)*, pages 1–9. IEEE, Nov 2012.

[251] Junyeol Song, Hyunkyo Jeong, Soojung Hur, and Yongwan Park. Improved indoor position estimation algorithm based on geomagnetism intensity. In *2014 International Conference on Indoor Positioning and Indoor Navigation (IPIN)*, pages 741–744. IEEE, Oct 2014.

[252] Jinseon Song, Soojung Hur, Yongwan Park, and Jeonghee Choi. An improved RSSI of geomagnetic field-based indoor positioning method involving efficient database generation by building materials. In *2016 International Conference on Indoor Positioning and Indoor Navigation (IPIN)*, pages 1–8. IEEE, Oct 2016.

[253] Qu Wang, Haiyong Luo, Fang Zhao, and Wenhua Shao. An indoor self-localization algorithm using the calibration of the online magnetic fingerprints and indoor landmarks. In *2016 International Conference on Indoor Positioning and Indoor Navigation (IPIN)*, volume 978, pages 1–8. IEEE, Oct 2016.

[254] Shervin Shahidi and Shahrokh Valaee. GIPSy: Geomagnetic indoor positioning system for smartphones. In *2015 International Conference on Indoor Positioning and Indoor Navigation (IPIN)*, pages 1–7. IEEE, Oct 2015.

[255] You Li, Peng Zhang, Xiaoji Niu, Yuan Zhuang, Haiyu Lan, and Naser El-Sheimy. Real-time indoor navigation using smartphone sensors. *2015 International Conference on Indoor Positioning and Indoor Navigation, IPIN 2015*, pages 13–16, Oct 2015.

[256] Vania Guimaraes, Lourenco Castro, Susana Carneiro, Manuel Monteiro, Tiago Rocha, Marilia Barandas, Joao Machado, Maria Vasconcelos, Hugo Gamboa, and Dirk Elias. A motion tracking solution for indoor localization using smartphones. In *2016 International Conference on Indoor Positioning and Indoor Navigation (IPIN)*, pages 1–8. IEEE, Oct 2016.

[257] Paolo Barsocchi, Antonino Crivello, Davide La Rosa, and Filippo Palumbo. A multisource and multivariate dataset for indoor localization methods based on WLAN and geo-magnetic field fingerprinting. *2016 International Conference on Indoor Positioning and Indoor Navigation (IPIN)*, pages 4–7, Oct 2016.

[258] Yuntian Brian Bai, Tao Gu, and Andong Hu. Integrating Wi-Fi and magnetic field for fingerprinting based indoor

positioning system. In *2016 International Conference on Indoor Positioning and Indoor Navigation (IPIN)*, pages 1–6. IEEE, Oct 2016.

[259] Andreas Ettlinger and Gunther Retscher. Positioning using ambient magnetic fields in combination with Wi-Fi and RFID. In *2016 International Conference on Indoor Positioning and Indoor Navigation (IPIN)*, pages 1–8. IEEE, Oct 2016.

[260] Xumeng Guo, Wenhua Shao, Fang Zhao, Qu Wang, Dongmeng Li, and Haiyong Luo. WiMag: Multimode Fusion Localization System based on Magnetic/WiFi/PDR. In *2016 International Conference on Indoor Positioning and Indoor Navigation (IPIN)*, pages 1–8. IEEE, Oct 2016.

[261] Anna Maria Vegni and Mauro Biagi. An indoor localization algorithm in a small-cell LED-based lighting system. In *2012 International Conference on Indoor Positioning and Indoor Navigation (IPIN)*, pages 1–7. IEEE, Nov 2012.

[262] Chi Zhang and Xinyu Zhang. LiTell: Robust Indoor Localization Using Unmodified Light Fixtures. *Proceedings of the 22nd Annual International Conference on Mobile Computing and Networking - MobiCom '16*, pages 230–242, 2016.

[263] Sven Heissmeyer, Ludger Overmeyer, and Andreas Muller. Indoor positioning of vehicles using an active optical infrastructure. In *2012 International Conference on Indoor Positioning and Indoor Navigation (IPIN)*, pages 1–8. IEEE, Nov 2012.

[264] Guillermo del Campo-Jimenez, Jorge M. Perandones, and F. J. Lopez-Hernandez. A VLC-based beacon location system for mobile applications. In *2013 International Conference on Localization and GNSS (ICL-GNSS)*, pages 1–4. IEEE, June 2013.

[265] Yohei Nakazawa, Hideo Makino, Kentaro Nishimori, Daisuke Wakatsuki, and Hideki Komagata. Indoor positioning using a high-speed, fish-eye lens-equipped camera in Visible Light Communication. In *International Conference on Indoor Positioning and Indoor Navigation*, pages 1–8. IEEE, Oct 2013.

[266] Yohei Nakazawa, Hideo Makino, Kentaro Nishimori, Daisuke Wakatsuki, and Hideki Komagata. LED-tracking and ID-estimation

for indoor positioning using visible light communication. In *2014 International Conference on Indoor Positioning and Indoor Navigation (IPIN)*, pages 87–94. IEEE, Oct 2014.

[267] Takayuki Akiyama, Masanori Sugimoto, and Hiromichi Hashizume. Light-synchronized acoustic ToA measurement system for mobile smart nodes. In *2014 International Conference on Indoor Positioning and Indoor Navigation (IPIN)*, pages 749–752. IEEE, Oct 2014.

[268] Md Shareef Ifthekhar, Nirzhar Saha, and Yeong Min Jang. Neural network based indoor positioning technique in optical camera communication system. In *2014 International Conference on Indoor Positioning and Indoor Navigation (IPIN)*, volume 2014, pages 431–435. IEEE, Oct 2014.

[269] Kazuki Moriya, Manato Fujimoto, Yutaka Arakawa, and Keiichi Yasumoto. Indoor localization based on distance-illuminance model and active control of lighting devices. In *2016 International Conference on Indoor Positioning and Indoor Navigation (IPIN)*, pages 1–6. IEEE, Oct 2016.

[270] W. Zhang and M. Kavehrad. A 2-D indoor localization system based on visible light LED. In *2012 IEEE Photonics Society Summer Topical Meeting Series, PSST 2012*, pages 80–81, 2012.

[271] Giulio Cossu, Marco Presi, Raffaele Corsini, Pallab Choudhury, Amir Masood Khalid, and Ernesto Ciaramella. A Visible Light localization aided Optical Wireless system. In *2011 IEEE GLOBECOM Workshops, GC Wkshps 2011*, pages 802–807, 2011.

[272] Kejie Qiu, Fangyi Zhang, and Ming Liu. Visible Light Communication-based indoor localization using Gaussian Process. *2015 IEEE/RSJ International Conference on Intelligent Robots and Systems (IROS)*, pages 3125–3130, 2015.

[273] Divya Ganti, Weizhi Zhang, and Mohsen Kavehrad. VLC-based Indoor Positioning System with Tracking Capability Using Kalman and Particle Filters. *2014 IEEE International Conference on Consumer Electronics (ICCE)*, pages 476–477, 2014.

[274] Jeffrey D. Poston, Javier Schloemann, R. Michael Buehrer, V. V. N. Sriram Malladi, Americo G. Woolard, and Pablo A. Tarazaga. Towards indoor localization of pedestrians via smart building vibration sensing. *Proceedings of 2015 International Conference on Localization and GNSS, ICL-GNSS 2015*, 2015.

[275] Yukitoshi Kashimoto, Manato Fujimoto, Hirohiko Suwa, Yutaka Arakawa, and Keiichi Yasumoto. Floor vibration type estimation with piezo sensor toward indoor positioning system. In *2016 International Conference on Indoor Positioning and Indoor Navigation (IPIN)*, pages 1–6. IEEE, Oct 2016.

CHAPTER 2

An Approach towards GIS Application in Smart City Urban Planning

Sourav Banerjee

Kalyani Govt. Engg. College

Chinmay Chakraborty

Birla Institute of Technology, Mesra

Debashis Das

Kalyani Govt. Engg. College

CONTENTS

2.1 INTRODUCTION

The Geographic Information System (GIS) is already a massive imaginative step in the field in these days of urban planning. Since its development in the 1970s, GIS has had a major impact on geographical independent investigation and the usefulness of business in governance and the region. GIS skills have increased over the past few decades, and now GIS has provided a wide range of data execution and dissection tools. Another delightful move was the relevance of GIS, Global Positioning System (GPS), and remote sensing methods such as satellite images, light detection and ranging, and data generation, which authoritatively remotely sensed data. Figure 2.1 [49] shows how GIS can support urban planning.

On the one hand, smart cities are showing considerable opportunities for development in the coming long period; while on the other hand, they are also offering different challenges [1]. Smart city ventures may be complex with privately owned and condo complexes upheld by a control, street, water, waste, and sewage framework, i.e. a virtual living and breathing city. A commercial success find may have been a mandatory requirement now for a common cultural entrepreneurship level to somehow enable heterogeneous representatives of the smart city social environment to integrate, coordinate, and work synergistically. A centralized data framework predicated mostly on GIS provides an IT system that coordinates each smart city's entire process -starting as well from categorization, arrangement as well as development to endorse -not as each partner but

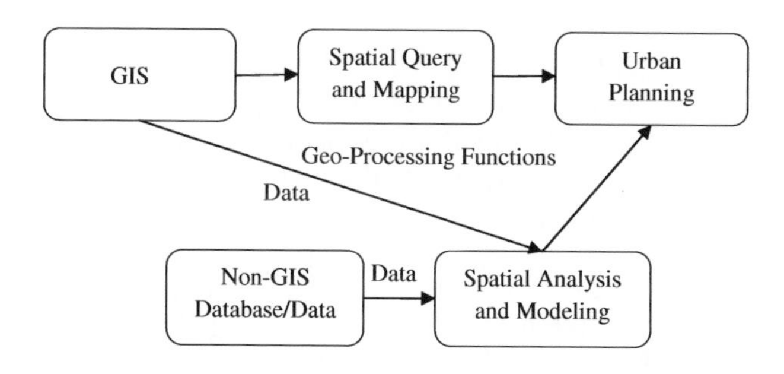

Figure 2.1 GIS and urban planning.

also every perspective. GIS is utilized for the identification of geological features, the examination of lands and sediments, geological data, and the production of regional geographical three-dimensional (3D) views. GIS can store information in a database and represent it visually in the form of mapped data. GIS technology is used to assess urban development and its extension path and to find suitable urban planning locations. Some considerations must be weighed to identify places ideal for urban development: lands must have adequate availability; lands must be more or less dry; land shortage or limited use value should be accessible at present and there must be an adequate water source. There are also many GIS applications, which are discussed in detail in this chapter.

The chapter consists of 12 sections. Section 2.1 provides the introduction and describes GIS and its application in urban planning. Section 2.2 describes GIS in transportation analysis and planning which are more broadly used in the transportation planning associations, mainly among metropolitan transportation agencies. Many advanced countries have managed highway maintenance which is enhancing a major issue. Section 2.3 describes the application of GIS in waste management planning. Waste management problems are becoming the spearhead of the global substantial program at an expanding persistence, as per population and implementation growth outcome in expanding allotments of waste. Section 2.4 describes the GIS application of regional planning. Section 2.5 describes the GIS application in resource management. The profitable, societal, and artistic possibilities of any dominion are primarily resolved by the plot and water assets. For this restriction, there is a requirement to confirm that these assets are functionally persuaded. Section 2.6 deals with the application of GIS in domain monitoring and assessment. The classic natural GIS customers are centered on natural problems. Section 2.7 describes GIS in socio-economic development. A vital and fast-developing application field for GIS is that of socioeconomic or population-related information. Section 2.8 describes the GIS application in emergency management. Section 2.9 discusses the implication of GIS in the domain of education. Section 2.10 and 2.11 present the role of machine learning (ML) and cybersecurity in GIS. Section 2.12 concludes the entire chapter.

Nowadays GIS is being rapidly recycled in urban planning in the advanced and developing countries. GIS tools assist creators to inspect problems more rapidly and thoroughly and formulate solutions. Many arranging divisions that had gotten mapping frameworks within the past have moved to GIS.

2.1.1 Objectives

The main intentions of this chapter are as follows:

- To provide a concept of the prefatory ideas about GIS application in urban planning. For example: what is GIS?, the prosperity of GIS, applications of GIS in the urban area, etc.
- To analyze, the application of GIS in urban planning and to evaluate the mechanizations applied in the specific area as an example: using internet GIS in urban planning.
- To display the improvement and constructive approach using GIS and recognize the principle discoveries.
- To propose some recommendations to enhance the application of GIS in urban planning.

2.1.2 Contribution

One natural consequence of the transformation process to the modern digital society which is entirely dependent on specific information was the growing political and economic significance of GIS, particularly throughout the last decade. This chapter describes how GIS can eventually find its applications completely different from many other cases of particularly sharp urban neighborhoods in developed countries than in the basic building nearby. The GIS could be a rising procedure that can be used to make the perfect use of resources; life appropriately could be an essential tool for transforming urban areas into urban communities of the shrewd. The smart city has overwhelming reference points both for the people and the government.

2.1.3 GIS in Urban Planning

At first, GISs high set-up and labor costs were a fence for the procurement of geospatial innovations within the urban arrangement zone [2]. At that time, a grid-based program known as the IMGRID of Sinton was effective. The software concentrated especially only on the visualization of systematic abilities only with the small assumption. Just as the cost of the machinery dropped significantly and also the user-friendliness of the GIS computer extended the program, the initial application of GIS in urban planning was drastically formed – so as the world aims to try to continue towards urbanization, it continues to create the need for a strong urban arrangement. GIS observes only its application as an explanatory and modeling device in the urban arrangement. It can be used in a wide complication cluster. This includes issues related to database structures and simple and composite explanatory models that cannot be distinguished. Also, GIS is productive in monitoring a range or conducting a practicality consideration of an area for a positive reason. The study of practicality of indeed smaller structures such as academies and health clinics is vitally important and can be carried out mostly with the help of GIS. The application of GIS increased with more elite gear to amass superior in areas where variations of an outline or interchange plans are required. Urban planning in GIS is the most advantageous planning application [3]. At the time of expansion, it can be used to evaluate urban progress and its direction. GIS data could then be used to help improve the urban site, and the disarray that could be produced by bottleneck usually stays away from benefits.

Smart cities are the gigantic, ever-evolving issue frameworks. Carefully driving the advancement of an area requires vigorous, shaded, and continually upgraded spatial data, as well as the ability to use that data to resolve issues. This challenge has put geographically data analytics and technical innovation at the center of urban planning. Figure 2.2 gives us knowledge about GIS applications in urban area planning [48]. Spatial information focuses on the way to the revision of quality of life and the development of faultless communities, while experts in GIS use spatial consideration for transforming this information into useful understanding and solutions. GIS applications cover major application areas in urban arranging and advancement: Infrastructure management (transport, public utilities, and storm-water/waste), regional planning, resource management, environmental assessment, socio-economic development, emergency management, education, etc. Urban planning GIS applications apply for area observation, regional potential, and site selection studies and analysis of viability. GIS can also be useful for spatial plan documentation and for authorizing advancement, construction, and installation in the consent process. The benefits of using GIS in urban planning are as follows: efficiency is increased and time/money is saved, generate revenue and support decision-making, improve accuracy, manage resources, automate tasks, increase government access, increase public involvement and actively promote larger public agency collaboration.

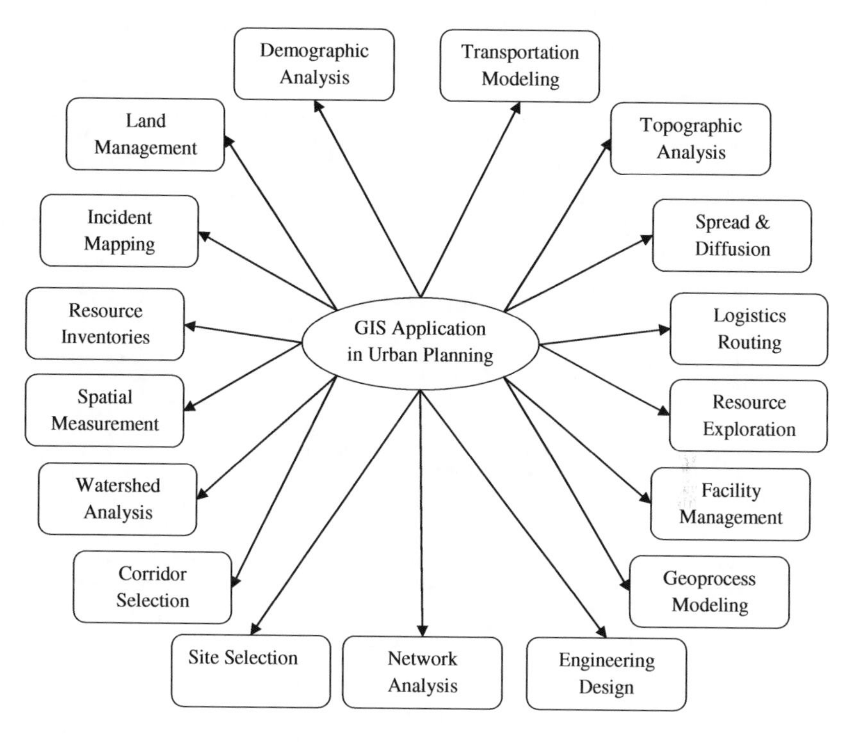

Figure 2.2 GIS application in urban planning.

2.1.4 GIS as an Integrated System for Smart Cities

Detailed research into savvy towns requires a coupled approach to the problem to the obligatory state of craftsmanship, an economical and comprehensive method of reasoning with bits of knowledge about feasible transportation, green vitality, natural quality, careful construction, reasonable accommodation, and strength for chance, accessibility of consumable water and nutrition and many other recognizable parking spaces. Figure 2.3 focuses on GIS applications in our smart city area [50]. Within the existing smart cities, geo-administration, geospatial information, and applications [4] have expanded within the framework of ventures largely driven by government offices (e.g. shrewd city of Dubai). In terms of expansion, online maps, and other software applications (for Google applications) can be used. In India, the use of Information Communication Technology in most Indian cities has started, but the comprehensive social structure that is driven by innovation in data and communication can be established through urban arrangements using GIS.

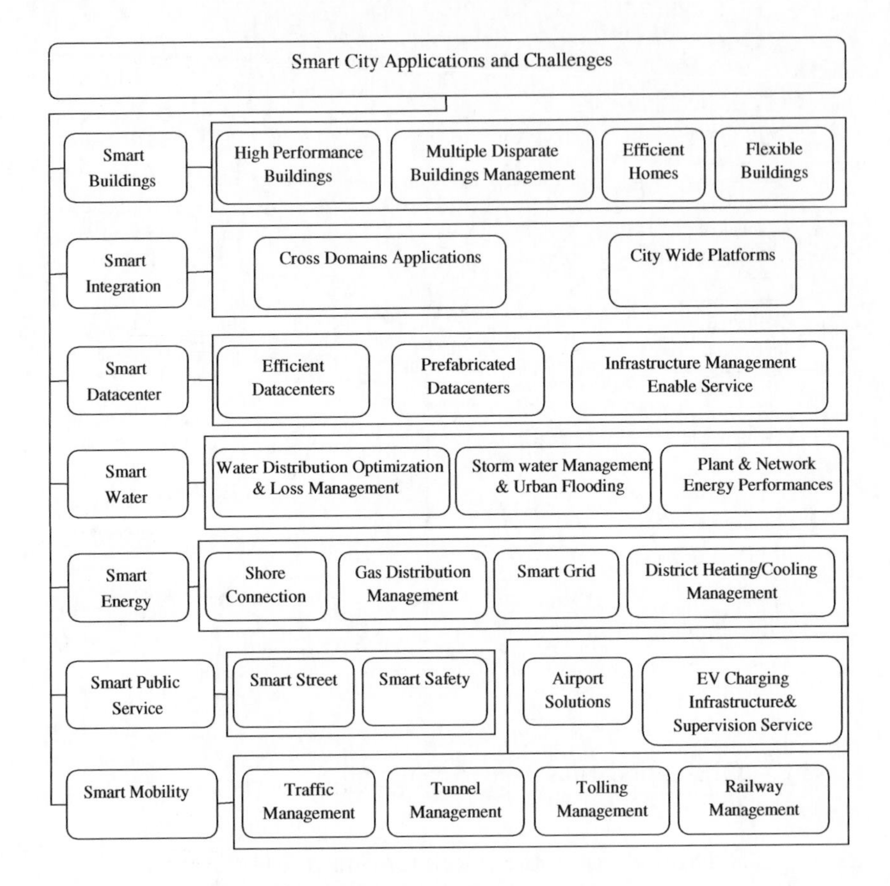

Figure 2.3 Smart city applications and challenges.

A centralized GIS-based information system provides an IT framework for maintaining and transferring information and apps throughout the life cycle of the city.

Acquire: Find the right sites for urban development, see legal boundaries, arrive at the correct assessment of your existing/new destinations.

Planning and Design: Recognize inadequacy and make the best arrangements. The computer-aided design, building information modeling is bringing more prominent analytics and cost-estimation capabilities to your foundation plan process.

Construct: The coordinated program of expansion and financial management with GIS is to oversee superior ventures. GIS can provide a single section point for all archives and records related to construction.

Sell: Get it how and where to advertise advances in the city, attract buyers and residents, and advance maintenance rates.

Maintain: Easily achieve resources that are different. Coordinate your resource stock with review history and cost-effectively manage your core business.

2.1.4.1 GIS Applications for Smart Cities

Site Selection and Land Acquisition: GIS can combine and coordinate distinctive data types to help shape better choices together, providing high-quality visualization tools that can advance understanding and enhance the choice for the location to distinguish proof, valuation, and ultimate determination. By analyzing area information proximity to the street arrangement, soil wealth, arrival utilization, soil capacity bearing, groundwater depth, and helplessness to disasters such as surges, and seismic tremors – genuine domain organizations can arrive at property valuation. It is possible to assess the merits of one location or area over another by analyzing, mapping, and modeling. *Environmental/Legal Compliance:* GIS makes assembly administrative needs less time-consuming and easier to achieve by giving the controllers to open a common stage for communication. GIS-based crop yields that are entirely graphical can help generate reports quickly that clearly describe how regulatory compliance basic requirements and by-laws are met. *Planning, Design, and Visualization*: Geo design will become the real key system for conceptualizing and arranging smart cities; it will assist in every arrangement from conceptualizing ventures to site analysis, planning, partner interest and collaboration, recreation, and evaluation. A 3D GIS could also be used to re-enact a business, environment, or basic circumstance practically. *Construction and Project Management*: GIS makes a big difference in the organization of all relevant extensive data, somewhat from sediment-specific information and geotechnical ponders to arrangements, completely natural ponders, drawings, maps of extent, stocks, and global resource gain control. *Sales and Marketing*: City engineers should win over all the long-planned businesses with GIS by attempting to make insightful deals devices and promoting reports that highlight an unused area's major financial real potential or future progress. *Facility Management (FM):* A GIS-based actual data legal framework provides an efficient political establishment for senior operations management by creating data coordinates that make a difference in making better allocation choices possible. GIS can be coordinated with the current administration system of the office and expanded. *Operations and Reporting:* GIS can track and analyze resources over time and space and provide an understanding through the map and easy-to-understand report visualization of data. It encourages operations to see that it incorporates maps, records, charts, and more based on live geographic information in a web site or web service.

2.1.5 GIS Software Types

During the device screening process, various types of GIS devices are available on the market and are worth considering [57]. The preceding collection is extracted from the GIS dictionary of the Environmental Systems Research Institute (ESRI).

- *Add-Ons/Extensions:* Refer to available device modules that attach to the currently existing software systems advanced resources and functions.
- *Computer-Aided Design:* Refer to the software that allows graphical and geographic details to be planned, created, and displayed.
- *Desktop Mapping:* Refers to desktop computer software, which varies from systems capable of viewing data only to complete data handling and evaluation.
- *GPS:* Refers to the data, software, and hardware collection capable of viewing, collecting and editing geographic data and basic mappings on or without a GPS recipient.
- *Image Processing:* Pertains to software that operates remotely sensed data to enhance the interpreting and thematic classification of a picture (e.g., aerial photographs, satellite imaging).
- *Internet Mapping:* Refers to applications that offer complex charts, data, and essential spatial features online.
- *Mobile GIS:* It involves the applications for the use of portable or smartphone wireless devices to map, GIS, and GPS.

- *Photogrammetry:* It refers to photographic images that extract useful information from software, in particular for creating accurate maps (for example, the distance between two objects).
- *Routing/Delivery GIS:* Refers to applications developing and managing distribution pathways.
- *Spatial Database:* It refers to spatial data and its associated explanatory information editing and arranging software for reliable space and recovery.
- *Web Service:* It involves apps not based on the hardware platform's operating system but available over the Web to be used in several certain software.

2.1.6 GIS Websites

To determine the GIS mechanism, web searches are especially helpful. In comparison, literature reviews of GIS software and web sites [58], such as those shown in Table 2.1, can help identify items that meet the needs of an organization. Several specific words include GIS, online mapping, and spatial analysis.

2.2 GIS IN TRANSPORTATION ANALYSIS AND PLANNING

The GIS makes use of more transport organizations, mainly among metropolitan transport organizations. Interstate nature conservation administration is becoming a vital issue in many advanced nations. Currently, numerous powers are shrewd to use GIS for inter-states and transportation administration as shown in Figure 2.4 [6], due to cost reduction and friendly GIS expansion. GIS provides transport organizers with a medium to store and analyze information on people's land use, travel, etc. Map/display and information integration are the most important essential to use GIS. The usage of GIS for transport applications includes road maintenance, representation of activities, mischance review, and course arrangement and natural evaluation of street plans. The main needs and challenges related to innovation in GIS administration are the development and maintenance of a database, the selection, and development of equipment and programs, the use of innovation to solve problems, subsidize, organize, and provide access and others. Standard

TABLE 2.1 GIS Analysis Software

GIS Software (Open Source)	Websites
ArcGIS (ESRI)	https://www.esri.com/en-us/home
Geomedia (Hexagon Geospatial)	https://www.hexagongeospatial.com/
Global Mapper (Blue Marble)	https://www.bluemarblegeo.com/products/global-mapper.php
Manifold GIS (Manifold)	http://www.manifold.net/
Bentley Map	http://www.bentley.com/
IDRISI (Clark Laboratories)	https://clarklabs.org/
AutoCAD Map 3D (Autodesk)	https://www.autodesk.com/products/autocad/included-toolsets/autocad-map-3d
Maptitude (Caliper Corporation)	https://www.caliper.com/maptovu.htm
SuperGIS (Supergeo Technologies)	http://www.supergeotek.com/
Tatuk GIS	https://www.tatukgis.com/Home.aspx

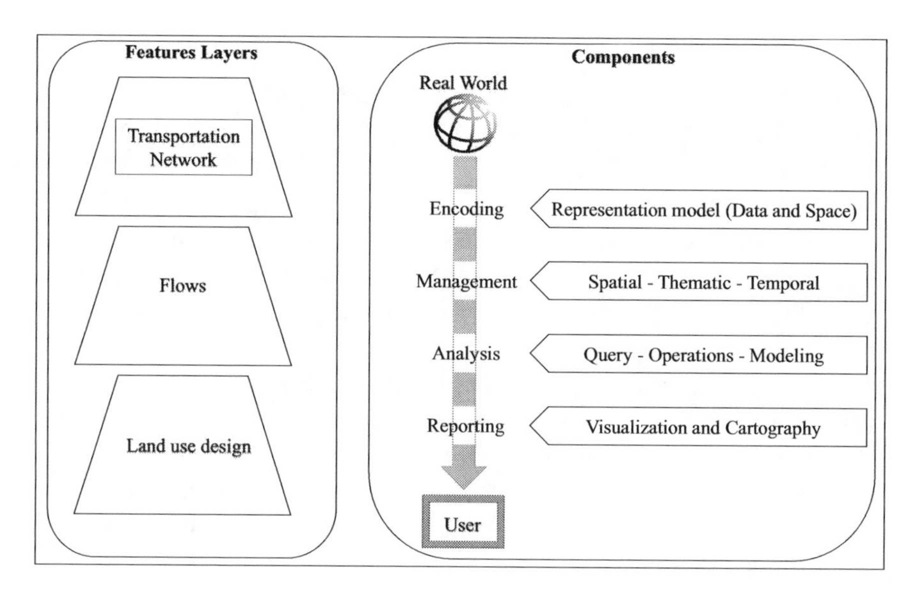

Figure 2.4 GIS and transportation.

GIS capabilities include topical mapping, insights, diagramming, lattice manipulation, bolster resolution system, modeling and calculation, and simultaneous access to various databases [5].

GIS application has transportation significance due to the fundamentally spatially conveyed nature of transportation-related information and requirements for various types of level examination arrangements, factual analysis, and spatial investigation and control. At the GIS stage, transport organizes the database by joining numerous sets of its characteristics and spatial information through its own fairly straight reference framework. The most advanced use of GIS is its ability to enter and analyze spatially disseminated information concerning its genuine spatial area surfaced based on a field of scope that allows for inconceivable examination with other database management frameworks. The main advantage of using the GIS is not necessarily the user-friendly visual approach and display, but also the spatial examination capacity and appropriateness to use standard GIS features such as topical mapping, charting and network-level investigation, synchronous access to a few layers of information and overlapping of the same, as well as the ability to interface with external programs and computer bolstering, information management, and user-specific capabilities [7].

2.2.1 Applications of GIS in Transportation Planning

Public transportation is definitely geographic and therefore GIS has an innovation mostly with the calculable huge potential for achieving significant efficiency and efficiency gains for the vast number of sophisticated transportation implementations, and also the opportunity to succeed to progress in traditional applications. GIS transport applications can be viewed as (a) data recovery; (b) data integrator; or (c) data analysis. The other applications are as follows: executive information system, pavement management system, bridge management, maintenance management, safety management, transportation system management, travel demand forecasting, corridor preservation, right-of-way, construction management, hazardous cargo routing, overweight/oversize vehicles permit routing, accident analysis, environmental impact, and the financial impact on the countryside and the investigation of value-capture, and others.

2.2.2 Internet GIS and Its Applications in Transportation

Two of the 1990s innovations, the Web and Geographic data frameworks, have changed how to transport experts get, share, circulate, and analyze information and data. Public transportation information [8] providers, counting government offices and private entities, are also finding the convenience of producing and distributing data on the World Wide Web and some have specifically designed their web pages. Developing technology-Internet GIS combines Internet and GIS, which provides better approaches for transportation experts to access, share, and transmit data on transportation. Web GIS could be a modern innovation linked to the handling of spatial information on the Internet. This may be a network-centered GIS device that operates the Web as major operations for entry and transmission of specific information and investigation devices to advance visualization and spatial information integration. Almost any data that can only be displayed on a sketch such as a road and travel levels, conditions, climate data, and the like can be decided to move using Web GIS. It also offers the planned Internet exchange of information and transportation examination.

2.2.3 GIS-T Analysis and Modeling

Geographic Transportation Information Systems (GIS-T) applications picked up numerous standard GIS capabilities (inquiry, geocoding, buffer, overlay, etc.) to back up information management, investigation, and visualization needs [8]. Transport, like many other areas, has expanded its possession of special examination strategies and models. Cases include most brief methods and steering calculations (e.g. traveling sales consultant issues, vehicle steering issues), spatial effective communication models (e.g. gravity demonstration), arrange stream issues (e.g. least collected stream issue, most severe live stream problem, make arrangements stream balance designs), office area issues (e.g. media issue, scope issue, maximum coverage issue, center issue), travel request models (e.g. four-step trip era, trip dispersion, modular component, activity-based travel request models, and later activity-based travel request models), and land-use interaction models. While essential transport evaluation strategies (e.g. most brief wayfinding) can be launched in the most commercial GIS computer program, other transport investigation strategies and models (e.g. travel request models) are opened in a few commercial programs bundles as they were specifically. Fortunately, the GIS component arrangement approach adopted by GIS computer program companies gives talented GIS-T clients a far better environment to advance their customized assessment strategies and tactics. Over the past two decades, many GIS-T applications have been executed at various transport organizations: planning, designing and managing infrastructure; analysis of transport safety; and travel requirements; monitoring of traffic; planning and operations for public transit; evaluation of the environmental impact; hazard migration and intelligent transportation systems. The social interaction between both the transport framework and its environment makes GIS innovative and creative.

2.3 GIS IN WASTE MANAGEMENT PLANNING

Issues of waste management are coming at an expanding recurrence to initiate the worldwide natural plan, as population and expansion of utilization result in the development of quantities of waste. The need for upgraded execution at more cost is not kept for advanced nations seeking to use frameworks for continuous composite isolated waste collection, treatment, and recovery. Under a different setting, it also uses its weight to the advancing countries' civil administrations, which seek, without doubt, to help create the trash collection and to open up well-being security for large populations of highly urbanized areas with a strong new framework and budget constraints [9–11].

2.3.1 Waste Storage System

The second functional element of solid waste management is waste storage at the source. Following buildings and other sources of waste generation, waste is usually placed in waste canisters on both sides of the roads. It is undesirable to store waste for a long time due to the microbiological decline of the waste. Unclaimed trash will pose a serious threat to human health because it is a source of termites, flies, and bacteria. Another disadvantage of such trash collection is the bad smell generated around the garbage area, which itself significantly reduces the current market value of the area due to non-esthetic conditions, leaving a bad intuition and posing a threat to the environment. The problems commonly observed in the area of the country or even the important issues were (a) at the normal interim, the trash is not lifted, (b) most of the time, the waste containers are either uncovered or not lying upright in an aggravating condition, and lying full of rubbish without being cleaned jointly, and (c) strong waste categories such as paper, glass, polythene, and food fabric were not partitioned. While on the other hand, the democratic professionals already had their reasons for the waste maintenance fumbled.

a. The ordinary citizens do not throw the canisters inside the waste, so it often lies outside and all around the containers, attempting to make the maximum range around the plastic bin look tainted.

b. The capacity to generate waste is lower compared with the sum of waste produced in the city.

There is also insufficient labor and equipment.

2.3.2 Elements of Solid Waste Management

The following functional elements are used in solid waste management: waste generation, treatment, separation, processing, and storage of waste at source, collection and transfer of waste, separation, processing, and processing of waste and waste disposal.

2.3.3 Significance of GIS in Waste Management

The issue of waste isn't an expansion of quantities, but more generally an insufficient management framework. And one of the main operational issues in solid waste management is trying to improve data translation

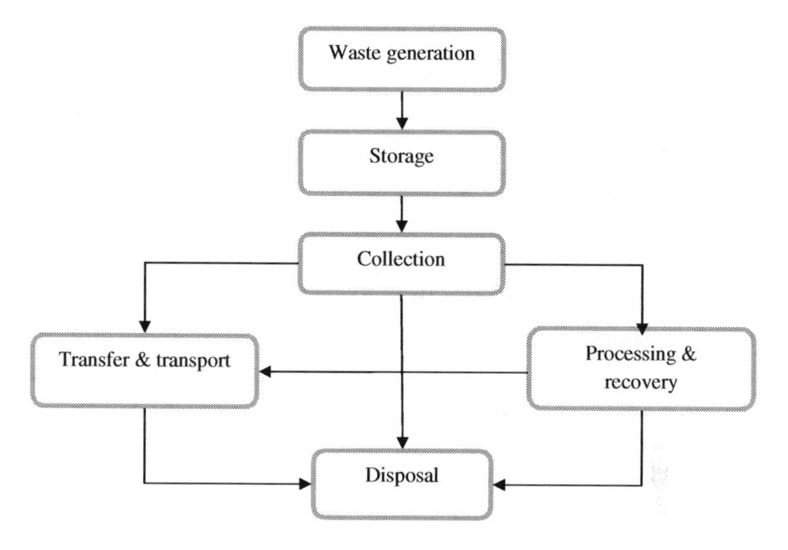

Figure 2.5 The interrelationship between the functional elements in a waste management system.

tools and methods [13]. All of the other health problems associated with solid waste management [14] have both spatial and non-spatial information components. In most cases, strong waste information and records are not controlled contrary to each other and are often confined. Due to inappropriate data or record management, evaluating system functionality and sometimes efficiency becomes onerous. Figure 2.5 [12] shows the functional elements of a waste management system.

2.3.4 GIS for Sustainable Waste Management

GIS is one of the leading advances in capturing, storing, and controlling, analyzing, and displaying spatial information. This information is essentially arranged within the framework of computerized maps in topical layers. The application of GIS with related advances created by the coordinate (e.g., Worldwide Positioning System-GPS and Inaccessible Detection) helps to record spatial information and to apply this information directly for investigation and cartographic representation. Unsegregated GIS innovation has been recognized as one of the most promising approaches to computerizing the waste planning and management method these days. The primary wide-ranging application of GIS braced waste management modeling lies in landfill siting and waste collection and transport advancement [15].

2.3.5 GIS Modeling for the Optimization of Waste Collection and Transport

Optimizing the routing basic framework for collecting or transporting cosmopolitan solid waste can be an influential figure in a strong, environmentally friendly, and cost-effective waste management system. The classic Traveling Salesman Problem (TSP) is often exceptionally alike [16]. However, the major issues of optimizing the direction of strong trash collection control systems may be an extremely unbalanced TSP due to the limitations of the street organization. Therefore, adjustments are required to the classic TSP algorithm, making the situation somewhat more complicated. The GIS can provide notable financial and natural reserve funds through travel time consumption, and separate use of fuel and poison discharges [17–20].

2.3.6 Application of GIS Technology in Waste Management

With the development of technology, especially IT, it is indeed important today that such a technology is used in any planning, including waste management planning. The implications of the planning information system have often been studied and discussed, but less frequently explored rigorously and coherently integrating both conceptual and empirical aspects, including the use and application of the urban and regional planning information system. In particular, the information system performs three key functions:

- Identify the information needed to plan, implement, monitor, and evaluate.
- They use data collection, processing, analysis, and dissemination methods that meet accuracy, timeliness, and cost standards.
- They provide a governance structure that brings constant dialog between users and information providers.

In organizations that have started using computer technology in information handling, no significant change has been made in the set-up of smart cities or regional administration, in the decision-making structure or the actual work of planning and decision-making agencies.

The advantages of waste management are as follows: the profitable practice maintains a clean and fresh environment, saves the earth and energy, reduces contamination of the environment; and waste management will help you earn money and create jobs. The disadvantages of waste management are as follows: the process is not always cost-effective, the life of the resulting product is short, often the sites are dangerous, practices are not carried out consistently, and the management of waste can cause more problems.

2.4 REGIONAL PLANNING

An expansionist arrangement could have been a land-use agency trying to arrange and allocate land-use exercises, foundations, and expansion of settlements over such an imperatively broader range of property than just a discreet smart city [21]. By far the most substantial aspect of regional arrangement would be to facilitate the relationship between humans, global economy, natural resources, and environment, as well as the district relationship. By constructing this regional framework based on GIS, individuals working with areal arrangements can not only make arrangements based on measurable information but also consider the issue for space. This study included numerous different models that made organizers make a difference in making choices from completely different angles. Based on territorial arrangement lists and common organization of geographic data, the GIS-based territorial arrangement framework was developed. Using physical geographic information and social financial measurable information as a source of information, the framework combining GIS innovation can spatialize the lists of territorial arrangements as well as imitate the 3D or 2D approaches resulting from the models. All the models were made in the framework as a dynamic link library. This shows a framework of five modules [21] that appeared in Figure 2.6.

2.5 GIS IN RESOURCE MANAGEMENT

For the most part, any country's financial and social reasons are chosen by the land and water assets it has [22]. All these completely normal assets are also crucial to a country's economy as they try and play an unfriendly role in business giving; then they are indeed a citation of even more crude components for a few more businesses, acting as a source of food and wages, pharmaceutical and vitality. On the show day, due to the ever-growing population of human creatures, the use of the resource display in the world has been exaggerated [23]. The use of natural resources has the direct effect of broadening the scope of

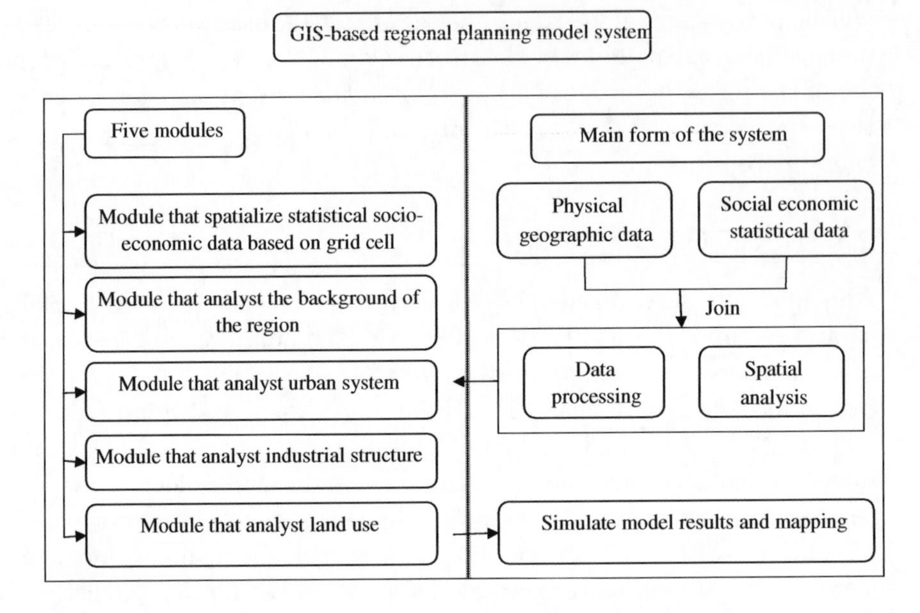

Figure 2.6 The framework of the regional planning model system.

living, changing climate patterns, and decreasing the financial and social interest guaranteed by their use. Because of these confines, there is a ghastly need to ensure that these resources are supervised satisfactorily. In the area of a normal resource supervised to induce this goal, a part of administration hones was created. Through these innovations, directors can create informative data and data that can be linked to create sound resolutions for achievable improvement. GIS is linked to creating specific information on how this raw data has been used to induce achievable common resource management within the decision-making process.

2.5.1 The Role of GIS in the Management of Natural Resources

Most land asset management features require information about the current range of attributes and how they have changed their dispensation in the past [24]. Such information can be composed through ground survey and/or aerial survey and satellite imagery application. By seizing these spatial data on a computer-based GIS and overlaying disparate data sets, the land asset planner and manager can inspect changes in property dispensation. This is necessary to assess the collision of previous planning determinations and to inventory existing policies.

2.5.2 Use of Remote Sensing and GIS in Natural Resource Management

Further detection and GIS are normally connected within the mapping process in common resource administration. It is possible to connect these innovations to advance a map distinction. Examples are land cover maps, vegetation maps, soil maps, and geology maps. To create a pre-field outline, the visual and computerized information collected is more often than not analyzed. The various information regarding components is analyzed.

2.5.3 Application of GIS in Natural Resource Management

The crucial use of GIS in distinctive feature inventory management is to challenge natural issues such as a surge, avalanche, soil scraped locations, monsoon season, and seismic tremor. GIS is addressing the current climate change issues, territorial misfortune, population development, contamination, etc. Application of GIS in normal resource management [25] is the arrangement for these issues. GIS has solved numerous common environmental issues. GIS would be an important tool linked again to the management of shared resources. A GIS helps to provide ingenious information when building works or any agrarian works in land administration. In characteristic resource management, GIS protects a wide range of biodiversity through the pre-information it provides, numerous natural living spaces and helps to arrange vegetation and fauna preservation. GIS provides hydrological information for the investigation of tipping point administration and Watershed analysis. Additional applications are briefly noted as below: management of facilities, analysis topography, and network analysis, modeling transport, design of engineering, analysis of demography, and modeling of geo processes. A few major GIS applications are discussed below.

2.5.3.1 Hazard and Risk Assessment

Within the exhaustion of a characteristic danger such as surge, avalanche, soil scrap spot, woodland fires, seismic tremor, and dry season, GIS in common asset management is connected. These characteristic calamities cannot be completely stopped, but these notices can be reduced by early arrangements and procedures. The GIS is linked in characteristic resource management to analyze, organize, supervise, and monitor common risks. GIS in common resource management provides spatial information about the calamities that have recently taken some time or may have emerged to protect early chance.

2.5.3.2 Change Detection

In common resource management, GIS gives almost land field data to change between periods. Reports of land alter taken note could be a supporting application in the assessment of land alters deforestation, urbanization, fracture of the environment, etc. The information obtained from GIS that can be worn out and around the area to think about a certain zone and observe it.

2.5.3.3 Natural Resource Inventory

Traditional asset stock can be a factual study of the characteristic asset condition. It provides different kinds of data on the environmental health condition and a pragmatic approach to the preservation program obtained in characteristic resource management through GIS. The GIS map data provides area data and current assets.

2.5.3.4 Environmental Monitoring

GIS provides graphical information in natural resource management that makes a difference in environmental monitoring. It selects highly subjective and quantitative relevant information about natural issues such as those for contamination, land corruption, and soil breakdown. GIS observes these issues in characteristic resource management and divinizes the long-term dangers.

2.6 GIS IN ENVIRONMENTAL MONITORING AND ASSESSMENT

When it comes to applications for natural modeling [26], the GIS program can be evaluated in many respects comparable to a programming dialect. Tragically, as it may be, the agent's natural GIS clients are focused on natural issues, and as it has barely had the extra specialized information that is essential for advancing unused GIS. First, recognizing the different types of natural observation and evaluation frameworks is valuable. Examination frameworks focus on the use of information control and display devices. The data frameworks are much more tired and stressed somehow on information capacity and administration.

GIS incorporates a primary role to play in natural observation, not astonishing. GIS is updated as a device for introducing information from assigned estimation stations (e.g. water quality sensors based on fields). Most new standard GIS program bundles do not have adequate equipment to take care of worldly information; the interface of orders must be expanded or assisted for a discussion of GIS and external computer program systems combination approaches [29–33].

2.6.1 Monitoring Systems in General

There are many different kinds of frameworks used for observation. Many of them robotize the collection and pre-handle the information. When planning a natural observation framework, consideration should be given to taking after components:

a. Satisfactory delays are likely to range from genuine real-time inspection to manual inspection after an investigation of the research facility;

b. The need for quality control may range from the display of unadulterated crude information to the preparation of absolute quality/quality control;

c. There is a regular discrepancy between manual and computerized systems for the examination of information. Any process is linked to genuine observation; the strategy includes taking steps after steps:

d. Communication assignment that collects and communicates information from a sensor (or other data source) to an accepting control system;

e. Pre-processing information using steps such as those for calibration, monitoring, and organization;

f. Store the information in a few database sorts;

g. Show the customers the information in an appropriate form.

2.6.2 Examples of Monitoring Systems

It is possible to use GISs at all Environmental Impact Assessment (EIA) stages [27]. EIA can be an assurance handle for recognizing and expecting collisions in the common environment. GIS will provide the EIA with a better approach to the analysis and control of spatial objects, and the investigation will provide an upgraded way of communicating, which can be an incredible appreciation of the open support process. The application of GIS within the EIA handle, where open cooperation is incredibly important, requires the development of applications that enable a higher understanding of spatial wonders. There are many reasons for this, the most common being:

a. Extended financial perspectives do not permit the collection of adequate information to "feed" investigative tools;

b. Information characteristics make simple extrapolation unpleasant;

c. A few more specifications are also related but it is not possible to specifically maintain the required information.

2.6.2.1 Real-Time Monitoring

The two most profitable parameters of the tidal ponds water interior are its oxygen and salinity as they have an overall impact on which species will flourish. An enormous deluge can cause saltiness to spread, while smaller ones can cause exhaustion of oxygen. The show status of the vital parameters can be continuously evaluated in near-real-time by using the GIS capabilities. Moreover, this approach allows a comparison with notable estimates and thus provides an amazing basic premise for a preliminary discussion.

2.6.2.2 Local Area Monitoring

A huge number of natural issues concerning the marine environment must be observed and evaluated within the development phase of the settled interface. In surveying the status of the environment, information such as hydrographic parameters, dregs transport, eel-grass development, musk scope, angle migration, water quality, and fowl life are considered. The EAGLE [28] scheme provides information on the ecosystem' show state as well as inter-atomic data with a framework for figure modeling. To adjust financial and natural interests, these components are used together to evaluate different development scenarios. EAGLE is built using the ArcView GIS from ESRI in conjunction with other specific programming tools. The system's main part runs on a PC hardware, using both local and wide area network technology for the project description, environmental challenges or legal requirements, monitoring programs, modeling, feedback loop, EAGLE system, and sub-systems.

2.6.3 Role of Remote Sensing and GIS in EIA

Topographic data framework (GIS) and Farther Detecting [34] play a critical role in the development of robotic spatial datasets and the establishment of spatial relationships. The EIA was carried out on the Man Waterway, Gujarat, India using GIS and Farther Detection software-Arc/ Info and ERDAS. Application studies of GIS in EIA found that while GIS is widely used, its usefulness is primarily limited to basic GIS capabilities such as map production, classic overlay, or buffering. This use abuses GIS's key advantage for EIA, its ability to conduct spatial examination and modeling. Some of GIS's use for EIA are complex methods of modeling representation, store information, and an estimate of the total collision.

The EIA is used:

a. Forms without a doubt that nearby organizations consider vital natural impacts arising from under – organizational undertakings.

b. Establish a strategy that gives the open the opportunity for substantial cooperation within the agency's thinking of the activity in question.

c. The EIA was rearranged to just be a comprehensive and detailed examination that analyzed thoroughly the discoveries of the potential natural collision of the given extent together tended to open concerns through the use of further detection and advances in GIS.

2.6.4 Advantages of Using GIS for EIA

In applying GIS for EIA [34], the individuals and organizations that opposed the survey recorded an arrangement of preferences or benefits.

a. The most critical advantage assessed by EIA professionals was the ability of GIS to conduct spatial examination and modeling, such as spatial issues, spatial resource, contamination investigation, imperative examination, and overlay operations.

b. The second most frequently cited advantage was the convenience of GIS to provide a clearer (or more attractive) introduction of the product, which is linked to GIS's ability to appear spatial information (from conventional advanced topical mapping to more creative and complex 3D visualization).

c. The power of GIS to store, manage and organize spatial information (indeed for exceptionally expanding data sets), as well as the ability to combine and control specific types of spatial data already associated with it, is the most sought-after focal point proposed by EIA experts.

d. Yet another vital advantage is the basic way in which GIS allows data to be adjusted and changed, and the unused results to be obtained change the basic conditions for the elective scenarios age, which is what EIA studies require.

e. There have also been other points of interest, although GIS is less commonly used, such as adaptability speed, accuracy, and unwavering quality.

2.6.5 Disadvantages in Using GIS for EIA

In contrast to the points of interest shown, there are issues related to GIS innovation that anticipate wider use or full potential.

a. The main critical disadvantages of using GIS for EIA are the time and the toll that should be passed on them.

b. The length of time required for setting up the GIS, building the database, and inputting and preparing important information are key imperatives and are probably the main imperative components to avoid the wider use of GIS in natural consultancies. There are two reasons for this.

c. The need for advanced information, the high preparation, and specialized skill prerequisites are other drawbacks indicated by the request.

d. Information infraction or deep thought of accuracy has been identified as a key GIS issue.

e. User-related errors were also said as a drawback when using computer-based GIS methods rather than more common manual strategies. Precision issues and quality controls of information are of particular importance in EIA, bearing in mind that natural articulations can be used as legitimate records [35].

f. Some consultancies described fragile GIS expository capabilities for specific purposes as an existing impediment to these systems [36].

2.7 GIS IN SOCIO-ECONOMIC DEVELOPMENT

GISs are connected by wide specialist differences to help them solve a wide range of spatially-based problems. Here, we focus on how GIS can be of productivity for professionals in socio-economic advancement who are forced with the errand of developing neighboring economies. Socioeconomic developers need strong decision-making tools to assist them in the examination of sport, show and spread, and make enlightened judgments about where to find modern businesses or increase existing ones. GIS devices can provide the basic stage for visualization, modeling, testing, and collaboration. One of the potential components that affect urbanization is often considered a driving constraint as well as a consequence of socio-economic development [36]. The population-related information is a critical and rapidly developing field of application for GIS [37]. This other information is also contrasted sharply here with all these relating either to material objects with certain locations, such as forests, geological infrastructures, or transportation systems. Within the request for computer frameworks, all these modern data sources will be monitored and analyzed, and GIS, specially equipped to take care of socio-economic information, will rapidly duplicate. Given these developments and the technology-and application-led growth of GIS to date, it is essential to establish a clear theoretical framework and to understand the specific socio-economic data type's issues. Currently, it is essential to look in a few details at the robotic frameworks used to handle socio-economic information.

2.7.1 Methodological Issues of Socio-Economic Data Integration in GIS Applications

Spatial population database design and modeling have been reviewed extensively by [38] whose own publications or articles evaluate most of the methodological issues endured in constructing world population registries and coordinating the spatial review of socio-economic, agrarian and natural information. Major issues can occur only on information about the quality and/or spatial information. Because of the exceptionally heterogeneous sources of socio-economic information, an array of integration problems can occur missing information on the position, inconsistent methodologies, and classifications, different units of space, various aggregation levels, gaps in the topic, and spatial data. While not initially seen as a device for socio-economic advancement in the community [39], GIS is rapidly becoming a particularly sought-after district innovation to offer. Modeling and spatial examination is another tool for arranging socio-economic advancement that applies to GIS.

2.7.2 Use of GIS by Economic Development Professionals

It is difficult to measure the region to which GIS is connected by nearby financial advancement experts [40] as usage information is compiled unsystematically. A 2004 study by the National Association of Development Officials Investigates establishment of 464 territorial improvement organizations may be an enlightening source. The study had a 63% countervailing rate. Although not all territorial advancement organizations in the Joined States used GIS, many stipulated that they would do so shortly. Only 11% of respondents announced they had not been arranged soon to receive GIS. The main applications of GIS were transported arrangements (81% of respondents), land-use arrangements (75%), and community and financial improvements.

2.7.3 Advantage of GIS in Economic Development Planning

a. The amount of time spent on long overwhelming errands is reduced by GIS. This allows time for other errands to be made, and it spares cash from districts as more workers do not need to be contracted.

b. Besides, finding individuals who are insinuating GIS becomes easier [40] as the industry and colleges currently offering early courses and certification programs are advancing.

c. The task was to select an area along the Interstate exit and choose whether the location would be suitable for a modern restaurant based on a wide range of variables.

2.7.4 Disadvantage of GIS in Economic Development Planning

If GIS is so extraordinary, why don't communities still apply this innovation at that point?

a. Initially, a particularly high startup [40] took a toll to create a GIS, and districts have exceptionally tight budgets.

b. Working information and skills representatives are thick to discover. It is both time- consuming and exorbitant to collect information, and it is accurate to guarantee the information.

2.8 GIS IN EMERGENCY MANAGEMENT

Emergency management as the teacher and caller use science, innovation, arrangement, and administration to negotiate with incomparable opportunities that can harm or kill large numbers of individuals, cause widespread damage to property, and disturb community life. GIS was arranged to support topographical requests and ultimately to make spatial choices [41]. GIS significance in crisis management evolves directly even from the inclinations of those wanting to join such entrepreneurialism set up to boost spatial free choice in an area with a solid prerequisite for addressing various vital spatial judgments. For this kind of reason, contemporary GIS applications in crisis management, together with an intrigued in advancing this slant, have been created a long time later.

The advanced and powerful overall improvement of geospatial technological innovation works emphatically and impacts all aspects of disaster management – such as relief (modeling hazards and helplessness in creating procedures), readiness (defining crisis response and clearing plans), reaction (implementing such plans), and recovery (evaluating harms, modifying, anticipating repetition, and teaching to the public) [42].

Considering that we cannot stop normal disasters, choosing potential dangers, and where they relate to our communities is essential. Figure 2.7 shows GIS in emergency management and related areas [41].

2.8.1 Implementing the Mission with GIS

- Work to ameliorate emergency impacts
  - Analysis of hazards and serious risk sculpting
- Try to educate about readiness for calamity
  - Fully interactive use of the Internet and public at large illustrations

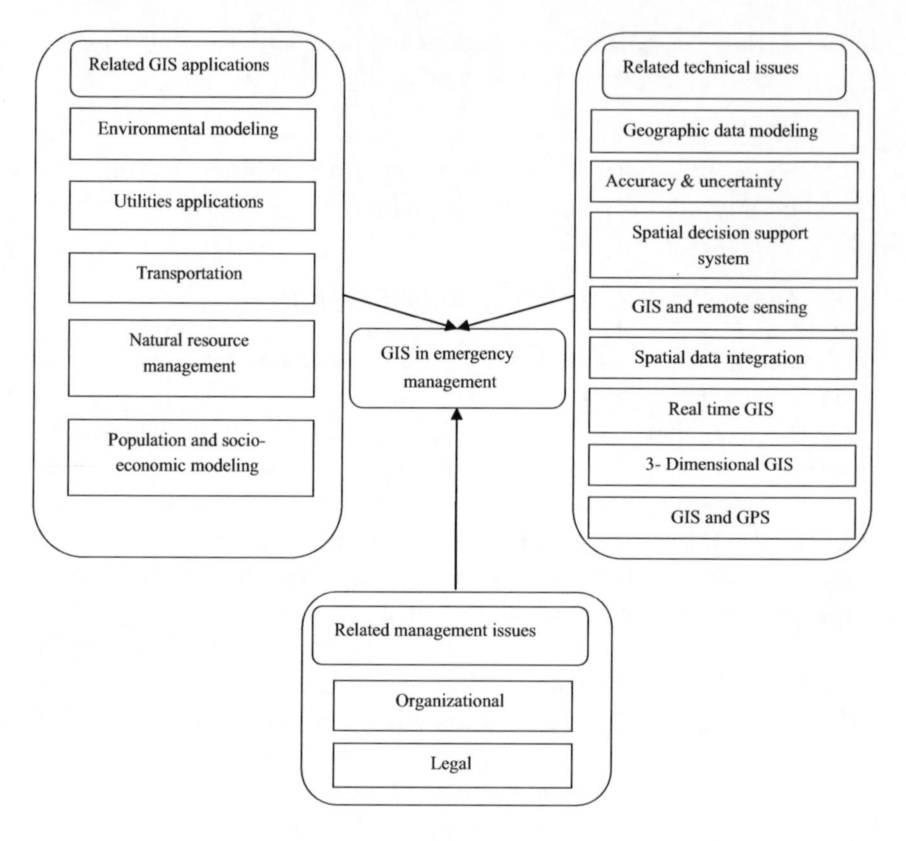

Figure 2.7 GIS in emergency management and related areas.

- Clear and direct actual improvement of both the city's emergency and disaster plans
  - Modeling and geospatial analysis
- Make arrangements or support emergency response and recovery
  - Applications for situational awareness and workflow recovery
- Try to collect, dissect and disseminate data on the occurrence
  - Geospatial maps and maps of occurrence

Emergency management services [51] code requires each ideological housing development within the commonwealth to prepare and support emergency operations to ensure that the damage and annihilation caused by the disaster are preserved and minimized [43].

2.8.2 Emergency Operations Plan

The emergency operations plan presents a collection of techniques and methods, allots parts and obligations to organizations to oversee an emergency occurrence or an arranged occasion over all stages of an occurrence, and communicates an all-encompassing approach.

2.9 GIS IN EDUCATION

GIS provides a solid decision-making toolkit [44] that can be linked in the instructive organization, instructional arrangement, and instruction. GIS provides chairmen with a way of imagining and supervising everything from observing campus security, mapping campus buildings, cable, and other foundations, running school buses, planning where and when to shut down schools and opening unused ones, and strategizing enrolment efforts. GIS also provides instructive policymakers with devices to see designs in instructive performance and where to stamp modern programs. GIS masters should know the basic standards of cartography at a basic level. This teaches essential standards [44] that must be taken into account by anyone who considers the outline to also effectively communicate data to everyone else. Numerous related disciplines are advised by GIS [45], and appropriate performance improvement electives are a great way of teaching individuals who want to finish GIS masters but are not locked in as GIS clients by and by. For those who need to supplement GIS instruction, there is a wide range of instructive alternatives available. These are the following choices: (a) formal university learning courses, (b) short lectures and occasional random conferences, (c) courses for vendors, (d) workshops for conferences, (e) workshops for professional, and (f) courses for self-study. Applying GIS is not just about using innovation and becoming a master in using GIS tools but perhaps cultivating a whole range of communication, information, media familiarity, and pivotal consideration skills.

2.9.1 Trials and Tribulations of GIS in K-12 Education

Audet authenticated numerous comments from Tinker [47] in their 1996 thinking about GIS and K-12 [46] learning. They offer a few key reviews for learning GIS in the classroom, counting instructors' ability to advance and gradually categorizing students' problem-solving styles. For K-12 understudies, there are many other positive benefits of GIS innovation. Spatial competence and geographical competence, characterized as the ability to recognize the area or topology of the focus and properties of the outline, are two of these points of interest. Finally, understanding scale and determination seems to be a fundamentally critical understudy errand, most promptly supported by GIS application [48]. Applications include office planning exercises that encompass room capacity, a framework for innovation, readiness for crises, and security on campus. Policymakers use geographic tools to make data-driven decisions at all levels through the Internet of Things (IoT) network [52]. GIS is connected at both the nearby and national level to imagine imperative usage, execution, and compliance information. These pivotal considerations include cases: (a) performance for students, schools, and districts, (b) ratios between teacher and student, (c) limited rates of English skills, (d) expenditure per pupil, (e) recipients of free/reduced lunch, (f) federal funding distribution, (g) the country, district, and school comparisons, (h) associations between the educational or community at large features.

2.10 ML IN GIS APPLICATIONS

ML is progressively advancing as a successful tool to predict non-linear time-series data such as weather and environmental data. For the air quality prediction authors used various ML models [53]. ML methods pledge the revolutionary growth of meteorology with a dataset of time series [54,59]. There are different ML use cases for the development of GIS applications such as land classification, agriculture, road networks, identification and tracking of an object, 3D modeling, and digital evaluation, geospatial attribute trending (census, twitter), image stitching and preprocessing. Venkatraman et al. [55] proposed and analyzed the framework for a successful malware classification of hybrid image-based solutions with profound learning architectures.

A major issue confronting technicians, the agencies of public services and the infrastructure is that without existing coordinate/projection systems there are lots of old maps. Artificial Intelligence (AI) and ML are mainly used for object detection in the GIS and geospatial fields. With the automatic identification of artifacts, you can retrieve photos by dramatically reducing the number of annotators in any digitalization project. The convergence of AI and GIS is on the increase with the trend of digital transformation. AI allows quick and cost-effective seizing and changes the maps or GIS databases. In the classification, prediction, and segmentation of images, it supports GIS strongly. Another use of

ML is to stitch orbital or satellite photographs together as one builds a sculpture. For economic models, GIS should be applied to collect and correlate economic and social data with store profits. However, GIS is a very excellent use of AI and ML, since space-time points are fixed and can work as a consistent reference point for data analysis that is quite detached.

2.11 CYBERSECURITY IN GIS

To prevent financial losses and reputational damage incurred by those intrusions, cybersecurity experts use various tools and techniques. Geographic information science provides tools that allow enterprises to evaluate and build stronger safeguards in compromised networks. The role of GIS in cybersecurity is still increasing with more enterprises exploring the importance of addressing geospatial challenges to avoid an emerging variety of hazards [56]. A clear viewpoint on the data flow through the network of an enterprise contributes to operational awareness of any interruption or malfunction of the system that can conflict with the activity. Spatial data connects an occurrence to specific sites, helping specialists to determine whether the problem is coming from a systematic attempt to compromise the machine and assess its impact. Cybersecurity is vital to the proper functioning of key infrastructures in our globally connected society. Cyber-attacks on power companies can follow many forms, for example transmitting false energy consumption details in particular areas. Algorithms for security detection can identify distribution load problems which may imply that providers get inaccurate information. When discrepancies occur, energy suppliers can assess if they have been hacked and respond accordingly. Cybersecurity professionals need to implement a security layer that prevents unauthorized access to remote sensing data and location details.

2.12 CONCLUSION

For a long time, GIS has been used to screen distinctive geological highlights for nature change. Engineers and scientists continue to abuse GIS observing properties to track changes in design or arrival behavior over a specified period. It makes different experts make educated choices about a range's progress condition and work out a plan. It can be used throughout a smart city's life cycle – from location selection, planning, and development to use and support. It is a perfect innovation that can scale across any scope, from a person's resource within a building to an essentially global setting that binds together all the perspectives of a keen city that arranges and improves. Working towards urban arrangement, the primary era had to rely on sociologists, designers, and financial specialists' consideration and supposition to reach their destinations. At that point, with GIS's approach, the situation changed with experts taking advantage of urban arrangement's most extreme benefits. GIS is not now a fair instrument for capturing and analyzing, but an important resource for spatial modeling, making choices, and a parcel of other disciplines. The above survey suggests that scale is a critical figure that recognizes the arrangement of GIS applications. The GIS meets the main requirement of competent spatial data handling capability in all levels of arrangement. The application of GIS for territorial financial arrangement has been overwhelmingly accentuated. The critical thinking of GIS only as an extremely high-profile, high-tech, one-source arrangement for numerous issues and the appeal of a centralized, all-encompassing database have reinforced the fundamental political back-up subsidizing GIS selection. Sociological contemplation is gaining ground over technical difficulties, and it could be a noticeable concern to arrange a bunch to get to the framework. It may be a matter of guessing to what extent GIS is changing the arrangement. It is still vague whether the pace of alteration is sufficient to be called an insurgency or whether GIS will be essentially absorbed into the current honey arrangement. GIS gets up for well-established, politically favorable, non-automated strategies that can jumble the handling of the decision-making by wastefulness and documentation need.

References

[1] Xhafa, S., Kosovrasti, A., (2015), Geographic Information Systems (GIS) in urban planning. *European Journal of Interdisciplinary Studies*, vol. 1, issue 1, pp. 85–92.

[2] Admin, Grind GIS (2015), What is a Geographic Information System? https://grindgis.com/what-is-gis/what-is-gis-definition/. Accessed 5 Jan 2019.

[3] GIS Contributor (2014), GIS in Urban Planning. https://www.gislounge.com/gis-urban-planning/. Accessed 5 Jan 2019.

[4] Admin, Grind GIS (2018), Importance of GIS in Planning. https://grindgis.com/gis/importance-of-gis-in-planning. Accessed 5 Jan 2019.

[5] Mackaness, W.A. (1994), Curriculum Issues in K-12. GIS/LIS '94, pp. 560–568. http://libraries.maine.edu/Spatial/gisweb/spatdb/gislis/gi94070.html.

[6] Paul, S. (2018), Urban Planning and GIS. https://www.geospatialworld.net/blogs/gis-in-urban-planning/. Accessed 21 Dec 2018.

[7] Kumar, A., (2015), GIS for smart cities, *Esri India*, vol. 9, issue 1. http://www.esriindia.com/~/media/esri-india/files/pdfs/news/arcindianews/Vol9/gis-for-smart-cities.pdf. Accessed 10 Jan 2019.

[8] Space and Naval Warfare Systems Center Atlantic (2013), Geographic Information System Software Selection Guide, https://www.dhs.gov/sites/default/files/publications/GIS-SG_0713-508.pdf. Accessed 05 Jan 2020.

[9] GIS Geography (2020), Commercial GIS Software: List of Proprietary Mapping Software. https://gisgeography.com/commercial-gis-software/. Accessed 15 Jan 2020.

[10] Rodrigue, J.P. (2017), *The Geography of Transport Systems*, New York: Routledge, p. 440. ISBN 978-1138669574. https://transportgeography.org/?page_id=6578. Accessed 10 Jan 2019.

[11] Gupta, P., Jain, N., Sikdar, P.K., Kumar, K. (2003): Geographical Information System in Transportation Planning. *Map Asia Conference*, New Delhi, India.

[12] Vonderohe, A.P., Travis, L., Smith, R.L., Tasai, V. (1993), Adoption of Geographic Information System for Transportation, Transport Research Board, National Research Council, Washington, DC, NCHRP Report 359.

[13] RenPeng, Z., Beimborn, E.A. (1998), Internet GIS and its Applications in Transportation. *TR News*.

[14] Gautam, A.K., Kumar, S. (2005), Strategic planning of recycling options by multi-objective programming in a GIS environment.

Clean Technologies and Environmental Policy, vol. 7, pp. 306–316, ISSN 1618-9558.

[15] Kanchanabhan, T.E., Abbas Mohaideen, J., Srinivasan, S., Lenin KalyanaSundaram, V. (2011), Optimum municipal solid waste collection using geographical information system (GIS) and vehicle tracking for Pallavapuram municipality. *Waste Management & Research*, vol. 29, pp. 323–339, ISSN 1096-3669.

[16] Vijay, R., Gupta, A., Kalamdhad, A.S., Devotta, S. (2005), Estimation and allocation of solid waste to bin through geographical information systems. *Waste Management & Research*, vol. 23, issue 5, pp. 479–484, ISSN: 0734-242X.

[17] Dr. Vohra notes for the course CE 647 Municipal Solid Waste Management.

[18] Upasna, S., Natwat, M.S. (2003), Selection of potential waste disposing sites around Ranchi Urban Complex using Remote Sensing and GIS techniques. *Map India.*

[19] Tchobanoglous, G., Theisen, H., Vigil, A.S. (1993), *Integrated Solid Waste Management: Engineering principles and management issues.* McGraw-Hill Inc., New York.

[20] Santos, L., Coutinho-Rodrigues, J., Current, J. (2008), Implementing a Multi-vehicle Multiroute Spatial Decision Support System for Efficient Trash Collection in Portugal. *Transportation Research, Part A: Policy and Practice*, vol. 42, issue 6, pp. 922–934, ISSN: 0965-8564.

[21] Dantzig, G.B, Fulkerson, D.R., Johnson, S.M. (1954), Solution of a large-scale traveling salesman problem. *Operations Research*, vol. 2, pp. 393–410, ISSN 0030-364X.

[22] Johansson, O.M. (2006), The effect of dynamic scheduling and routing in a solid waste management system. *Waste Management*, vol. 26, pp. 875–885, ISSN 0956-053X.

[23] Kim, B.I., Kim, S., Sahoo, S. (2006), Waste collection vehicle routing problem with time windows. *Computers and Operations Research*, vol. 33, pp. 3624–3642, ISSN 0305-0548.

[24] Sahoo, S., Kim, S., Kim, B-I., Kraas, B., Popov, A. (2005), Routing Optimization for Waste Management. *Interfaces*, vol. 35, issue 1, pp. 24–36, ISSN: 0092-2102.

[25] Tavares, G., Zsigraiova, Z., Semiao, V., Carvalho, M. (2008), A case study of fuel savings through optimization of MSW transportation routes. *Management of Environmental Quality*, vol. 19, issue 4, pp. 444–454, ISSN: 1477-7835.

[26] Xu, R., Wang, L., Wang, Z., Guo, Y. (2010), A GIS-based system for regional planning and its application in Chengdu, China, *The 2nd International Conference on Information Science and Engineering*, Hangzhou, pp. 3424–3427. doi: 10.1109/ICISE.2010.5691857.

[27] Harahsheh, H. (2001), *Development of Environmental GIS Database and its Application to Desertification Study in Middle East.* Chiba: Chiba University (2001).

[28] Swe, M. (2005), Application of GIS and remote sensing in Myanmar. http://www.aprsaf.org. Accessed 1 Jan 2019.

[29] Deane, G.C. (1994), The role of GIS in the management of natural resources, *Aslib Proceedings*, vol. 46, Issue 6, pp. 157–161, https://doi.org/10.1108/eb051360 .

[30] Uiz, (2017), 5 Ways to Use of GIS in Natural Resource Management. http://uizentrum.de. Accessed 2 Jan 2019.

[31] Ji, W., Mitchell, L. C. (1995), Analytical model-based decision support GIS for wetland resource management. In Lyon J G, McCarthy J (eds) *Wetland and Environmental Applications of GIS.* New York, Lewis Publishers, pp. 31–45.

[32] Abel, D.J., Kilby, P.J., Davis, J.R. (1994), The systems integration problem. *International Journal of Geographical Information Systems*, vol. 8, issue 1, pp. 1–12.

[33] Federa, K., Kubat, M. (1992), Hybrid geographical information systems. *EARSel Advances in Remote Sensing*, vol. 1, issue 3, pp. 89–100.

[34] Goodchild, M.F., Haining, R., Wise, S. (1992), Integrating GIS and spatial analysis: problems and possibilities. *International*

Journal of Geographical Information Systems, vol. 6, issue 5, pp. 407–423.

[35] Maguire, D.J. (1995), *Implementing Spatial Analysis and GIS Applications for Business and Service Planning.* Cambridge, UK: GeoInformation International, pp. 171–191.

[36] Steyaert, L.T., Goodchild, M.F. (1994), Integrating geographic information systems and environmental simulation: a status review, *Scanning Electron Microscopy*, pp. 333–355.

[37] Larsen, L., GIS in Environmental Monitoring and Assessment, pp. 999–1007. https://www.geos.ed.ac.uk/~gisteac/gis_book_abridged/files/ch71.pdf. Accessed 10 Jan 2019.

[38] Maucha G. (2014), The EAGLE data model. https://lcluc.umd.edu/sites/default/files/lcluc_documents/27_Eagle_Maucha_0.pdf, pp. 1–17. Accessed 29 Dec 2019.

[39] Mondal, P., Role of Remote Sensing and GIS in Environmental Impact Assessment (EIA). http://www.yourarticlelibrary.com/environment/role-of-remote-sensing-and-gis-in environmental-impact-assessment-eia/27476. Accessed 16 Jan 2019.

[40] Epstein, E. (1991), Legal aspects of GIS. In Maguire, D., Goodchild, M., Rhind, D. (eds) *Geographical Information Systems: Principles and Applications.* Harlow: Longman. vol. 1, pp. 489–502.

[41] Kidner, D., Dorey, M., Sparques, A. (1996), GIs and visual impact assessment for landscape planning. *Proceedings of the Conference GIS Research UK (GISRUK) '96*, University of Kent, Canterbury, pp. 89–95.

[42] Batty, M. (2013), *The New Science of Cities.* Cambridge, MA: MIT Press.

[43] National Center for Geographic Information and Analysis (NCGIA)/University of California. Santa Barbara, CA 93106, USA. Tel: (805) 893-8224. Fax: (805) 893-8617.

[44] Goski, A.M. (2005), Economic Development and GIS. http://www.umich.edu/~econdev/gis/index.html. Accessed 19 Jan 2019.

[45] Reid N., Carroll M.C., Smith B.W., Frizado J.P. (2009), GIS, and economic development. In Gatrell, J.D., Jensen, R.R. (eds), *Planning and Socioeconomic Applications. Geotechnologies and the Environment*, Springer, Dordrecht, vol. 1, pp. 5–28.

[46] Drabek, T.E., Hoetmer, G.J., (eds), (1991), *Emergency Management: Principles and Practice for Local Government.* Washington, DC: International City Management Association.

[47] Cova, T.J. (1999), *GIS in Emergency Management, Chapter in Geographical Information Systems: Principles, Techniques, Applications, and Management*, John Wiley & Sons, New York, pp. 845–858.

[48] Bouzige, J.B. (2017), Smart City Summit: Managing the Smart City. https://www.ekimetrics.com/news/smart-city-summit-managing-smart-city?page=7. Accessed 2 Feb 2018.

[49] The Role of GIS in Emergency Planning, City of Philadelphia. https://www.dvrpc.org/Committees/IREG/Presentations/2011-06.pdf. Accessed 19 Jan 2019.

[50] Kerski, J.J. (2018), Why GIS in Education Matters. https://www.geospatialworld.net/blogs/why-gis-in-education-matters/. Accessed 19 Jan 2019.

[51] Savery, J.R. (2005), BE VOCAL: Characteristics of successful online instructors. *Journal of Interactive Online Learning*, vol. 4, issue 2, pp. 141–152.

[52] Audet, R.H., Abegg, G.L. (1996), Geographic information systems: Implications for problem-solving. *Journal of Research in Science Teaching*, vol. 33, issue 1, pp. 21–45.

[53] Geospatial World (2010), The History and Application of GIS in K-12 Education. https://www.geospatialworld.net/article/the-history-and-application-of-gis-in-k-12-education/. Accessed 27 Jan 2019.

[54] Amit, B., Chinmay, C., Anand, K., Debabrata, B. (2019), *Emerging Trends in IoT and Big Data Analytics for Biomedical and Health Care Technologies.* Elsevier: Handbook of Data Science Approaches for Biomedical Engineering, Ch. 5, pp. 121–152, ISBN:9780128183182.

[55] Athira, V., Geetha, P., Vinayakumar, R., Soman, K.P. (2018), DeepAirNet: Applying recurrent networks for air quality prediction, *Procedia Computer Science*, vol. 132, pp. 1394–1403, https://doi.org/10.1016/j.procs.2018.05.068.

[56] Chinmay C. (2019), *Computational Approach for Chronic Wound Tissue Characterization.* Elsevier: Informatics in Medicine Unlocked, 17, pp. 1–10, https://doi.org/10.1016/j.imu.2019.100162

[57] Chakraborty, C., Gupta, B., Ghosh, S.K., Das, D, Chakraborty, C. (2016), Telemedicine supported chronic wound tissue prediction using different classification approach. *Journal of Medical Systems*, vol. 40, issue 3, pp. 1–12.

[58] Venkatraman, S., Alazab, M., Vinayakumar, R. (2019), A hybrid deep learning image-based analysis for effective malware detection. *Journal of Information Security and Applications*, vol. 47, pp. 377–389. https://doi.org/10.1016/j.jisa.2019.06.006.

[59] USC Dornsife (2019), Online GisProgramme, https://gis.usc.edu/blog/gis-and-cybersecurity/. Accessed 26 Dec 2019.

CHAPTER 3

A Review of Checkpointing and Rollback Recovery Protocols for Mobile Distributed Computing Systems

Houssem Mansouri

Ferhat Abbas Sétif University 1

Al-Sakib Khan Pathan

Independent University

CONTENTS

3.1 INTRODUCTION

Mobile computing field has shown a great level of improvement in the past decade. Nowadays, hundreds and thousands of mobile devices are used for various purposes, and their resources today allow running advanced distributed applications connecting several devices in a network fashion. Such environment has opened new opportunities, but the issue of fault tolerance remains as a great challenge. A system is called *smart* when it has some kind of ability to understand its own standing and the context of a situation based on which it can decide which way to go forward or continue operations. Fault tolerance and recovery is one of the core aspects required for mobile distributed computing systems (MDCS). When a setting like Internet of Things (IoT) is considered, this is even more important to introduce appropriate smart mechanisms of operation.

One of the fault tolerance techniques is *checkpointing*, which is a way to roll back (or, go back) a system to the most recent *failure-free* status. A checkpoint is basically a snapshot of the system's state which could be recorded when the distributed application is running normally. Then, in case of a failure, that snapshot could be used to go back to the state when the system showed no error or failure.

When a failure occurs in MDCS, it is not enough to roll back only the failed process, because this is often dependent on other processes or other processes may depend on this one. In fact, in such a setting, the devices are mobile and they can interact/communicate among themselves. Hence, to get back the original *fault-free* or *failure-free* state, we need to take into account many other interrelated processes.

In fact, MDCS environment introduces new challenges for fault-tolerant computing, generally and for the protocols of checkpointing, especially. Compared with traditional distributed environments, wireless networks often have a low bandwidth for which speed would be relatively slower and throughput would be constrained. In addition, mobile hosts (MHs) possess limited resources, are usually more exposed to failures, can roam, and get disconnected frequently while operating. Traditional checkpointing protocols cannot be directly applied for mobile environments. Indeed, it is necessary to reconsider these protocols to adapt them to fit for the mobile context.

After this, the main system model is presented in Section 3.2. Section 3.3 adds the background and definitions. Section 3.4 elaborates the checkpointing techniques in this area. In Section 3.5, a literature survey is done. We classify the proposed protocols in Section 3.6. Lastly, Section 3.7 concludes the chapter.

3.2 THE SYSTEM MODEL

There is a common system model that is used for almost all works in this field. Figure 3.1 depicts the main model on which almost all works that came in sequence are based on, since the first one by Acharya and Badrinath [1] up to the recent one [2]. There are mainly two types of entities: MHs and Fixed Hosts (FHs) like *Mobile Support Stations (MSSs)* and *Static Hosts (SHs)*. The fixed network is formed with all the FHs and a communication path connecting them. An MSS communicates with MHs within its cell via a wireless medium and connects the wireless network.

This distinctive characteristics of MHs are (compared to FHs) [1,3] as follows:

- MHs can change cells and get frequently disconnected [3].
- MHs are vulnerable to physical damage and can be broken/ exhausted [4].

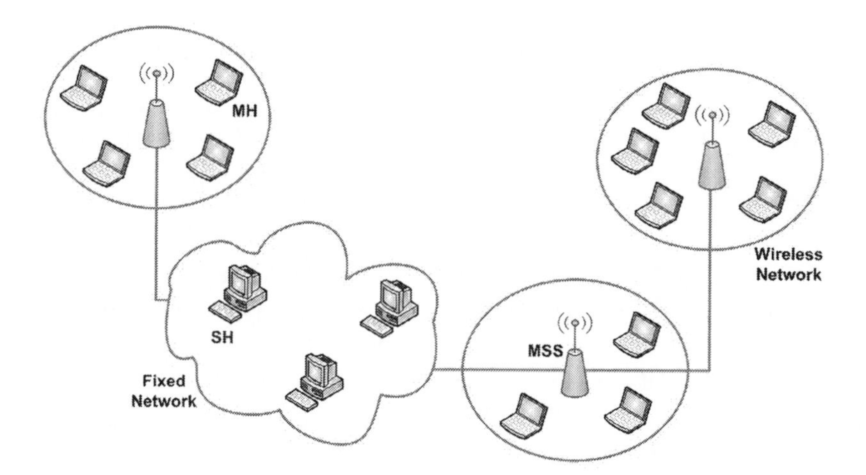

Figure 3.1 System model.

- As *mobility* is an integral feature, the storage is also considered unstable for an MH [5].
- Run by batteries, the power source is really constrained and a replacement is often needed [4].
- Due to resource limitation, MHs have to use low bandwidth wireless channels to MSSs [6].

3.3 BACKGROUND AND DEFINITIONS

- *Local Checkpoint*: In MDCS, the local states of all processes are saved at different points of time. Such locally saved status information is called the local checkpoint, which could be denoted by C_i, k [7].
- *Checkpoint Interval*: For any process (P_i), the ith checkpoint interval means all the computations performed by the process between ith and $(i + 1)$th checkpoints (this is excluding the $(i + 1)$th one) [8].
- *Global Checkpoint*: This could be represented as $\{C_1, x, C_2, y, \ldots, C_n, z\}$, which is basically a set of local checkpoints of all contributing processes. This also includes the status information of communication channels [1,5] involved in the operation.
- *Orphan Message*: A message is termed *orphan* when there is a record in the local checkpoint of a destination process that an event has happened but then its *send event* itself is missing [9].
- *Consistent Global Checkpoint*: When the local checkpoint of a process shows reception of a message, the local checkpoint of the associated sending process must also record sending of that message [10,11].
- *Distributed Checkpointing*: Even if a failure occurs (which is still normal), bringing the system to a consistent global checkpoint is the main objective of a checkpointing protocol [12,13]. Various issues are taken into consideration for distributed checkpointing.

3.4 CHECKPOINTING TECHNIQUES

Figure 3.2 shows the categories of checkpointing techniques. Let us discuss about each of them.

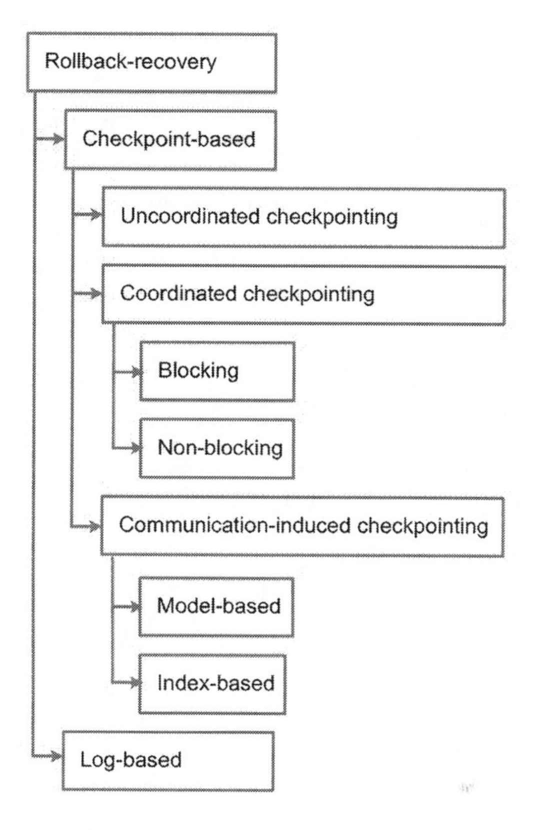

Figure 3.2 Various techniques for checkpointing in MDCS environment.

3.4.1 Checkpoint-Based

When a failure occurs, this technique tries to restore the most recent consistent state of checkpoints (called *recovery line*) of the system [14]. There are three variations of this technique [15].

3.4.1.1 Uncoordinated Checkpointing

In uncoordinated checkpointing, no coordination is required among the processes running in the system. A process makes independent decision when to take its own checkpoint [14–16].

3.4.1.2 Coordinated Checkpointing

In this technique, checkpoints are taken by the processes in a way to achieve consistent global state. The processes would coordinate among themselves for this. A two-phase *commit* structure can be used for this technique [14,15,17,18]. Two types of coordinated checkpointing are as follows:

- *Blocking Checkpointing*: In this approach, during the execution of checkpointing mechanism, the processes temporarily stop the computation or their execution to save checkpoint [14,19].
- *Non-Blocking Checkpointing*: In this approach, while executing checkpointing process, there is no need to temporarily halt the mobile distributed application's computation.

3.4.1.3 Communication-Induced Checkpointing

In this technique, only a subset of interacting processes takes checkpoints. This set has a specific technical term, "*minimum set*" [14,15] which is of two types [8]:

- *Model-Based*: In this checkpointing technique, it is required to prevent some patterns of communications and checkpoints (some models are used) that may cause inconsistent states.
- *Index-Based*: This technique uses an assigned sequence number (i.e., index) to each local checkpoint.

3.4.2 Log-Based

In this technique, all the non-deterministic events' determinants are logged (on stable storage) by each process (that is running in the system). When failure occurs, the affected processes can recover by replaying the corresponding non-deterministic events in the order they were saved.

3.5 LITERATURE SURVEY – EARLY STAGE TO CURRENT TREND

For our work, we have tried to cover the area taking into consideration all works in sequence to make it as comprehensive as possible till the time of writing of this chapter. Many works are very similar but have significant difference(s) in the idea. Some have concrete proofs of efficiency while some are mainly theoretical contributions. In this section, we talk about all those works; however, the sequence of exact publishing year may not be maintained because of the required flow of text and the interlinked ideas of various approaches.

Krishna, Vaidya, and Pradhan [6] are the earliest authors who introduced the problem of checkpointing for distributed systems in mobile computing in 1993. Acharya and Badrinath [1] are the first authors who proposed the first checkpointing protocol for MDCS in 1994, satisfying the constraints of mobility. The majority of the proposed protocols that have come later are based on this work.

Manivannan and Singhal [20] develop a quasi-synchronous checkpointing protocol that uses a coordination of communication-induced checkpoints for progression of recovery line. The overhead involved in the recovery is very low since messages are logged and replayed using selectively pessimistic message logging approach at the receiver.

Prakash and Singhal [21] describe a synchronous checkpointing protocol which makes it optional for each host to save a local checkpoint. Also, it uses a non-blocking approach (i.e., the underlying computation is not blocked or stopped). The recovery algorithm here is a compromise between two diverse recovery strategies: fast recovery with high communication and storage overheads and slow recovery with little communication overheads.

Neves and Fuchs [22] design a protocol targeted especially for mobile environments. Even without the need of exchanging messages using a local timer, authors propose a simple mechanism which is used to keep the checkpoint timers approximately synchronized. This mechanism can record consistent recoverable global checkpoint, which is a positive aspect of this work.

Quaglia, Ciciani, and Baldoni [23] show a comparative analysis of several particularly interesting communication-induced checkpointing protocols. Such comparison has been carried out by a simulation study while varying both the mobility assumptions and the disconnection rate of MHs. Results show that index-based checkpointing performs better.

The concept of *mutable checkpoint* is introduced by Cao and Singhal in [24]. The objective of this new issue is to avoid the overhead of transferring huge volume of data via wireless network to the stable storage in MSS, since a mutable checkpoint can be saved anywhere even in the main memory or the local disc of MHs. Also, the authors present a technique to minimize the number of mutable checkpoints.

Yang et al. [25] consider a system with two distinct parts: an FH network and the wireless network. Then, they propose two checkpointing algorithms using a two-phase-cut approach. Their mechanism basically adopts a two-tier system model as required for their protocol design.

The two-phase-cut algorithms employ an efficient message passing mechanism with low overhead.

Cao and Singhal [26] argue that there was no non-blocking type protocol that would compel only a minimum number of processes to save the local checkpoints. Based on the finding, the authors propose a new checkpointing algorithm, which forces only a minimum number of processes to take checkpoints and reduces the blocking time during the checkpointing process.

A hybrid protocol is proposed by Higaki and Takizawa [27]. Here, MHs take checkpoints in an asynchronous manner while the FHs do that in a synchronous manner. The authors also show that the hybrid checkpointing protocol implies less total processing time than the synchronous one.

Ssu et al. [28] present a checkpointing protocol which uses a time-based technique to reduce the number of message, which is integrated with leasing enabled storage managers to control disk-space effectively. The checkpointing mechanism uses also a three-level storage hierarchy to improve checkpointing performance.

Juang and Chen [29] propose an efficient non-blocking coordinated checkpointing protocol, where only a few MHs need to roll back once and then can immediately continue the operation. The algorithm only needs $O(n_r)$ messages to rollback the system to a consistent state, where $O(n_r)$ is the total number of *MHs* in the system related to the failed *MH*.

Park and Yeom [30] describe an optimistic message-logging based asynchronous scheme. The drawback of this is that the erratic MHs and loose network connection may really hamper effective coordination. To avoid the overhead of message logging by MHs and any additional information carried on each message between a MH and the MSS, the tasks of logging and dependency tracking are fully performed by MSSs.

Gass and Gupta [31] design an efficient communication-induced protocol. The exception of this is that it does not take basic checkpoints but forces checkpoints (when needed); also no extra information needs to be piggybacked with the application messages.

Cao and Singhal [9] develop a relatively new protocol for checkpointing based on mutable checkpoint according to their previous result in [24] and [26]. The proposed non-blocking algorithm avoids the avalanche effect and forces only a minimum number of processes to take their checkpoints on the stable storage.

Morita and Higaki [32] present an extended three-phase hybrid coordinated checkpoint and log-based protocol for MDCSs. The order information of events in a mobile station and message content is separately transmitted to an access point without additional control messages.

Morita and Higaki [33] achieve a hybrid checkpoint protocol where MSSs take local checkpoints asynchronously and MHs do the same synchronously. In addition, mobile stations use message logs to get a consistent state.

Manabe [34] shows a coordinated checkpointing protocol. This protocol could be shown as an optimal solution among the Chandy and Lamport's [5] distributed snapshot algorithms, when certain assumptions are considered so that the message overhead is independent of the number of MHs, and the number of checkpoints is minimized under two assumptions.

Park, Woo, and Yeom [35] devise a recovery protocol using checkpointing and message logging based on movement scheme. The recovery information is kept at the visited MSSs while the host could be mobile.

Lin and Dow [36] describe a two-phase protocol: coordinated checkpointing phase among the MSSs and communication-induced checkpointing phase between each MSS and its MHs. The failure recovery algorithm in this investigation integrates the advantages of coordinated and communication-induced checkpointing algorithms.

Park, Woo, and Yeom [37] present a region-based checkpointing and message logging technique. Here, a group of cells is assigned to a manager who deals with the recovery of MHs in a region. The proposed scheme considers the reduction on both of the failure-free operation cost and the failure-recovery cost.

Lin, Wang, and Kuo [38] design a coordinated checkpointing protocol based on time, which reduces the number of checkpoints with a limited number and the size of messages transmitted via the wireless network. The protocol uses the accurate timers in the MSSs to adjust the timers in the MHs.

Park, Woo, and Yeom [39] show an efficient asynchronous optimistic logging checkpointing protocol. Here, logging and dependency tracking tasks are performed by the MSSs using the volatile log space. The MHs only carry minimum information so that the mobility of MHs can properly be traced by the MSSs.

Lin et al. [40] propose an enhanced coordinated checkpointing protocol based on time. This protocol reduces the number of checkpoints to nearly minimum, by integrating an improved timer. The protocol performs very well in terms of minimizing the number and size of messages transmitted via the wireless network. Tracking and computing the dependency relationship between processes are performed in the MSSs.

Kumar, Mishra, and Joshi [41] present a non-blocking type minimum-process coordinated checkpointing protocol to minimize the number of unusable checkpoints and also the required bandwidth. The advantage is that it allows processes to send or receive messages during checkpointing.

Ahmed and Khaliq [42] present a low overhead protocol that uses the resources of MHs efficiently, and requires minimal amount of information to be exchanged over the network with minimal checkpointing overhead and few exchange of checkpoint messages.

Ni, Vrbsky, and Ray [43] develop a new idea called a proxy coordinator. By this, they minimize the overhead of coordinating messages (for mobile participant) to only $O(1)$. Also this non-blocking checkpoint algorithm does not request all processes to participate in a checkpoint event.

Jiang and Li [44] design a checkpointing protocol based on message logging. The MH-to-MH communication is done with minimal contention of the network (i.e., wireless medium) and a relatively lower latency for message, which benefits from decreasing contention and message transmission latency.

Singh and Cabillic [45] present a coordinated checkpointing algorithm. To achieve reduced energy consumption alongside faster recovery, this algorithm uses anti-messages with selective logging by calculating the optimal checkpointing interval.

Chen and Lyu [46] introduce an equi-number checkpointing strategy. The analytical results show that the program execution time is an exponential function of n without checkpointing and is a linear function of n with checkpointing.

Lin, Wang, and Kuo [47] describe another coordinated checkpointing protocol based on time. This protocol minimizes the number of transmitted messages as well as their size. The protocol is non-blocking because inconsistencies can be avoided by the piggybacked information in every message.

Park, Woo, and Yeom [48] propose a scheme which uses message logging and independent checkpointing. The recovery information remains at the MSS where the information was first saved. The performance of the proposed scheme provides various levels of failure-free operation cost and recovery cost by adjusting movement factors.

Agbaria and Sanders [49] show an efficient checkpointing protocol. This is an interesting contribution, because it could be characterized as a robust adaptation of a classical distributed checkpointing protocol without adding any constraint to the system.

Ahn, Min, and Hwang [50] design a causal message logging protocol that employs the idea of independent checkpointing. It includes a failure-free mechanism which needs to only locate the mobility agent. Moreover, the protocol allows other failed processes to progress their recoveries and reduce the garbage collection overhead.

Li, Wang, and Chen [51] present a coordinated checkpointing algorithm which has a low cost. Here, the information about checkpoint dependency (among MHs) is recorded in the corresponding MSS. During checkpointing, the algorithm blocks the minimum number of MSSs for the identification procedure.

Li and Wang [52] describe a min-process coordinated checkpointing algorithm which uses the computation power and power of the MSSs. The algorithm reduces the time latency for a global checkpointing procedure significantly. Furthermore, it only forces the minimum number of MHs to take their checkpoints and minimizes the number of blocked MSSs.

Kumar et al. [53] propose a minimum-process synchronous checkpointing protocol, where they optimize the number of useless forced checkpoints as compared to [28]. An attempt has been made to reduce the period during which the processes may be forced to take additional checkpoints due to non-intrusiveness.

Li and Shu [54] propose a coordinated checkpointing scheme. This mechanism minimizes the delay involved in the process of saving a global checkpoint. To track and record the dependency of checkpoints, a piggyback technique is employed.

Li, Jiang, and Zhang [55] design a novel checkpointing protocol, in which two key problems, message order and duplicate message, are effectively solved. This approach provides a better performance in terms of the fail-free overhead and recovery overhead than the traditional approaches.

Kumar, Kumar, and Chauhan [56] develop a hybrid checkpointing protocol, in which synchronous checkpoints are taken by FHs while MHs take checkpoints independently which advantageously increases the performance of the wireless network.

Tantikul and Manivannan [57] present an efficient communication-induced checkpointing mechanism. Based on the log-based protocol, the work proposes log messages only selectively, which could reduce the message logging overhead.

Men and Yang [58] coin the concept of computing checkpoint for designing an efficient consistent non-blocking type coordinated checkpointing protocol which could reduce the coordination overhead and the number of checkpoints.

A hybrid coordinated checkpointing protocol is proposed by Kumar, Kumar, and Chauhan in [59]. Here, for a fixed number of times, all processes should coordinate among themselves to take a checkpoint. A mobile node with low activity or doze mode operation may not be disturbed during the checkpointing process.

Brzeziński, Kobusińska, and Szychowiak [60] introduce a new class of consistency models called session guarantees. The proposed protocol combines logging and checkpointing in order to save space in the stable storage and spends less time during rollback operations.

Gupta, Rahimi, and Liu [61] design a single-phase non-blocking coordinated checkpointing protocol. This protocol produces a consistent checkpoint, avoiding the overhead of taking any temporary, tentative, or mutable checkpoint unlike some other significant related works.

Gupta et al. [62] show an approach for designing non-blocking single-phase synchronous checkpointing protocol with a low overhead. The presented non-blocking approach uses minimum interaction (only once) between the initiator process and the system of n processes and there is no synchronization delay.

George, Chen, and Jin [63] present an efficient movement-based checkpointing and logging checkpointing protocol. In this case, a checkpoint is saved only when a threshold of *mobility handoffs* is exceeded. The value of *mobility handoffs* is governed by the failure rate, log arrival rate, and the mobility rate of the application and MH.

Li and Shu [64], based on their protocol proposed in [54], present a coordinated checkpointing scheme that aims to reduce the delay involved in the process of global checkpointing. The idea is to collect and store process dependency information when processes exchange computation messages.

Rao et al. extend the work done by [9] and develop a proxy-based coordinated checkpointing protocol [65]. In this work, MHs seamlessly save checkpoints on their respective proxies running on the middleware.

The scheme manages storage and processing overhead from low-power mobile devices and delegate it to their respective proxies running on MSS.

Awasthi and Kumar [66] describe a minimum-process coordinated checkpointing protocol. The identifying feature of this protocol is that it uses some probabilistic approach to minimize the number of unusable checkpoints and blocking times. But a few processes may be blocked during the execution of the checkpointing protocol.

Aliouat, Mansouri, and Badache [67] propose a non-blocking coordinated checkpointing protocol, which is designed to meet efficiency and optimization in recovery time using pessimistic message logging.

Imran et al. [68] design a proxy-based uncoordinated checkpointing protocol. This scheme takes storage and processing overhead from low-power MHs and delegates to their respective proxies. The simulation results show that this protocol is reliable.

Men et al. [69] present a low-overhead non-blocking protocol which minimizes the number of checkpoints and coordinating messages. This protocol also takes advantage of the piggyback technique and the communication vector by reducing the checkpoint latency.

Singh [70] shows a combination of time-based and index-based approaches. In this work, an index-based checkpointing protocol is mainly proposed which uses time-coordination for consistent checkpointing. The local timers are synchronized through piggybacking control information on application messages.

Kanmani, Anitha, and Ganesan [71] propose a protocol based on the timeouts of the coordinator process. The proposed protocol minimizes the number and overhead of coordination message significantly.

Chowdhury and Neogy [72] describe a coordinated non-blocking checkpointing protocol. In this case, an initiator MSS sends checkpoint requests to all other MSSs. Those MSSs in turn request some selective MHs. Since the simplicity of pessimistic logging makes it attractive for practical applications, this concept is used here.

Biswas and Neogy [73] design a coordinated checkpointing protocol that can globally ensure a consistent set of checkpoints that specifically focus on lessening power consumption and effective use of the limited available memory and bandwidth.

Kumar [74] presents a hybrid blocking minimum-process coordinated checkpointing protocol executed for a limited number of times where no useless checkpoints are taken and an effort has been made to optimize the blocking of processes.

Men and Xu [75] analyze how the performance and effectiveness metrics are affected by checkpointing and handoff schemes. The simulations show that there does not exist a single recovery scheme that performs well for all types of mobile environments.

Gupta, Chauhan, and Kumar [76] survey and classify the checkpointing recovery techniques for distributed mobile computing systems into user triggered and transparent checkpointing.

Gao, Deng, and Che [77] develop an index-based protocol that uses time-coordination for consistently checkpointing in MDCS setting. In this case, there is no need to send extra coordination messages. However, the synchronization of timers needs to be dealt with.

Gupta, Chauhan and Kumar [78] describe a minimum-process checkpointing protocol in which no useless checkpoint is taken. This protocol reduces the blocking of processes by allowing the processes to do their normal computations, send messages, and receive selective messages during their blocking period.

Ci et al. [79] show how to handle the recovery information in an efficient way. The design of the mechanism is made in such a way that when an MH goes out of a particular range, the new MSS needs only a segment of the recovery information (i.e., just a limited migration would do the purpose).

Singh, Bhat, and Kumar [80] propose an index-based checkpointing protocol. For taking consistent global checkpoint, the mechanism uses time for indirect coordination (to initiate the creation of checkpoint), irrespective of the message sending rate.

Chowdhury and Neogy [81] present a rollback recovery protocol based on independent checkpointing and message logging. The novelty of this protocol is that it uses mobile agents. The algorithm uses the number of hops that an MH takes to decide whether or not to shift the checkpoints and message logs for faster recovery upon possible failure.

Pamila and Thanushkodi [82] design a new log management checkpointing protocol. The existing pessimistic schemes are compared with this mechanism and a trade-off analysis is performed between the costs required to manage log and for profit in terms of improved failure recoverability.

Lim [83] develop a new checkpointing scheme based on checkpoint-related software agent called checkpointing agent. The checkpointing agent is used to make consistent global checkpoints on behalf of the application processes. The proposed scheme provides the capability of semantics-aware checkpointing by paying only a minimal cost.

Basu et al. [84] study some existing checkpointing protocols to identify their relative merits and demerits. The authors try to reduce the recovery cost as well. The proposed algorithm has three subparts – a checkpointing scheme, a failure recovery scheme, and authentication scheme.

Kumar, Kumar, and Chauhan [85] describe a hierarchical coordinated checkpointing protocol, which is a non-blocking type. Also, it requires minimum message logging, steady storage, and number of processes to take checkpoint compared with other works.

Biswas and Neogy [17] propose a checkpointing protocol where MHs save checkpoints based on mobility and movement patterns. Hence, the concept of migration of checkpoint is introduced before planned disconnection so that checkpointing can be completed without any delay, resulting in enhanced fault tolerance in the proposed scheme.

Awasthi, Misra, and Joshi [86] present a quasi-synchronous checkpointing protocol that does not require any synchronization message. Also, it reduces the checkpointing overheads at MHs but requires logging for mobile nodes.

Singhal [87] shows some recent results in checkpointing protocols for MDCS. Specifically, the author presents results with a classification of checkpointing protocols.

Kumar, Chauhan, and Kumar [88] develop a minimum-process non-blocking coordinated checkpointing protocol. However, this mechanism incurs some computation overhead on the MHs and suffers from low memory. The proposed algorithm has low communication and storage overheads.

Garg and Kumar [9] design a non-blocking coordinated checkpointing protocol which requires only a minimum number of processes to take permanent checkpoints. The protocol reduces the message complexity as compared to Cao-Singhal algorithm [8], while keeping the number of useless checkpoints unchanged.

Kumar and Garg [89] describe a hybrid coordinated checkpointing protocol with probabilistic approach. This protocol has the ability to reduce useless checkpoints and blockings of processes. Concurrent initiations of the proposed protocol do not cause its concurrent executions.

Kumar and Khunteta [90] present a minimum-process coordinated checkpointing protocol in which no unusable checkpoint is taken; even there is no blocking of processes.

Gahlan and Kumar [91] present a review of some protocols. The authors conclude that the time taken by checkpointing should be minimal using an anti-message logging technique. The method captures the transitive dependencies during the normal execution by piggybacking dependency vectors onto computation messages.

Kumar and Garg [16] discuss various issues related to the checkpointing for MDCS. They also present a survey of some checkpointing protocols.

Khunteta and Kumar in their work in [19] survey and compare some algorithms of checkpointing which are designed specifically for mobile distributed systems.

Garg and Kumar [92] analyze some existing coordinated checkpointing protocols on the basis of various parameters like number of processes required to checkpoint, blocking time, concurrent execution, synchronization message overhead, piggybacked information messages onto computation messages, and the number of useless checkpoints.

Kumar, Chauhan, and Kumar [93] present a survey of some checkpointing algorithms. They also propose an efficient checkpointing protocol to reduce the checkpointing overheads. This protocol does not have any overhead for synchronization message for its use of time to coordinate indirectly.

Kanmani, Anitha, and Ganesan [94] propose a new non-blocking min-process checkpointing protocol that reduces the large overheads of a previous non-blocking protocol. This new protocol reduces the number of processes taking checkpoints and also diminishes the dependency array information.

Gupta, Kumar, and Solanki [95] design a minimum-process coordinated checkpointing protocol. This protocol is *domino-effect-free*. At most two checkpoints of a process are needed for it on stable storage, and it requires only a few processes to take checkpoints.

Garg and Kumar [14] review and compare different checkpointing schemes: independent, coordinated, communication-induced checkpointing and message logging.

Garg and Kumar in [96] talk about the different aspects of checkpointing for distributed systems in an MDCS environment.

Kumar, Gupta, and Solanki [18] develop a minimum-process checkpointing protocol, where no useless checkpoints are taken and an effort has been made to optimize the blocking of processes. The authors try to reduce the loss of checkpointing effort when any process fails to take its checkpoint in coordination with others.

Kumar [97] presents a minimum-process coordinated checkpointing protocol, where no useless checkpoints are taken. This protocol reduces the blocking time to bare minimum by computing the exact minimum dependency set information in the beginning.

Khunteta and Kumar [15] review and compare various approaches of checkpointing for MDCS taking a set of properties into consideration. These include the assumption of piecewise determinism, performance overhead, and storage overhead.

Khunteta and Kumar [98] describe a minimum-process coordinated checkpointing protocol which ensures that unnecessary checkpoints are not taken and blocking of processes is not done. Also, it makes an effort so that anti-messages are logged only for a very few messages at the time of the checkpointing operation.

Kumar and Garg [99] propose a minimum-process coordinated checkpointing protocol. In this protocol, the authors tried to optimize the number of useless checkpoints and blocking of the processes. They use some probabilistic approach.

Kumar and Gahlan [100] design a minimum-process coordinated checkpointing protocol for non-deterministic MDCS setting where unnecessary checkpoints are not taken. Authors try to reduce the checkpointing time and blocking time of processes by limiting the checkpointing tree which may be formed in [26].

Biswas and Neogy [101] propose a recovery scheme which applies two measures: recovery time and cost which depends on the number of MSSs from which information needs to be acquired as well as the distance between them.

Marzouk and Jmaiel [102] address two issues. First, they talk about the existing solutions that use checkpointing and mobility in distributed applications. Second, they propose policies to allow dynamic selection of checkpointing and mobility techniques based on the execution environment.

Panghal, Rana, and Kumar [103] design an efficient non-blocking minimum-process synchronous checkpointing protocol which is suitable for MDCS. It compels only those MHs which are directly or transitively dependent and to take their checkpoints during the checkpointing operation.

Kumar and Garg [104] develop a coordinated checkpointing protocol in which at the first phase, only soft checkpoints would be taken by all concerned MHs. A soft checkpoint is somewhat similar to a mutable checkpoint as presented in [9]. In this mechanism, if a process fails to save a checkpoint in the first phase, the MHs need to discard their soft checkpoints only.

Lim in [105] suggest minimization of the usage of wireless communications. The distributed application in this work is able to adjust the strictness of checkpointing. It means the maximum rollback distance that sets the value of how many recent local checkpoints that can be rolled back.

Gupta, Liu, and Koneru [106] present a non-blocking synchronous checkpointing approach which determines the globally consistent checkpoints. In this approach, checkpoints can be taken by only those processes that have sent some messages after their last checkpoints.

Sachan and Maheshwari [107] propose a recovery scheme that optimizes the costs of both failure-free and failure recovery operations. The algorithm restricts the number of MSSs from which the recovery information of a failed image could be retrieved.

Dey and Biswas [108] propose an algorithm based on movement-based secure independent checkpointing and logging for MDCS. In this scheme, an MH takes a checkpoint only when its handoff count exceeds a predefined threshold value.

Nagbal and Kumar [109] propose a minimum-process coordinated checkpointing algorithm, which avoids taking unnecessary checkpoints as well as process blocking during checkpointing at the cost of logging anti-messages of very few messages.

Biswas and Neogy design in [110] an algorithm based on the message logging and independent checkpointing. The authors also present a cryptographic method for securing checkpoints and logs. The algorithm ensures protection of checkpoint against security attack against both nodes and links.

The non-blocking algorithms in [111] is an efficient time coordinated checkpointing algorithm that uses time to indirectly coordinate to minimize the number of coordinated message and involves a minimum number of processes to record a checkpoint. Also, the cost of coordinating messages of these algorithms is the smallest.

The coordinated checkpointing algorithm proposed in [112] requires only a minimum number of processes to take a checkpoint. This scheme designed by the generalization of Chandy and Lamport's algorithm [5], uses a delayed checkpointing approach.

Li et al. propose in [113] a checkpointing message logging scheme considering the visit time of the process to the base station. This scheme also suffers from some drawbacks. In fact, this scheme works efficiently only if the stay times of the process in the base station are long enough.

Sharma et al. show in [114] a non-blocking proxy based synchronous checkpointing approach that has a simple data structure, and it forces minimum number of processes to take checkpoint. In the proposed scheme, one MSS and a lot of proxy MSSs work as per the workload.

In [115], checkpointing protocol with hypercube topology is proposed. A *d* dimensional hypercube consists of *2d* process, and a *d+1* dimensional hypercube can be created by connecting two *d* dimensional hypercubes.

A hybrid checkpointing algorithm is proposed in [116], which works in two phases. This algorithm can deal with the constraints related to mobility and not dependent on any topology, structure, or routing protocol.

In [117], a checkpointing algorithm is proposed for mobile computing systems. This algorithm is efficient and optimized in terms of incurred time-space overhead during the process of checkpointing and also during the period of normal execution of the application.

In [118], the authors present an adaptive, coordinated, and non-blocking type checkpointing algorithm which could ensure fault tolerance feature in cluster-based Mobile Ad Hoc Network (MANET). In this mechanism, checkpoints are taken by only a minimum number of MHs in the cluster. The experimental results show that the proposed scheme performs well compared to other related works.

In [119], the authors propose an efficient minimum-process non-intrusive snapshot algorithm that is suitable for mobile distributed applications running in a Vehicular Ad Hoc Network (VANET), and this is the first work of this kind.

In [120], the authors present a similar mechanism as of [118] with some extensions. The proposed algorithm in [121] synchronizes the intra- and intercluster checkpointing processes in such way that each checkpoint taken locally in any cluster is a segment of the consistent global state.

In [2], the authors propose a new optimization approach based on dependency matrices offering the advantage of distribution to reduce the execution/blocking time to the strict necessity against the coordination message overhead.

3.6 CLASSIFICATION OF THE PROTOCOLS

Table 3.1 shows a classification of the checkpointing protocols designed for the MDCS; this classification is done according to the techniques of checkpointing used. The protocols that are mentioned in more than one

TABLE 3.1 Classification of Various Works Covered in This Survey

Checkpoint-Based					
	Coordinated		Communication-Induced		
Uncoordinated	**Blocking**	**Non-Blocking**	**Model-Based**	**Index-Based**	**Log-Based**
[1][22][27][30][33]	[18][25][26]	[2][7][9][17][21][24][27][28]	[31]	[20]	[17][30][32][33]
[35][37][48][50][55][56]	[46][52][54]	[29][32][33][34][36][38][40]	[36]	[23]	[35][37][39][44]
[57][63][68][70]	[61][64][66]	[41][43][45][47][53][56][58]	[42]	[49]	[45][48][50][55]
[79][81][86][93]	[74][78][84]	[59][60][62][65][67][69][71]	[57]	[70]	[57][59][63][68]
[108]	[89][95][97]	[72][73][83][88][94][98][101]	[87]	[77]	[79][81][82][85]
	[99][100][105]	[103][104][106][107][109]	[116]	[80]	[86][90][108]
	[112]	[111][115][118][119][121]			[111][121]

class are hybrid protocols – this type of protocol uses more than one technique of checkpointing to save a global consist checkpoint. These protocols represent a percentage of 30% among the total number of the proposed protocols. The log-based checkpointing protocols represent more than 27% among the total number of the proposed schemes, but more than 86% of these protocols are hybrid protocols. This means that other checkpoint-based techniques are used in these protocols along with the log-based technique. The uncoordinated checkpoint-based protocols represent a percentage of 22% among the total number of the proposed protocols, but those using only uncoordinated techniques do not exceed 3%.

3.7 CONCLUSIONS

Mobile environments pose new challenging problems in the design of checkpointing protocols for distributed computing systems because of their inherent restrictions. We see that the checkpointing protocols designed for traditional setting cannot be applied directly to these systems. By this time, more than one hundred protocols for checkpointing and rollback recovery have been published in the existing literature since 1993 up to 2020. The majority of these protocols are based on the first protocol proposed by Acharya and Badrinath (1994). Some of these protocols are tuned or modified from the traditional schemes while others have been designed from scratch. It is to be noticed in Table 3.1 that more than 47% of the proposed schemes are non-blocking coordinated checkpoint-based protocols. The checkpoint-based approach can reinstate the system state to its most recent consistent checkpoint when a failure case occurs. This approach is therefore relatively less restrictive plus easier for implementation. The coordinated approach needs coordination among the processes for their checkpoints to generate the global consistent system state. The positive point in this is getting only one permanent checkpoint which can be stored. When a failure or fault occurs, processes roll back to the last checkpointing state. The non-blocking approach does not need blocking of the running of distributed application when the rollback recovery protocol is running.

We have reviewed in this work more than one hundred papers published in the literature which address the issue of checkpointing in an MDCS environment. We have also classified the works.

References

[1] A. Acharya and B.R. Badrinath, "Checkpointing distributed applications on mobile computers," in *Proceedings of the 3rd IEEE International Conference on Parallel and Distributed Information Systems*, Austin, TX, USA, pp. 73–80, 1994.

[2] H. Mansouri and A.-S.K. Pathan, "Checkpointing distributed computing systems: An optimization approach," *International Journal of High Performance Computing and Networking*, vol. 15, no 2/3, pp. 202–209, 2019.

[3] B.R. Badrinath and A. Acharya, T. Imielinski, "Impact of mobility on distributed computations," *Operating Systems Review, ACM SIGOPS*, vol. 27, no. 2, pp. 15–20, 1993.

[4] G.H. Forman and J. Zahorjan, "The challenges of mobile computing," *IEEE Computing*, vol. 27, no. 4, pp. 38–47, 1994.

[5] K.M. Chandy and L. Lamport, "Distributed snapshots: Determining global states of distributed systems," *ACM Transactions on Computer Systems*, vol. 3, no. 1, pp. 63–75, 1985.

[6] P. Krishna, N.H. Vaidya and D.K. Pradhan, "Recovery in distributed mobile environments," in *Proceedings of IEEE Workshop on Advances in Parallel and Distributed Systems,* Princeton, NJ, USA, pp. 83–88, 1993.

[7] R. Garg and P. Kumar, "A non-blocking coordinated checkpointing algorithm for mobile computing systems," *International Journal of Computer Science*, vol. 7, no. 3, pp. 41–46, 2010.

[8] M. Elnozahy, L. Alvisi, Y-M. Wang, and D.B. Johnson, "A survey of checkpointing protocols in message-passing systems," *ACM Computing Surveys*, vol. 34, no. 3, pp. 375–408, 2002.

[9] G. Cao and Singhal, "Mutable checkpoints: A new checkpointing approach for mobile computing systems," *IEEE Transaction on Parallel and Distributed Systems*, vol. 12, no. 2, pp. 157–172, 2001.

[10] S. Alagar and S. Venkatesan, "Causal ordering in distributed mobile systems," *IEEE Transactions on Computers, Special Issue on Mobile Computing*, vol. 46, no. 3, pp. 353–361, 1997.

[11] L. Lamport, "Time, clocks, and the ordering of events in a distributed system," *Communication of ACM*, vol. 21, no. 7, pp. 558–565, 1978.

[12] R.H.B. Netzer and J. Xu, "Necessary and sufficient conditions for consistent global snapshots," *IEEE Transactions on Parallel and Distributed Systems*, vol. 6, no. 2, pp. 165–169, 1995.

[13] F. Mattern, "Virtual time and global states of distributed systems," in *Proceedings of the International Workshop on Parallel and Distributed Algorithms*, Amsterdam, North-Holland, pp. 215–226, 1989.

[14] R. Garg and P. Kumar, "A review of checkpointing fault tolerance techniques in distributed mobile systems," *International Journal on Computer Science and Engineering*, vol. 2, no. 4, pp. 1052–1063, 2010.

[15] A. Khunteta and P. Kumar, "An analysis of checkpointing algorithms for distributed mobile systems," *International Journal on Computer Science and Engineering*, vol. 02, no. 04, pp. 1314–1326, 2010.

[16] P. Kumar and R. Garg, "Checkpointing based fault tolerance in mobile distributed systems," *International Journal of Research and Reviews in Computer Science*, vol. 1, no. 2, pp. 83–93, 2010.

[17] S. Biswas and S. Neogy, "A mobility-based checkpointing protocol for mobile computing system," *International Journal of Computer Science and Information Technology*, vol. 2, no. 1, pp. 135–151, 2010.

[18] P. Kumar, P. Gupta and A.K. Solanki, "Dealing with frequent aborts in minimum-process coordinated checkpointing algorithm for mobile distributed systems," *International Journal of Computer Applications*, vol. 3, no. 10, pp. 7–12, 2010.

[19] A. Khunteta and P. Kumar, "A survey of checkpointing algorithms for distributed mobile systems," *International Journal of Research and Reviews in Computer Science*, vol. 1, no. 2, pp. 127–133, 2010.

[20] D. Manivannan and M. Singhal, "A low-overhead recovery technique using quasi-synchronous checkpointing," in *Proceedings of*

16th IEEE International Conference on Distributed Computing Systems, Hong Kong, pp. 100–107, 1996.

[21] R. Prakash and M. Singhal, "Low-cost checkpointing and failure recovery in mobile computing systems," *IEEE Transactions on Parallel and Distributed Systems*, vol. 7, no. 10, pp. 1035–1048, 1996.

[22] N. Neves and W. K. Fuchs, "Adaptive recovery for mobile environments," *Communication of the ACM*, vol. 40, no. 1, pp. 68–74, 1997.

[23] F. Quaglia, B. Ciciani and R. Baldoni, "Checkpointing protocols in distributed systems with mobile hosts: A performance analysis," in *Proceedings of 12th PPS and 9th Symposium on PDP*, Orlando, FL, USA, pp. 742–755, 1998.

[24] G. Cao and M. Singhal, "Low-cost checkpointing with mutable checkpoints in mobile computing systems," in *Proceedings of 18th IEEE International Conference on Distributed Computing Systems*, Netherlands, pp. 464-471, 1998.

[25] Z. Yang, C. Sun, A. Sattar, and Y. Yang, "Consistent global states of distributed mobile computations," *Proceedings of 1998 International Conference on Parallel and Distributed Processing Techniques and Applications*, USA, pp. 1–9, 1998.

[26] G. Cao and M. Singhal, "On the impossibility of min-process nonblocking checkpointing and an efficient checkpointing algorithm for mobile computing systems," in *Proceedings of 27th IEEE International Conference on Parallel Processing*, Minneapolis, MN, USA, pp. 37–44, 1998.

[27] H. Higaki and M. Takizawa, "Checkpoint-recovery protocol for reliable mobile systems," in *Proceedings of 17th IEEE International Symposium on Reliable Distributed Systems*, West Lafayette, IN, USA, pp. 93–99, 1998.

[28] F-K. Ssu, B. Yao, W. K. Fuchs, and N. F. Neves, "Adaptive checkpointing with storage management for mobile environments,"*IEEE Transactions on Reliability*, vol. 48, no. 4, pp. 315–324, 1999.

[29] T-Y.T. Juang and Y-S. Chen, "An efficient rollback recovery algorithm for distributed mobile computing systems," in *Proceedings of IEEE International Conference on Performance, Computing and Communication,* USA, pp. 354–360, 2000.

[30] T. Park and H.Y. Yeom, "An asynchronous recovery scheme based on optimistic message logging for mobile computing systems," in *Proceedings of 20th IEEE International Conference on Distributed Computing Systems*, Taiwan, pp. 436–443, 2000.

[31] R.C. Gass and B. Gupta, "An efficient checkpointing scheme for mobile computing systems," in *Proceedings of the 13th International Conference on Computer Applications in Industry and Engineering*, Honolulu, HI, USA, pp. 323–328, 2001.

[32] Y. Morita and H. Higaki, "Checkpoint-recovery for mobile computing systems," in *Proceedings of IEEE International Conference on Distributed Computing Systems Workshop*, Mesa, AZ, USA, pp. 479–484, 2001.

[33] Y. Morita and H. Higaki, "Checkpoint-recovery for mobile intelligent networks," in *Proceedings of 14th International Conference on Industrial and Engineering Applications of AI and Expert Systems*, Budapest, Hungary, pp. 455–464, 2001.

[34] Y. Manabe, "A distributed consistent global checkpoint algorithm for distributed mobile systems," in *Proceedings of 8th IEEE International Conference on Parallel and Distributed Computing Systems*, South Korea, pp. 125–132, 2001.

[35] T. Park, N. Woo and H.Y. Yeom, "An efficient recovery scheme for mobile computing environments," in *Proceedings of 8th IEEE International Conference on Parallel and Distributed Computing Systems*, South Korea, pp. 53–60, 2001.

[36] C-M. Lin and C-R. Dow, "Efficient checkpoint-based failure recovery technique in mobile computing systems," *Journal of Information Science and Engineering*, vol. 17, no. 4, pp. 549–573, 2001.

[37] T. Park, N. Woo, and H.Y. Yeom, "Efficient recovery information management schemes for the fault tolerant mobile computing systems," in *Proceedings of 20th IEEE Symposium on RDS*, New Orleans, LA, USA, pp. 202–205, 2001.

[38] C-Y. Lin, S-C. Wang, and S-Y. Kuo, "An Efficient Time-Based Checkpointing protocol for mobile computing systems over wide area networks," in *Proceedings of the 8th International Conference on Parallel Processing*, Paderborn, Germany, pp. 59c76, 2002.

[39] T. Park, N. Woo and H. Y. Yeom, "An efficient optimistic message logging scheme for recoverable mobile computing systems," *IEEE Transactions on Mobile Computing*, vol. 1, no. 4, pp. 265–277, 2002.

[40] C-Y. Lin, S-C. Wang, S-Y. Kuo, and I-Y. Chen, "A low overhead checkpointing protocol for mobile computing systems," in *Proceedings of 9th Pacific Rim IEEE International Symposium on Dependable Computing*, Japan, pp. 37–44, 2002.

[41] L. Kumar, M. Mishra, and R. C. Joshi, "Low overhead optimal checkpointing for mobile distributed systems," in *Proceedings of 19th IEEE International Conference on Data Engineering*, Bangalore, India, pp. 686–688, 2003.

[42] R.E. Ahmed and A. Khaliq, "A low-overhead checkpointing protocol for mobile networks," in *Proceedings of IEEE Canadian Conference on Electrical and Computer Engineering*, Montréal, Canada, pp. 1779–1782, 2003.

[43] W. Ni, V.S. Vrbsky, and S. Ray, "Low-cost coordinated nonblocking checkpointing in mobile computing systems," in *Proceedings of 8th IEEE International Symposium on Computers and Communication*, Turkey, pp. 1427–1434, 2003.

[44] T-Y. Jiang and Q-H. Li, "A low-cost recovery protocol for mobile computation," in *Proceedings of the 4th IEEE International Conference on Parallel and Distributed Computing, Applications and Technologies*, China, pp. 644–647, 2003.

[45] P. Singh, and G. Cabillic, "A checkpointing algorithm for mobile computing environment," *Lecture Notes in Computer Science, Springer,* vol. 2775, pp. 65–74, 2003.

[46] X. Chen, and M.R. Lyu, "Performance and effectiveness analysis of checkpointing in mobile environments," in *Proceedings of 22nd IEEE International Symposium on Reliable Distributed Systems*, Florence, Italy, pp.131–140, 2003.

[47] C-Y. Lin, S-C. Wang, and S-Y. Kuo, "An efficient time- based checkpointing protocol for mobile computing systems over mobile IP," *International Journal of Mobile Networking and Applications*, vol. 08, no. 6, pp. 687–697, 2003.

[48] T. Park, N. Woo, and H.Y. Yeom, "An efficient recovery scheme for fault-tolerant mobile computing systems," *International Journal of Future Generation Computing Systems*, vol. 19, no. 1, pp. 37–53, 2003.

[49] A. Agbaria and W. Sanders, "Distributed snapshots for mobile computing systems," in *Proceedings of 2nd IEEE Annual Conference on Pervasive Computing and Communications*, Orlando, FL, USA, pp. 177–186, 2004.

[50] J-H. Ahn, S-G. Min, and C-S. Hwang, "A causal message logging protocol for mobile nodes in mobile computing systems," *Future Generation Computer Systems,* vol. 20, no. 3&4, pp. 663–686, 2004.

[51] G. Li, H. Wang, and J. Chen, "A low-cost checkpointing scheme for mobile computing systems," in *Proceedings 5th International Conference on Advances in Web-Age Information Management*, Dalian, China, pp. 97–106, 2004.

[52] G-H. Li and H-Y. Wang, "A novel min-process checkpointing scheme for mobile computing systems," *International Journal of Systems Architecture*, vol. 51, no. 1, pp. 45–61, 2005.

[53] P. Kumar, L. Kumar, R.K. Chauhan, and V.K. Gupta, "A non-intrusive minimum process synchronous checkpointing protocol for mobile distributed systems," in *Proceedings of the 7th IEEE International Conference on Personal Wireless Communications*, New Delhi, India, pp. 491–495, 2005.

[54] G. Li and L-C. Shu, "A low-latency checkpointing scheme for mobile computing systems," in *Proceedings of 29th IEEE Annual International Conference on Computer Software and Applications*, Edinburgh, Scotland, pp. 491–496, 2005.

[55] Q-H. Li, T-Y. Jiang, and H-J. Zhang, "A transparent low-cost recovery protocol for mobile-to-mobile communication," *International Journal of Software*, vol. 16, no. 1, pp. 135–144, 2005.

[56] L. Kumar, P. Kumar, and R.K. Chauhan, "Logging based coordinated checkpointing in mobile distributed computing systems," *Journal of Research*, vol. 51, no. 6, pp. 485–490, 2005.

[57] T. Tantikul and D. Manivannan, "A communication-induced checkpointing and asynchronous recovery protocol for mobile computing systems," *Proceedings of the Sixth International Conference on Parallel and Distributed Computing Applications and Technologies (PDCAT'05)*, Dalian, China, pp. 70–74, 5–8 Dec. 2005.

[58] C. Men and X. Yang, "Using computing checkpoints implement consistent low-cost non-blocking coordinated checkpointing," in *Proceedings of the 5th Springer International Conference on Parallel and Distributed Computing: Applications and Technologies*, Singapore, pp. 570–576, 2005.

[59] P. Kumar, L. Kumar, and R.K. Chauhan, "A hybrid coordinated checkpointing protocol for mobile computing systems," *IETE Journal of Research*, vol. 52, no. 2&3, pp. 247–254, 2006.

[60] J. Brzeziński, A. Kobusińska, and M. Szychowiak, "Checkpointing and checkpointing protocol for mobile systems with MW session guarantee," in *Proceedings of the 20th IEEE International Symposium on Parallel and Distributed Processing*, Rhodes Island, Greece, pp. 431–438, 2006.

[61] B. Gupta, S. Rahimi, and Z. Liu, "A new high performance checkpointing approach for mobile computing systems," *International Journal of Computer Science and Network Security*, vol. 6, no. 5B, pp. 95–104, 2006.

[62] B. Gupta, S. Rahimi, R.A. Rias, and G. Bangalore, "A low-overhead non-block checkpointing algorithm for mobile computing environment," in *Proceedings of the 1st International Conference on Advances in Grid and Pervasive Computing*, Taichung, Taiwan, pp. 597–608, 2006.

[63] S.E. George, I-R. Chen, and Y. Jin, "Movement-based checkpointing and logging for recovery in mobile computing systems," in *Proceedings of the 5th ACM International Workshop on Data Engineering for Wireless and Mobile Access*, Chicago, IL, USA, pp. 51–58, 2006.

[64] G. Li and L-C. Shu, "Design and evaluation of a low-latency checkpointing scheme for mobile computing systems," *The Computer Journal*, vol. 49, no. 5, pp. 527–540, 2006.

[65] I. Rao, N. Imran, P-W. Lee, and E-N. Huh, T-C. Chung "A proxy based efficient checkpointing scheme for fault recovery in mobile grid system," in *Proceedings of the 13th International Conference on High Performance Computing*, Bangalore, India, pp. 448–459, 2006.

[66] L.K. Awasthi and P. Kumar, "A synchronous checkpointing protocol for mobile distributed systems: Probabilistic approach," *International Journal of Information and Computer Security*, vol. 1, no. 3, pp. 298–314, 2007.

[67] M. Aliouat, H. Mansouri, and N. Badache, "Un protocole efficace non-bloquant et optimal d'établissement coordonné de points de reprise dans les systèmes mobiles," in *Proceedings of the* 7th *International Conference on New Technologies of Distributed Systems*, Marrakesh, Morocco, pp. 193–205, 2007.

[68] N. Imran, I. Rao, Y-K. Lee, and S. Lee, "A proxy-based uncoordinated checkpointing scheme with pessimistic message logging for mobile grid systems," in *Proceedings of the 16th ACM International Symposium on High Performance Distributed Computing*, California, USA, pp. 237–238, 2007.

[69] C. Men, L. Cao, L. Wang, and Z. Xu, "Low-overhead non-blocking checkpointing scheme for mobile computing systems," *International Journal of Tsinghua Science and Technology*, vol. 12, no. S1, pp. 110–115, 2007.

[70] K. Singh, "On mobile checkpointing using Index and time together," *Journal of World Academy of Science, Engineering and Technology*, vol. 32, pp 144–151, 2007.

[71] P. Kanmani, R. Anitha, and R. Ganesan, "Coordinated checkpointing with avalanche avoidance for distributed mobile computing system," in *Proceedings of the IEEE International Conference on Computational Intelligence and Multimedia Applications*, Sivakasi, Tamil Nadu, India, pp. 461–463, 2007.

[72] C. Chowdhury and S. Neogy, "A consistent checkpointing-recovery protocol for minimal number of nodes in mobile computing system," in *Proceedings of the 14th International Conference on High Performance Computing*, Goa, India, pp. 599–611, 2007.

[73] S. Biswas and S. Neogy, "A low overhead checkpointing scheme for mobile computing systems," in *Proceedings of the 15th IEEE International Conference on Advanced Computing and Communications*, Guwahati, India, pp. 700–705, 2007.

[74] P. Kumar, "A low-cost hybrid coordinated checkpointing protocol for mobile distributed systems," *International Journal of Mobile Information Systems*, vol. 4, no. 1, pp. 13–32, 2008.

[75] C. Men and Z. Xu, "Performance analysis of rollback recovery schemes for the mobile computing environment," in *Proceedings of the 3rd IEEE International Conference on Internet Computing in Science and Engineering* Harbin, China, pp. 436–443, 2008.

[76] S.K. Gupta, R.K. Chauhan, and P. Kumar, "Backward error recovery protocols in distributed mobile systems: A Survey," *Journal of Theoretical and Applied Information Technology*, vol. 4, no. 4, pp. 337–347, 2008.

[77] Y. Gao, C. Deng, and Y. Che, "An adaptive index-based algorithm using time-coordination in mobile computing," in *Proceedings of IEEE International Symposium on Information Processing*, Moscow, Russia, pp. 578–585, 2008.

[78] S.K. Gupta, R. K. Chauhan, and P. Kumar, "A minimum-process coordinated checkpointing protocol for mobile computing systems," *International Journal of Foundations of Computer Science*, vol. 19, no. 4, pp. 1015–1038, 2008.

[79] Y-W. Ci, Z. Zhang, D-C. Zuo, Z-B. Wu, and X.-Z. Yang, "Area difference based recovery information placement for mobile computing systems," in *Proceedings of the 14th IEEE International Conference on Parallel and Distributed Systems*, Melbourne, Victoria, Australia, pp. 478–484, 2008.

[80] A.K. Singh, R. Bhat and A. Kumar, "An index-based mobile checkpointing and recovery algorithm," in *Proceedings of 10th*

International Conference on Distributed Computing and Networking, Hyderabad, India, pp. 200–205, 2009.

[81] C. Chowdhury and S. Neogy, "Checkpointing using mobile agents for mobile computing system," *International Journal of Recent Trends in Engineering*, vol. 1, no. 2, pp. 26–29, 2009.

[82] J.C.M.J. Pamila and K. Thanushkodi, "Log management support for recovery in mobile computing environment," *International Journal of Computer Science and Information Security*, vol. 3, no. 1, pp. 1–6, 2009.

[83] S. Lim, "A new distributed checkpointing scheme for the mobile computing environment," in *Proceedings of the 9th IEEE-RIVF International Conference on Computing and Communication Technology*, Danang, Vietnam, pp. 1–4, 2009.

[84] S. Basu, S. Palchaudhuri, S. Podder, and M. chakrabarty, "A Checkpointing and recovery algorithm based on location distance, handoff and stationary checkpoints for mobile computing systems," in *Proceedings of IEEE International Conference on Advances in Recent Technologies in Communication and Computing*, India, pp. 58–62, 2009.

[85] S. Kumar, P. Kumar, and R.K. Chauhan, "Hierarchical non-blocking coordinated checkpointing algorithms for mobile distributed computing," *International Journal of Computer Science and Security*, vol. 3, no. 6, pp. 518–524, 2010.

[86] L.K. Awasthi, M. Misra, and R.C. Joshi, "A weighted checkpointing protocol for mobile distributed systems," *International Journal of Ad Hoc and Ubiquitous Computing*, vol. 5, no. 3, pp. 137–149, 2010.

[87] M. Singhal, "Recent results in checkpointing and failure recovery in distributed systems and wireless networks," in *Proceedings of the IEEE International Symposium on Parallel and Distributed Processing, Workshops and Phd Forum*, Atlanta, GA, USA, pp. 1–10, 2010.

[88] S. Kumar, R.K. Chauhan, and P. Kumar, "A low overhead minimum process global snapshot collection algorithm for mobile

distributed systems," *International Journal of Multimedia and its Applications*, vol. 2, no. 2, pp. 12–30, 2010.

[89] P. Kumar and R. Garg, "Soft-checkpointing based coordinated checkpointing protocol for mobile distributed systems," *International Journal of Computer Science Issues*, vol. 7, no. 3, pp. 40–46, 2010.

[90] P. Kumar, and A. Khunteta, "Anti-message logging based coordinated checkpointing protocol for deterministic mobile computing systems," *International Journal of Computer Applications*, vol. 3, no.1, pp. 22–27, 2010.

[91] P. Gahlan and P. Kumar, "Review of some checkpointing algorithms in distributed and mobile systems," *International Journal of Engineering Science and Technology*, vol. 2, no. 6, pp. 1594–1602, 2010.

[92] R. Garg and P. Kumar, "A review of fault tolerant checkpointing protocols for mobile computing systems," *International Journal of Computer Applications*, vol. 3, no. 2, pp. 8–19, 2010.

[93] S. Kumar, R.K. Chauhan, and P. Kumar, "Low cost coordinated checkpointing algorithm for mobile distributed systems," *International Journal of Wireless Communication*, vol. 2, no. 7, pp. 15–23, Oct. 2010.

[94] P. Kanmani, R. Anitha, and R. Ganesan, "A token ring minimum process checkpointing algorithm for distributed mobile computing system," *International Journal of Computer Science and Network Security*, vol. 10, no. 7, pp. 162–166, 2010.

[95] P. Gupta, P. Kumar, and A.K. Solanki, "Review of some minimum-process synchronous checkpointing schemes for mobile distributed systems," *International Journal on Computer Science and Engineering*, vol. 2, no. 4, pp. 1406–1410, 2010.

[96] R. Garg and P. Kumar, "Low overhead checkpointing protocols for mobile distributed systems: A comparative study," *International Journal of Engineering Science and Techniques*, vol. 2, no. 7, pp. 3267–3276, July 2010.

[97] P. Kumar, "A minimum-process global state detection scheme for mobile distributed systems," *International Journal of*

Engineering Science and Technology, vol. 2, no. 7, pp. 2853–2858, 2010.

[98] A. Khunteta and P. Kumar, "A non-blocking minimum-process checkpointing protocol for deterministic mobile computing systems," *Journal of Theoretical and Applied Information Technology*, vol. 17, no. 1, pp. 1–10, July 2010. http://www.jatit.org/volumes/research-papers/Vol17No1/1Vol17No1.pdf.

[99] P. Kumar and R. Garg, "An efficient synchronous checkpointing protocol for mobile distributed systems," *Global Journal of Computer Science and Technology*, vol. 10, no. 6, pp. 6–10, 2010.

[100] P. Kumar amd P. Gahlan, "A low-overhead minimum process coordinated checkpointing algorithm for mobile distributed system," *Global Journal of Computer Science and Technologies*, vol. 10, no. 6, pp 30–36, 2010.

[101] S. Biswas and S. Neogy, "A handoff based checkpointing and failure recovery scheme in mobile computing system," in *Proceedings of the 2011 IEEE International Conference on Information Networking,* Barcelona, Spain, pp. 441–446, 2011.

[102] S. Marzouk and M. Jmaiel, "A Survey on software checkpointing and mobility techniques in distributed systems," *Journal of Concurrency and Computation: Practice and Experience*, vol. 23, no. 8, pp. 1–17, 2011.

[103] K. Panghal, M.K. Rana, and P. kumar, "Minimum-process synchronous checkpointing in mobile distributed systems," *International Journal of Computer Applications*, vol. 17, no. 4, pp. 1–4, 2011.

[104] P. Kumar and R. Garg, "Soft-checkpointing based hybrid synchronous checkpointing protocol for mobile distributed systems," *International Journal of Distributed Systems and Technologies,* vol. 2, no. 1, pp. 1–13, 2011.

[105] S. Lim, "A tunable checkpointing algorithm for the distributed mobile environment," *International Journal of Computer Science Issues*, vol. 8, no. 6, pp. 1–9, 2011.

[106] B. Gupta, Z. Liu, and S. Koneru, "A low-overhead non-block checkpointing and recovery approach for mobile computing

environment," *Advances and Applications in Mobile Computing*, Chapter 3, InTech, ISBN: 978-953-51-0432-2, pp. 47–62, 2012.

[107] A.K. Sachan and P. Maheshwari, "A failure recovery scheme in mobile computing system based on checkpointing and handoff Count," *International Journal of Advanced Research in Computer Engineering & Technology*, vol. 1, no. 5, pp. 111–118, 2012.

[108] P. Dey and S. Biswas, "Handoff based secure checkpointing and log based rollback recovery for mobile hosts," *International Journal of Network Security & Its Applications*, vol. 4, no. 5, pp. 25–41, 2012.

[109] M. Nagpal and P. Kumar, "Anti-message logging based checkpointing algorithm for mobile distributed systems," *International Journal of Computer Applications*, vol. 69, no. 14, pp. 21–27, 2013.

[110] S. Biswas and S. Neogy, "Improving recovery probability of mobile hosts using secure checkpointing," in *Proceedings of the 2nd International Conference on Advances in Computing, Communication and Informatics*, India, pp. 984–989, 2013.

[111] J. Surender, S. Arvind, K. Anil, and S. Yashwant, "Low Overhead Time Coordinated Checkpointing Algorithm for Mobile Distributed Systems," *Lecture Notes in Electrical Engineering book series (LNEE)*, volume 131, Springer, New York, NY, 26 February 2013. https://link.springer.com/chapter/10.1007/978-1-4614-6154-8_17

[112] A. Kiehn, P. Raj, and P. Singh, "A causal checkpointing algorithm for mobile computing environments," in *Proceedings of 15th International Conference on Distributed Computing and Networking*, Coimbatore, India, pp. 134–148, 2014.

[113] X. Li, M. Yang, C.G. Men, Y.T. Jiang, and K. Udagepola, "Access-pattern aware checkpointing data storage scheme for mobile computing environment," in *Proceedings of the 11th International Conference on Mobile Systems and Pervasive Computing*, Niagara Falls, Canada, pp. 330–337, 2014.

[114] P.K. Sharma, P. Kumar, and S. Jangra, "Proxy MSS based synchronous checkpointing approach for mobile distributed systems," *International Journal of Innovations in Engineering and Technology*, vol. 3, no. 4, pp. 194–202, 2014.

[115] P. Sharma and A. Khunteta, "An efficient checkpointing using hypercube structure (self-adjusted) in mobile ad hoc network," in *ProceedingsIEEE Intl Conference on Recent Advances and Innovations in Engineering*, pp. 1–5, 2014.

[116] P.K. Jaggi, A.K. Singh "Opportunistic rollback recovery in mobile ad hoc networks," in *Proceedings of the IEEE International Conference on Advance Computing,* pp. 860–865, 2014.

[117] H. Mansouri, N. Badache, M. Aliouat, and A.-S.K. Pathan, "A new efficient checkpointing algorithm for distributed mobile computing," *Journal of Control Engineering and Applied Information*, vol. 17, no. 2 pp. 43–54, 2015.

[118] H. Mansouri, N. Badache, M. Aliouat, and A.-S.K. Pathan, "Adaptive fault tolerant checkpointing protocol for cluster based mobile ad hoc networks," *Procedia Computer Science*, vol. 73, pp. 40–47, Oct. 2015.

[119] H. Mansouri and A.-S.K. Pathan, "An efficient minimum-process non-intrusive snapshot protocol for vehicular ad hoc networks," in *Proceedings of the 13th International Conference of Computer Systems and Applications*, Agadir, Morocco, pp. 83–92, 2016.

[120] H. Mansouri and A.-S.K. Pathan, "Checkpointing distributed application running on mobile ad hoc networks," *International Journal of High Performance Computing and Networking*, vol. 11, Issue 2, pp. 95–107, 2018.

[121] H. Mansouri and A.-S.K. Pathan, "A resilient hierarchical checkpointing algorithm for distributed systems running on cluster federation," in S.M. Thampi, G.M. Perez, R. Ko, D.B. Rawat (eds.), "*Security in Computing and Communications*," Springer, pp. 99–110, 2020.

CHAPTER 4

Softwarized Network Function Virtualization for 5G: Challenges and Opportunities

Deborsi Basu and Raja Datta
Indian Institute of Technology Kharagpur

Uttam Ghosh
Vanderbilt University

CONTENTS

4.1 INTRODUCTION

Software Defined Network (SDN) is one of the major emerging technologies used in the development of fifth generation (5G) mobile communication systems. The 5G technology which is expected to be in the market by the end of 2020 is within a rapid growing phase, and researchers are continually incorporating new concepts and technologies in it. The raising demands of mobile data, high-speed internet connection, uninterrupted communication, low call droppings, new applications, and upgraded facilities are creating a huge load on the existing 3G, 4G/long-term evolution (LTE) networks. On the other hand, existing mobile networks are also struggling with many limitations such as complicated controlling protocols, heterogeneous configuration interfaces, comparatively high latency, and poor network performances [1,2]. In the upcoming 5G network these challenges will be dealt with various improvements like very high traffic handling capacity, high data rate, network flexibility, network security, efficient spectrum utilization, adaptability and vendor independence, maximum network coverage area, low latency, high scalability, high network performance, etc. in a resource restricted environment [3,4]. SDN technology plays an important role in providing various facilities and improvements inside a 5G network. Software defined networking is considered to be an approach for network management and controlling the network configurations programmatically in order to improve network performance [5,6]. In other ways it can be said that the main goal of SDN is to make a network open and programmable. If any network controlling organization wants a specific type of network behavior or try to implement a new program or feature inside the existing network architecture, they can develop and install the networking functions easily. This network functions can be anything like Traffic Engineering, Quality of Service (QoS), Packet Routing, Switching, Load Balancing, etc. The static architecture of traditional networks is decentralized and complex in nature and requires more flexibility and easy monitoring for troubleshooting. So, intelligent networking can be implemented using SDN by disassociating the forwarding process of network packets from the control plane to the data plane. Controllers which are called the brain of SDN are installed in the control plane, and the whole network intelligence is confined into it. This intelligent network centralization also has its own drawbacks like network security [6], network scalability, single-point of controller failure [7,8], etc.

To cope up with these drawbacks the idea of Network Function Virtualization (NFV) has been integrated with SDN to provide a better network softwarizing environment. The fundamental idea of NFV is to logically virtualize different network services running on dedicated network appliances such as network firewalls, switches/routers, and load balancers. The application-specific virtualize service instances run on commodity hardware. The preliminary objective of NFV is to reform the way networks are designed, and services are implemented and provided to the users. Any service provider can simplify a wide range of network services, optimize network functions, and maximize parametric efficiencies by introducing new profit generating modules. The NFV approach moves away from dependence on a variety of hardware platform to the use of small number of standardized platform types, with virtualization techniques used to provide the needed network functionalities. The NFV approach is applicable to any data plane packet processing and control plane function in fixed and mobile network environments. In 5G, NFV will enable various other features like network slicing, distributed cloud infrastructure, edge computing, and network softwarization. It will introduce more flexibility and programmability for the needs of tomorrow [9].

Based on recent researches, a detailed analysis has been made on SDN-based architectures that have been used in 5G and in areas where challenges have been faced. The advantages of using SDN and the areas where this technology can do remarkable improvements have also been explained. This analysis is helpful for future implications of configuration of network elements and in redesigning the network applications.

This chapter is also helpful for those who want to explore the area of Network Virtualization and SDN in the upcoming 5G technology. It gives a simple idea and builds up the basic fundamental concept that is an important prerequisite before going inside this interesting area.

Chapter Organization: The rest of the chapter is organized as follows: Section 4.2 elaborates the basic overview on SDN and its historical background, and it also explains the concept of NFV and its development and major importance in the upcoming 5G networks as well as in future day's evolutionary telecommunication networks. Section 4.3 drafts the idea of an SDN-/NFV-enabled 5G network architecture. In Sections 4.4 and 4.5, the challenges based on current day's network architectures and the probable solutions based on Softwarized NFV have been elaborated respectively. Finally, Section 4.6 concludes the chapter with future scope and research directions using this technology.

4.2 BACKGROUND

4.2.1 An Overview of SDN

This section gives a basic and simplistic overview about SDN, including the source and importance of SDN concept in modern day network.

SDN is all about restructuring, transferring, and modifying the existing network architecture in an efficient manner.

Let's consider a network in Figure 4.1, where different users are connected and enjoy different network services. This can be any network like a simplified mobile network, and the network topology is shown in Figure 4.1.

In between the two users, switches and routers exist, and data is traveling via different routers from one user to the other. Suppose there exist two paths, Path A and Path B, in between User I and User II through which a communication has been made. The routing protocol used is the Open Shortest Path First (OSPF) protocol. Depending on the best path between the two users the data packets are routed. So, every switch knows how to route a received packet based on the routing table it has. It does not have a global view of the network as a whole.

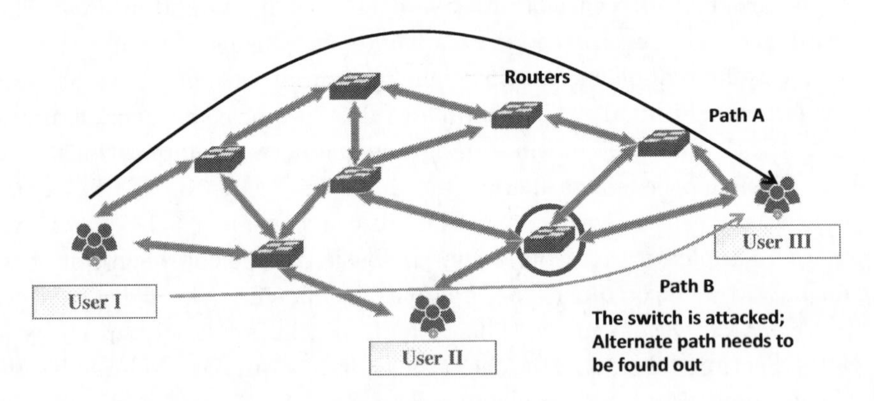

Figure 4.1 Overview of the current network.

Now consider one reason where a switch marked in red circle and is shown in Path B. It may be attacked or may face some problem and fail to route the received packets. So, an alternate path is required that will avoid the node and route the data to the destination successfully.

The distributed control makes things difficult to happen. Here, the routing technology is such that there is no centralized control over the network. If one switch is affected, the other switches are also affected through that path. This is one of the major limitations of the existing network. The present network also has vendor-specific architecture of switches that limit dynamic configuration according to application-specific requirements. Switches are required to configure according to the installed operating system (OS). A centralized control is not feasible in traditional networks as well.

Considering a network viewpoint, the switches along the routing path have been configured with thousands of lines of codes providing packet forwarding rules, mobility management (MM), traffic handling capacity, etc. They are costly as well. A simple inner architecture of the switches can be seen in Figure 4.2. Specialized packet forwarding hardware is at the bottom level. Millions of logic gates are integrated inside the hardware section consisting storage of approximately 10 GB of random access memory (RAM). Network OS (NOS) is at the middle position and works as an interface between the upper application layer and the hardware layer. Various application functionalities are handled by the application

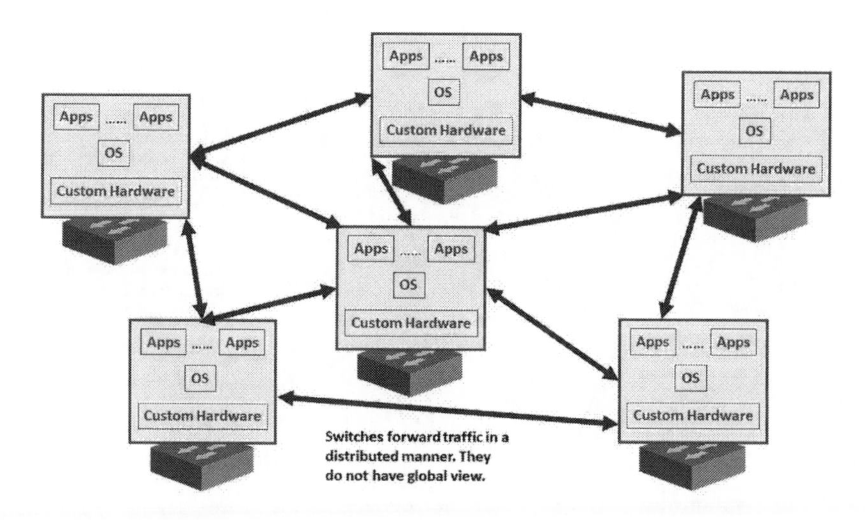

Figure 4.2 Snap of a real-time scenario of the current network.

layer. So, a switch is having all these sections combined together and has individual specifications depending on the requirement.

Now using the concept of SDN, the aim will be to make this network efficient to overcome the existing challenges and limitations that prevail. So, the main fundamental idea in SDN is to curb these problems by separating the application and OS layers from the hardware layer of every switch.

As can be seen in Figure 4.3, the Application layer along with the OS layer has been separated out from the Hardware layer in the switches for the case of an SDN network.

In the basic SDN architecture shown in Figure 4.4, there are network devices like switches or routers that lie at the bottom of the model in the

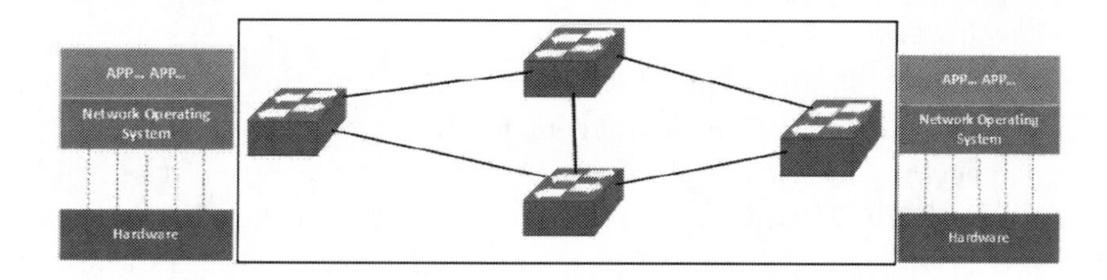

Figure 4.3 Current network architecture with SDN.

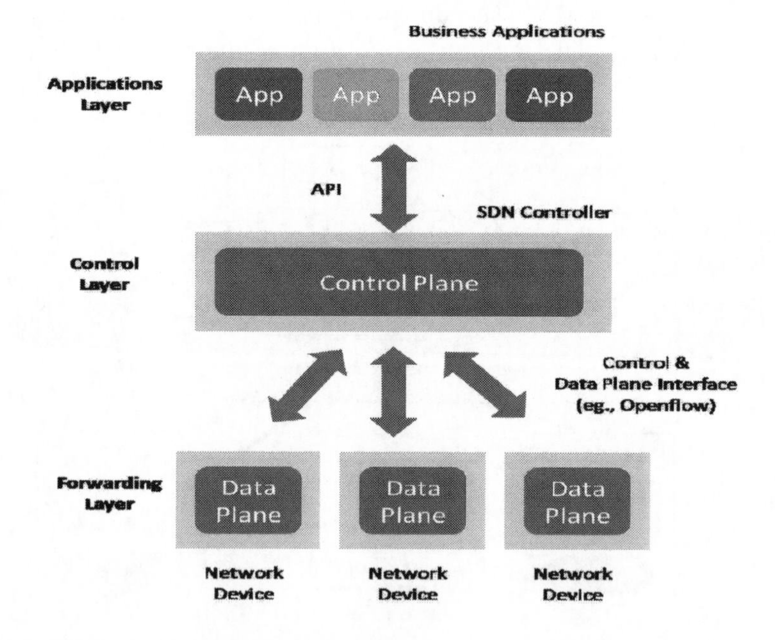

Figure 4.4 The basic SDN architecture.

forwarding layer. The main function of this layer is to forward the received packet from the source to the destination. This layer is termed as *Data Plane* or *Forwarding Plane*. The middle control layer consists of a NOS and is also known as *control plane*. At the top the *Application Plane* resides, which performs functionalities like security, routing, traffic engineering, and other applications. The main fundamental routing protocol that makes the communication between Data Plane and Control Plane is known as *OpenFlow Protocol*. The concept of SDN is made possible through the process of NFV. So, all the network functions have been performed over a virtualized network. The NFV gives a logical view of the SDN and based on that these functions are performed. So, in short, the main concept of SDN is to separate control logic from hardware switches, define the control logic in a centralized manner, control the entire network including individual switches, and make communication between the application, control, and data planes.

The main components or attributes of SDN:

a. Hardware switches

b. Controller

c. Application

d. Flow rules

e. Application programming interfaces (APIs)

4.2.2 Basic SDN Architecture

4.2.2.1 SDN Control Plane Architecture

The SDN control layer maps the service requests of application layer into specific commands and directives to data plane switches and provides applications with info about the data plane topology and activity. The SDN controllers have been formed with servers or cooperating sets of servers in the control plane (Figure 4.5).

Control Plane Functions:

a. *Shortest Path Forwarding*: Uses routing information collected from switches to establish preferred routes.

b. *Notification Manager*: Receives, processes, and forwards to an application events, such as alarm modifications, security alarms, and state changes.

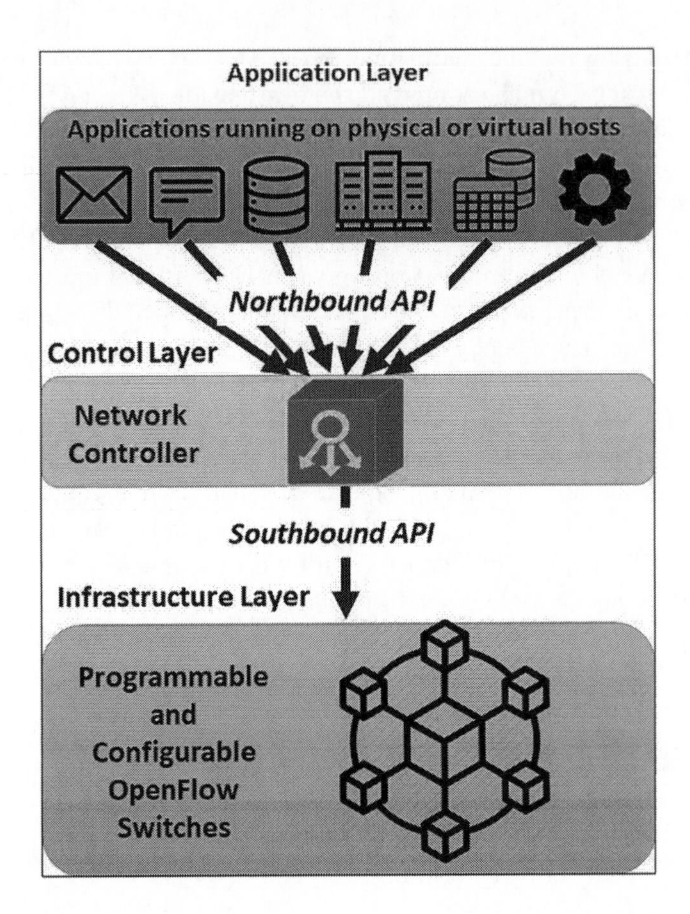

Figure 4.5 SDN architecture concept.

c. *Security Mechanisms*: Provides segregation and security enforcement in between application and facilities.

d. *Topology Manager*: Builds and maintains switch interconnection topology information.

e. *Statistics Manager*: Collects data on traffic through the switches.

f. *Device Manager*: Configures switch parameters and attributes and manages flow tables.

4.2.2.2 SDN Application Plane Architecture

The basic entities of application plain are services which define, monitor, and control network behavior and different network resources.

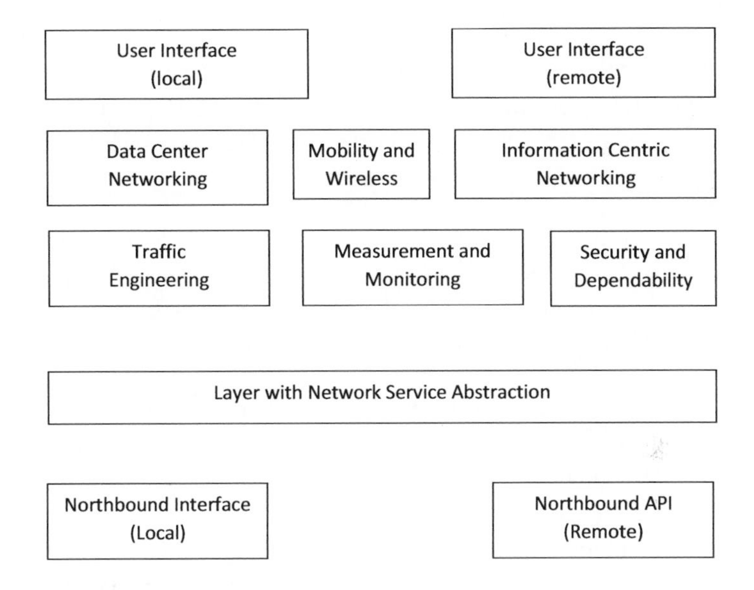

Figure 4.6 SDN application plane functions and interfaces.

The Application Control Interface (ACI) is responsible for the interaction between SDN control plane and SDN applications. It customizes the behavior of the properties of network functions. The abstraction of network resources has been done by ACI considering the programmable nature of the SDN control plane. The exposure of information and data models are done via ACI. Figure 4.6 shows the basic SDN application plane functions and architecture.

4.2.2.3 Operating Principle of SDN

With the advent of SDN–NFV the applications become network aware unlike the traditional networks where the network was application aware:

- Previously in traditional networks (i.e., non-SDN/NFV) network requirements are defined implicitly by applications only. They include several manual processing steps until sufficient resources, network functions, and control policies are there to back the application up.

- Traditional networks also suffer with flexible end user requirements which are absolutely dynamic in nature. Network throughput, delay, availability of network services, etc. are few of the essential user demands. Network service providers avoid the data packet headers with priority requests. They prefer to get generalized data traffic which can be routed as per the current network scenarios. Here SDN plays a vital role to satisfy both the parties by providing a software-controlled platform. This makes the realization of dynamic network behavior easy.

- Therefore, Telecommunication Service Providers (TSPs) are now starting to tackle the user requirements with utmost priority. That takes care of additional cost and proper classifications of services. SDN drives the user ability to totally specify their needs.

- Non-SDN traditional networks keep the information and network states hidden from the applications which are using them. Using SDN concept all the related network services and applications can keep a track of network states and adapt any change accordingly. The control plane has been decoupled from the data plane and logically centralized in nature. The centralized SDN controller translates the application rules to low level and summarizes the network states.

- Depending upon the performance requirements, scalability, and/or service reliability reasons, the SDN controllers can be placed in a centralized or distributive fashion. It is not necessary every time to make the controller physically centralized. Several independent instances of controllers can control network and servers' applications collaboratively.

- All the major controlling decisions are made on an up-to-date global view depending on different network states rather than making local decisions in an isolated behavior in every network hop. In SDN, the control plane acts as a single centralized NOS to resolve resource conflicts and process scheduling as well. Abstracting the device details at low level like optical vs. electrical transmission is taken care of within SDN.

All the SDN data paths have been controlled by the SDN controllers subject to limit their capabilities and thus does not compete with other elements of the control plane. This simplifies traffic scheduling and resource allocation. This improves the quality of network services and enables the network to run with complex policies with greater resource utilization. Different kinds of common information models (e.g., OpenFlow) are used to well understand the controller mechanism at different network levels.

4.2.3 An Overview of NFV

The main idea behind network virtualization is to design various networking functions and parameters using software defined programs [10,11]. Network services are the integrated part of the hardware until these applications have been decoupled from its hardware counterpart using NFV. Various network functions like Encryption, Firewall, Domain Name Service (DNS), Network Address Translation (NAT), Broadband Remote Access Servers (BRAS), etc. have been virtualized and incorporated into the virtual servers using NFV. Apart from using dedicated hardware, service providers can install inexpensive switches, routers, storage, servers, and virtual machines to perform network functions. This integrates multiple network functions inside a single physical server, hence reducing cost and minimizing complexity. If a customer requests a new network service, the service provider can simply set up a new virtual machine to perform that requested function (Figure 4.7).

The NFV framework mainly consists of three things:

- Softwarization
- Virtualization
- Orchestration and Automation

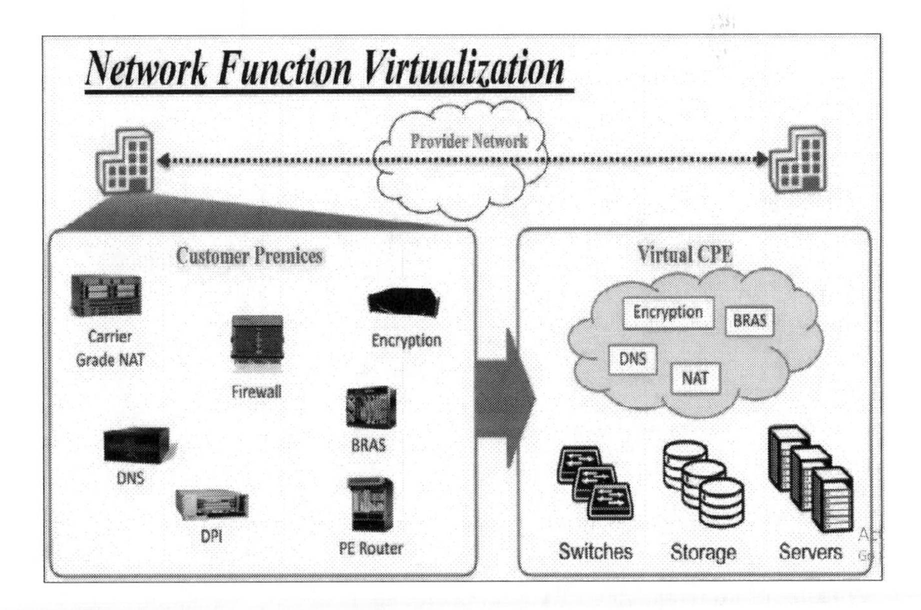

Figure 4.7 Realization of virtualized network functions.

NFV replaces the costly and dedicated hardware with Generic servers that use software to provide a bunch of different virtualized network functions (VNF). Virtualization is all about software packages that perform a specific function. The VNF is equivalent to the physical network and is capable to perform any network tasks like it can work as a switch or virtual switch, router or virtual router, virtual Home Location Register (HLR), virtual Master Switching System (MSC), virtual Short Message Service Center (SMSC), and many more. The orchestration interlinks the resources and networks required to produce cloud-based services and application. The NFV Orchestration (NFVO) uses a range of virtualized software and standard hardware (Figure 4.8).

The major components of NFV framework [12,13]:

1. VNFs are the form of network functions implemented using software. That same can be installed on a network function virtualization infrastructure (NFVI).

2. NFVI consists of all the hardware and software components, and it builds the environment where VNFs have been deployed. The infrastructure for NFV can span several geographical

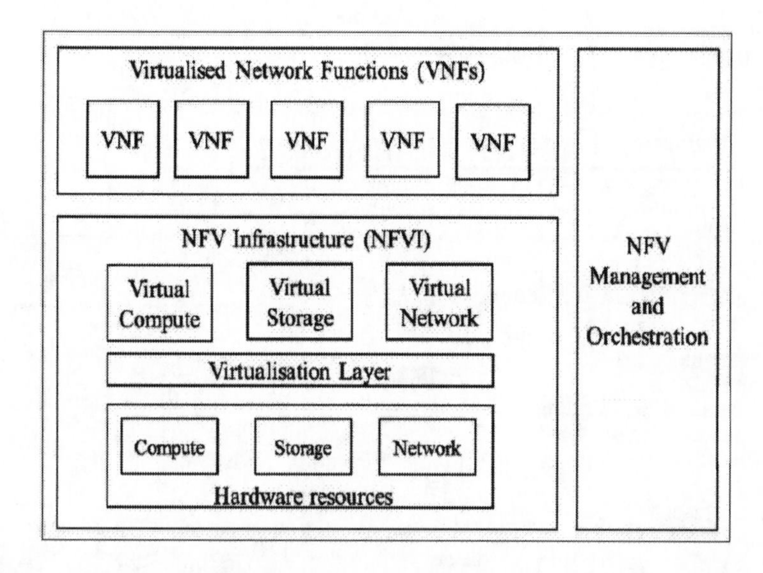

Figure 4.8 NFVO functions govern VNF managers and virtualized infrastructure managers in the NFV Management and Orchestration (MANO) section as well as the virtualization layer of the NFV Infrastructure (NFVI). (European Telecommunication Standards Institute, ETSI.)

locations. The network which is responsible to provide connectivity between these locations has been considered as part of the NFV infrastructure.

3. NFV-MANO Architectural Framework (NFV management and orchestration (MANO) architectural framework) is the host of all functional blocks. The MANO of NFVI and VNFs are done by these functional blocks by using data repositories, reference points, and structural interfaces.

The NFV platform works as the building block for both NFVI and NFVI-MANO. Virtual and physical processing along with storage resources define the role for NFVI. The role of NFV-MANO includes NFVI and VNF managers and virtualized OS on a hardware controller. Managing the carrier grade components and platform elements, helping them to recover from system failures and providing them sufficient security have been taken care of by NFV.

4.2.3.1 Background and Motivation for NFV

The term NFV has been first coined in the global summit of major telecommunication network operators (Nos) and service providers in 2012 in Germany. The main agenda of the meeting was to improve network operations in the high-volume multimedia era. This discussion results in publishing the first white paper on NFV titled *Network Function Virtualization: An Introduction, Benefits, Enablers, Challenges & Call for Action* [14]. In this white paper, the group listed out the major objectives of NFV, considering the standard IT virtualization technology to combine high-volume servers, routers, switches, storage, etc. which are compatible with respect to industry applications. Those appliances can be placed at network nodes, data centers, and end user premises.

The chapter also highlights that due to the addition of large and growing variety of proprietary hardware appliances the following negative consequences arise:

- Upgraded network services may require additional & modified hardware appliances, and placing them inside existing network environments is becoming extremely difficult.
- With the addition of a new hardware setup, the capital expenditure will also increase.

- Whenever new types of hardware appliances are required, NOs will be running out of skills required to design, manage, operate, and integrate complex hardware devices.
- The hardware-based devices reach the end of life quickly, resulting in a quick design-integrate-deploy cycle to replace or upgrade them with little or sometimes no revenue benefit.
- As technology accelerates to meet all the service demands of end users which fuel the network centric IT environment, it is becoming essential to develop a new virtualized hardware platform. This will inhibit the addition of a new revenue earning network service for TSPs.

The NFV brings the concept of platform independent network services by incorporating a variety of hardware platforms to a small number of standardized platform types with the help of network virtualization techniques. This approach can be applicable to any kind of control plane functions or data plane packet processing, forwarding in both static and dynamic network infrastructures.

4.3 THE RELATIONSHIP OF SDN/NFV WITH 5G

Software Defined Networking and NFV are highly correlated but independent of each other. Either of them can be implemented as a standalone architecture without considering the other one. Although the impact of their combined effect is quite remarkable as per the QoS is concerned, both of them are complementary to each other. Non-SDN mechanisms can also be implemented using virtualization concepts which are already in use in many data centers. The approach to separate the control plane and data forwarding plane using SDN can enhance network performance and simplify software compatibility with existing deployments. It facilitates operations and maintains software installation procedures. The NFV is capable enough to provide the required infrastructure using which the SDN software can be operated. Additionally, NFV is related closely with SDN objectives to use the generic switches and servers (Figure 4.9).

1. Portability or Interoperability:
2. Performance Trade-Off
3. Compatibility with existing platforms and co-existence of legacy

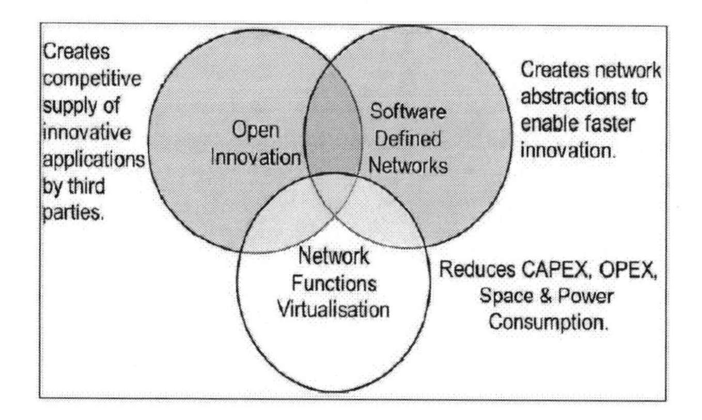

Figure 4.9 Relationship between software defined networking and network function virtualization challenges for network function virtualizations.

4. MANO
5. Automation
6. Security & Resilience
7. Network Stability
8. Simplicity
9. Integration

4.3.1 Basic 5G Network Architecture

The rapid growth of demands and services of the users from the next-generation telecommunication networks makes the researchers think about the next big modification and upgradation inside the telecom industry. Hence the role of 5G comes into action. The primary goal of present generations of mobile networks is to simply offer quick and dependable mobile data services to the network users. With the advent of 5G, the domain of wireless network services is growing rapidly. That facilitates the increasing demands of end users. The 5G network framework is coherent, dynamic, and flexible in nature, which comprises advanced technologies supporting a wide range of network applications. 5G uses the intelligent Radio Access Network (RAN) architecture, keeping it free from complex network infrastructures and base station proximity. 5G is aiming to enable virtual RAN with flexible interfaces developing additional access points for processing and routing data.

- *Spectrum Distribution of 5G Network*: 5G-NR (New Radio) is being allocated with multiple frequency ranges. In between the range of 30–300 GHz inside the radio spectrum millimeter wave belongs, and 24–100 GHz of frequency spectrum range has been used in 5G technology across the globe [15].

  Along with the millimeter wave frequencies the underutilized Ultra High Frequency (UHF) in the range of 300 MHz to 3 GHz are also being used for 5G. Due to shorter wavelengths (1–10 mm) the millimeter wave can only be used for densely populated regions with small coverage areas. The same becomes inactive for long distance communication. Among these upper and lower frequency boundaries dedicated to 5G, individual carrier has started to carve out their individual portions from the allocated 5G frequency spectrum.

- *MEC*: It stands for Multi-Access Edge Computing and belongs to one of the most important elements of the 5G network architecture. The concept is to bring the network services and applications from centralized servers to the network edge and closing the gap with the end users [15,16]. The quick and error-free content delivery between the user and host is possible through MEC.

  The MEC technology is an integral part of 5G but not exclusively. Using MEC network functions 5G networks will be having low latency, high bandwidth, and quick access to the RAN information that feather was absent in the predecessors of 5G. The NOs (network operators) require new approaches to converge the RAN and core network.

  Third Generation Partnership Project (3GPP) and 5G-Public Partnership Project (5G-PPP) specifications for 5G networks are ideal platforms for MEC deployment. The collaborative traffic routing of 5G network along with MEC has been defined in the 5G applications for edge computing. The distributive high-volume computational power of connected devices in Internet of Things (IoT) has been handled by MEC architecture utilizing proper latency and network bandwidth.

- *NFV in 5G*: NFV separates the software from hardware using SND concept and installs different network functions like load balancers, routers, and firewalls with virtualized instances running as software. This helps in the rapid installation of new network services and also eliminates additional capital investments, thereby providing reliable revenue generating services to the customers faster.

Inside 5G network architecture NFV virtualizes appliances to make it easily shareable among users. Network slicing technology is one of such techniques that enable multiple virtual networks to run simultaneously. NFV is also capable to address other 5G challenges through virtualized storage, computing, and network resources that are modified depending on applications and customer segments.

- *RAN Architecture in 5G*: The RAN in 5G uses the concept of NFV and follows the network disaggregation promoted by alliances such as Open-RAN (O-RAN). This creates an open platform of flexibility and new opportunities for the TSPs to compete in the telecom market. It also provides open interfaces and an open source of development that enables easy installation and scale up new technology as per the requirements. Multi-vendor cooperation for easy and smooth deployment of network services with off-the-shelf hardware is the primary objective of the O-RAN alliance. Capacitive user experience and controlling dynamic demand modes with increasing QoS of end users are done by network de-segregation and virtualized network components. As far as massive IoT or simple IoT application has been concerned, the virtualization of RAN provides a cost-effective hardware and software realization to help in an economical growth [17].

- *eCPRI*: Introduction of cost benefit interfaces such as Evolved Common Public Radio Interface (eCPRI) comes with functional split of services and network disaggregation. eCPRI is a protocol, which can be used in fronthaul transport networks. It will be included in standard Ethernet frames and User Datagram Protocol (UDP) frames. CPRI has been around for quite some time. But now, evolved CPRI (eCPRI) is becoming an important technology to understand for 5G. CPRI is an interface that sends data from the Radio Resource Units (RRUs) to the baseband unit. It's a serial interface, which provides a high-speed connection, and it's also a way to translate all those radio signals back to the computing function [18]. With increasing numbers of radio frequency (RF) carriers in 5G, the subsequent cost of RF interfaces increase. With the advent of eCPRI interfaces a significant reduction in cost can be seen as only a few numbers of interfaces can be used to test multiple carriers in 5G. In order to standardize the interfaces of 5G O-RAN fronthaul architectures including Data Unit (DU) eCPRI interface has been taken as a benchmark. The older version of

CPRI for 4G used to be vendor specific but eCPRI for 5G brings for flexibility inside this [19].

In order to meet the ever-growing need of 5G mobile networks the specifications of eCPRI support increased the network efficiency. Unlike CPRI units, eCPRI provides more network flexibility in placing and positioning of functional split in the Physical Layer (PHY layer) of a cellular base station unit. The scope of the eCPRI specification (as shown in Figure 4.10) aims to execute error-free flexible data packet transmission via a supportive packet-based network like Ethernet or internet protocol (IP). The protocol layer defined by eCPRI provides User Plane data-specific services to the layers above the protocol stack. The intercommunication in between eCPRI-Radio Equipment (RE) Control (eCPRI-eREC) and eCPRI-RE using internal radio base station interfaces via packet-based transport network. The eCPRI network faces three different types of control information flows such as eCPRI User Plane Data, Control and Management Plane (C&M Plane) data, and Synchronization Plane Data. eCPRI specifications above the Transport and Network Layer standards are used for C&M and Synchronization.

- *Network Slicing*: Network slicing is the key ingredient to enable the full potential of 5G network architecture. The networking and storage function can be included inside the 5G network by creating an end-to-end virtual network. Multiple logical network services can be run on a single server simultaneously using the concept of virtual network slicing. Slicing approach adds a new dimension in the NFV application platform [20,21]. The dynamic partitioning of network resources among multiple users or tenants can be done

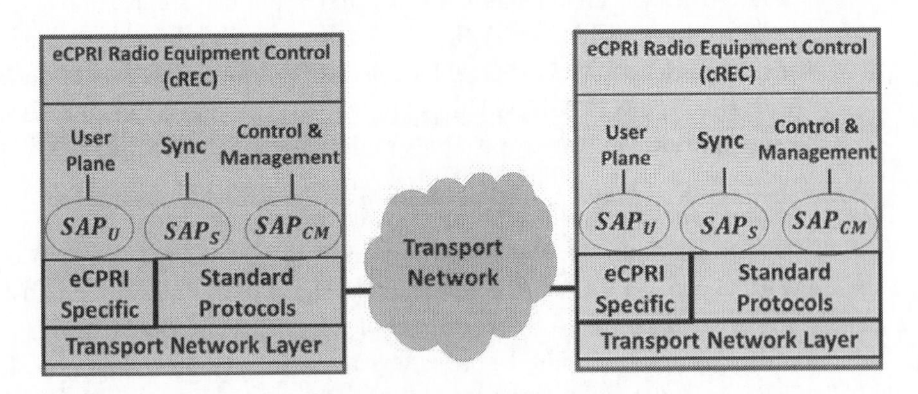

Figure 4.10 System and interface definition.

by NOs in 5G use cases depending upon latency, throughput, and availability demands.

The applications of Network Slicing are extremely necessary in IoT, where the user demands are always very high but the bandwidth requirement varies with time. IoT is having multiple application-specific verticals where the demands are also different. So, network slicing has become an important design consideration for 5G network architecture. Customization of the verticals in order to optimize deployment costs, resource utilization, and flexible network configuration are done using slicing approach. Moreover, network slicing makes the new 5G services quick time to market by expedited trials.

- *Beamforming*: It is another path breaking technology addiction in the development of 5G. Previously the conventional Base Stations used to transmit signal overall direction without having any prior knowledge of the existence of the reception. With the advent of Multiple Input and Multiple Output (MIMO) and massive MIMO systems many small antennas can be integrated within a small place to form directed beams. Signal processing algorithms can be used to find out the most efficient packet delivery path. Based on that data packet will be transferred from source to destination in a pre-allocated sequence [22].

The free space data transmission of 5G has been positively affected with its higher frequency and smaller antenna size. The free space propagation loss is proportional to the smaller antenna aperture, and similarly the diffraction loss is affected by higher frequencies, and hence the lack of wall penetration is significantly higher. However, smaller antenna size makes larger amount of arrays to be assembled together inside a small space. Dynamic re-assignments of beams have been performed by these smaller antennas several times per millisecond. Integrating a large number of arrays inside a smaller section narrow beams can be generated using the concept of massive MIMO, which effectively will be helpful to provide higher throughput with effective user mobility tracking (Figure 4.11).

4.3.1.1 The Core Network Architecture of 5G

The increasing network throughput demand of the 5G network is highly dependent on the core network architecture. As per 3GPP the

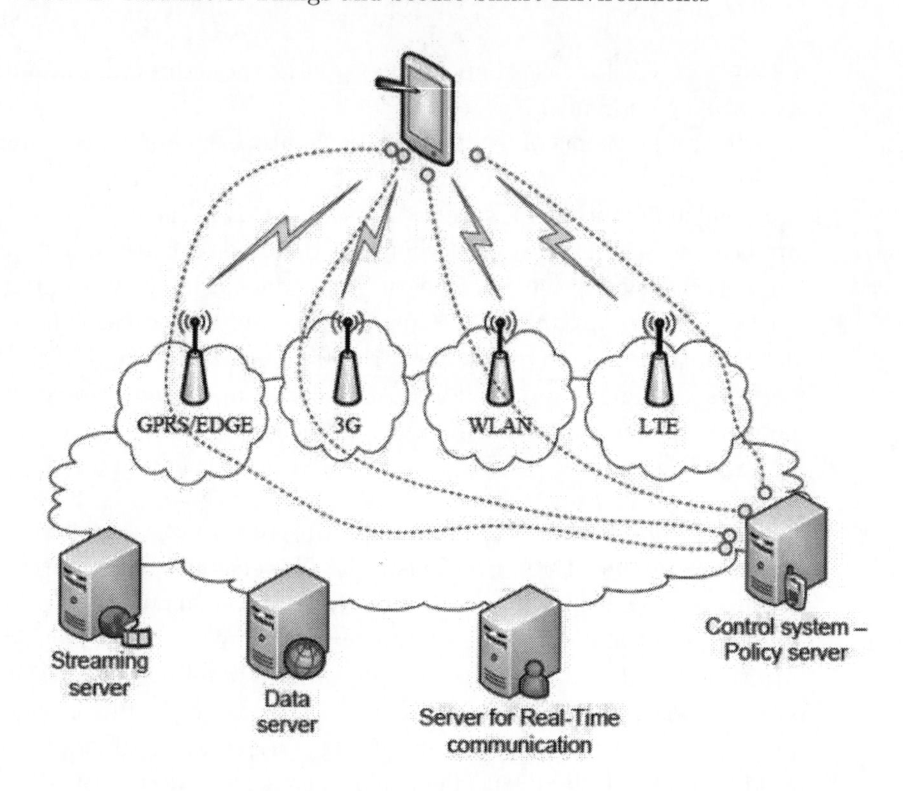

Figure 4.11 Basic 5G network architecture.

new 5G core has been supported with the cloud aligned service-based architecture (SBA) that covers all the sphere of 5G networks including security, authentication, data traffic aggregation from end devices, and information authentication. NFV is acting as an active tool in the internal design of 5G. The concept of virtualized software defined networking is capable enough of being deployed using the infrastructure of MEC, which is alike the architectural principles of 5G.

The functionalities of the components of 5G core network architectures (as shown in Figures 4.12 and 4.13) are explained below [23–26]:

- **Access and Mobility Function – AMF**

  Like the working of MM Entity (MME – Mobility Management Entity) inside the 4G network, AMF does the same for 5G network

  - Disconnects the RAN Control Plane (RAN CP) interface (N2)
  - Non-Access Stratum (NAS) signaling function

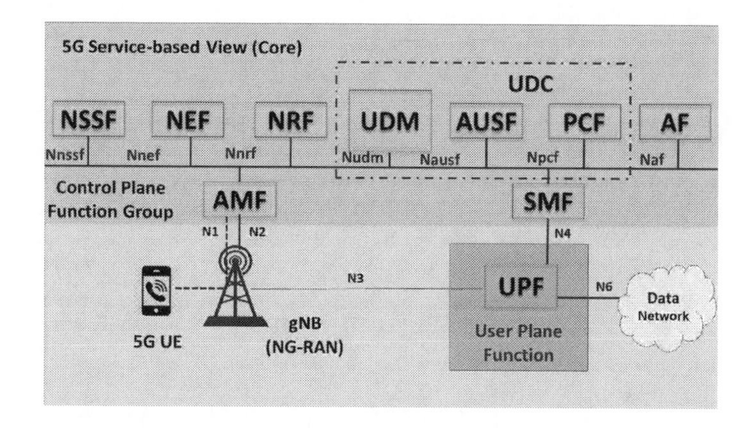

Figure 4.12 5G network core architecture. (VIAVI Solution Inc. Home – 5G Architecture.)

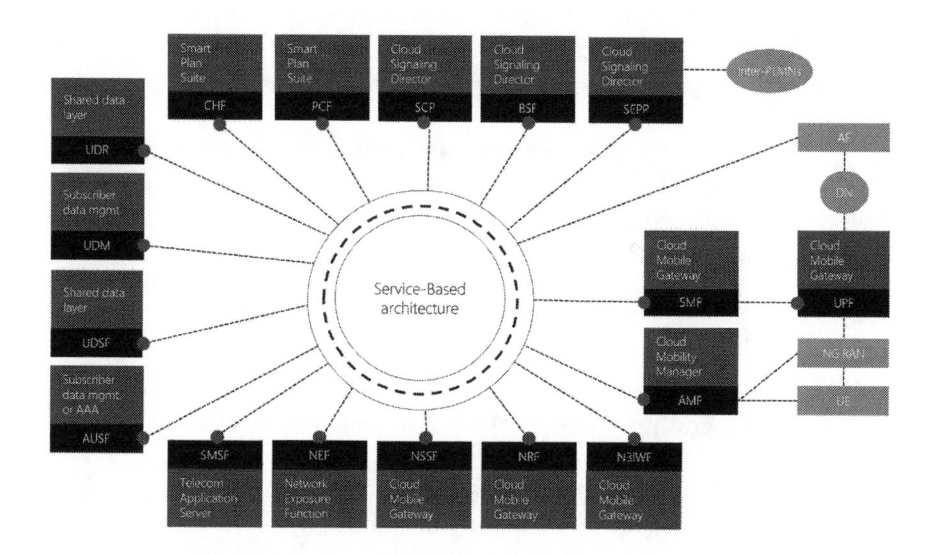

Figure 4.13 5G network core service-based architecture.

- Integrity protection and NAS ciphering
- NAS layer termination and MM
- NAS forwarding of Session Management (SM) layer
- Authenticates User Equipment/User Entity (UE)
- Security context management
- Management of registration content

  - Connection management
  - Management of reliability content
  - MM (mobility management)
  - Application of policies related to mobility from Policy Control Function (PCF) (e.g., mobility restrictions)

- **SM Function – SMF**

  The SMF executes the functions related to SMs like 4G MME does for Packet Gateway – Controller (PGW)-C and Software Gateway – Controller (SGW)-C.

  - IP address allocation to UEs
  - SM using NAS signaling
  - QoS and policy information has been sent to RAN via the AMF by SMF
  - Notification for downlink data
  - The User Plane Functions (UPFs) have been selected and controlled for routing the network traffic. Selecting UPF close to the edge of the network, the UPF section functions enable Mobile Edge Computing (MEC)
  - Policy and charging related to different functions are done. It also acts as the interface of all communications related to user plane services
  - Lawful intercept of the control plane

- **User Plane Function – UPF**

  One of the most important interfaces between SGW and PGW is UPF. The impact of UPF in the context of Control and User Plane Separation (CUPF) architecture are

  - EPC (Evolved Packet Core) PGW-U + EPC PGW-U → UPF in 5G
  - The following functions are performed by UPF
  - Routing and forwarding of data packets
  - Inspecting data packets to maintain QoS. Sometimes UPF integrates the Deep Packet Inspection (DPI) for classifying and identifying the category of data packets

  - It connects to the Point of Presence (PoP) inside the internet. UPF sometimes integrates NAT functions with Firewall
  - Dynamic anchor for Inter Radio Access Technology (RAT) and Intra-RAT signaling and handovers
  - Lawful intercepts of the User Plane
  - Traffic movement maintenance and traffic stat reporting

- **Policy Control Function – PCF**
  In 5G the functions of PCF are analogical with the functions of Policy and Chaining Function (PCRF) in 4G-LTE Networks.

  - Managing and defining policy rules like MM, roaming, and network slicing for control plan services
  - The policy decisions of User Data Repository (UDR) have been accessed as subscription information
  - New QoS and charging features are being supported

- **Authentication Server Function – AUSF**
  Like Home Subscriber Station (HSS) in 4G-LTE, AUSF takes care of the authentication functions.

  - Launches and maintains the Extensible Authentication Protocol (EAP) authentication server
  - Storing the session keys

- **Unified Data Management – UDM**
  The UDM performs parts of the 4G HSS function.

  - Generation of Authentication and Key Agreement (AKA) credentials
  - Subscription management
  - User identification
  - Access authorization

- **Application Function – AF**
  Just like EPC AF functions, this follows the same.

  - Application influence on traffic routing.
  - Accessing NEF
  - Interaction with the policy framework for policy control

- **NF Repository Function – NRF**
  The main job of NRF is to do function registration and discovery
  - Service registration and discovery function allow Network Functions to discover each other
  - Maintains NF profile and available NF instances
- **Network Exposure Function – NEF**
  Secure and flexible services and features for a 5G core network have been provided by NEF
  - Secure provision of information from an external application to 3GPP network
  - Exposes capabilities and events
  - Control plane parameter provisioning
  - Translation of internal/external information
  - Packet Flow Description (PFD) management. A PFD is a tuple of protocol, server-side IP, and port number
- **Network Slice Selection Function – NSSF**
  The fundamental task of NSSF is to properly guide the traffic through a network slice
  The following functions are being performed:
  - Selecting of the Network Slice instances to serve the UE
  - Determining the AMF set is used to serve the UE
  - Determining the allowed Network Slice Selection Assistance Information (NSSAI)

4.3.1.2 5G Geographical Architecture Adoption

In various geographical regions the deployment of stand-alone 5G infrastructure has been instantiated by the respective authorities. The advanced technological regions like Europe, North America, and few countries in Asia have already began minor deployments and other countries are closely following them. Approximately 55 active networks have been considered to be in action by the end of 2020. Large proliferation of carriers is enabling the rollout difficult for various countries in Europe. In order to solve these difficulties European Commission has created an action plan for 5G in Europe. Their main aim is to initiate the deployment of 5G roadmap to all the countries positively by the end of 2020 and start their functionalities as early as possible.

Every industrial nation is investing heavily to make the practical as well as the financial implication of the 5G conversion from the existing networks. China, Japan, and India are few of them. New infrastructures for both hardware and software level have been designed and manufactured by all big telecom industries creating a bonanza for them. The rapid deployment has been initiated as well. China is expecting to have 10,000 base stations by the end of 2020. The largest telecom service provider in India already upgraded their entire network in such a way to make it compatible with 5G.

4.3.2 SDN-/NFV-Enabled 5G Architecture

In this section a basic overview has been provided regarding the impact of network softwarization and virtualization of 5G systems.

Design implications of network architecture

The drastic change in network architecture is one of the most fundamental things to be considered while analyzing the concept of softwarization and virtualization of 5G systems. In the context of network architectural design perspective there are three different categories depending upon the placement of network functionalities. They are explained below:

1. *Full-Cloud Based Migration Architecture*: This process architects things inside a centrally placed cloud in which all network functionalities and signaling have been controlled. That may be an operator owned cloud. This kind of network modeling brings certain advantages like easy integration of new technologies, network service availability, and interoperability with other network services and protocols, and all these can be placed together at a cloud level. An example of such kind has been shown in Figure 4.14, where virtualized eNodeBs (logical eNBs) are directly connected to the central cloud controlled by an operator. Full-cloud based topological migration has been explained in detail in the following works and explained in [27–32]. The fundamental concept of cloud-based migration is to move all the user/control (U/C) - plane management and signalling decisions in a central cloud. The centralized architecture is always cost effective as it avoids the distributive computational and storage capabilities [27]. Basta et al. [27] explained the concept of data flows (incoming/outgoing signaling, data traffic) between

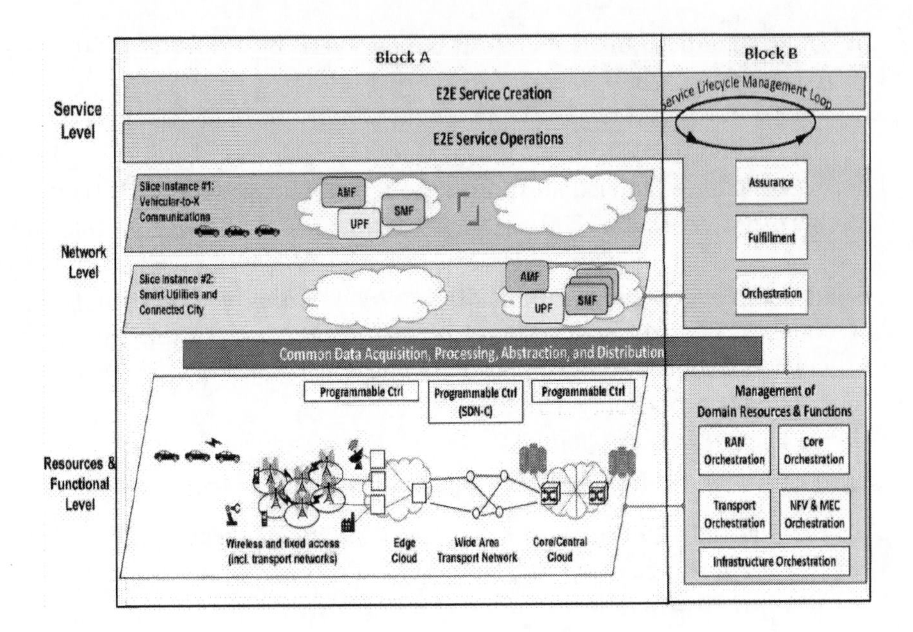

Figure 4.14 SDN-/NFV-enabled 5G architecture.

base stations (BSs) and central cloud using SDN controllers (i.e., intra-cloud communication). Works in [31,32] define the concept of multi-level controller placement (e.g., three levels). The job of the first controller is to access network selection and is termed as **device controller**. The second one is **edge controller** and is responsible for security, routing control, addressing, handover, authentication, etc. The last one does the C-plane instantiation management and U-plane load balancing and overall management of the two planes which is called orchestration controller. However, due to C-plan traffic operation load [33,34] and high traffic signal handling load on PGW [35], central cloud facility becomes a bottleneck in terms of capacity, latency, and reliability in between the connectivity of eNB and central cloud. Indeed, the network flexibility lies within the service chaining management but not in terms of placements of various functions within the network. As all the network functions are centralized in the cloud the flexibility of this architecture has been managed at the data plane level.

2. *C-Plane Cloud Migration Architecture*: The following configuration explains the functionality split of the User/Control plane as explained in [27,36], and the same has been represented in

Figure 4.12, and the central cloud controls the virtual eNB or simply the eNBs. The eNBs are attached with the User Plane entities distributed over the network. The virtualized user plane components have been migrated according to network traffic load and service requirements by the C-Plane cloud migration architecture [37]; thus, the reduction in bottleneck effect in centralized cloud has been done. The critical functions like GTP tunneling, which includes forwarding of data packet in U-plane to cloud, are still the major negative aspects. Customized hardware or middle boxes with programmable platform to handle dynamic network functions are included inside the cloud architecture.

3. *Scenario-Based Migration Architecture*: The fundamental idea of this kind of infrastructure is to place the network functionalities in both mobile and cloud infrastructures; these deployments are done depending upon network load conditions and service demands [27,38]. As can be seen in Figure 4.14, edge clouds are capable of running network functionalities [39], and in order to achieve higher network scalability, flexibility edge placements are required. Services dealing with delay sensitive applications or cases with additional network functionalities are required, and this architecture enables the same (e.g., support of non-3GPP access points). Yousaf et al. [40] proved that approximately 48% of network resource savings can be achieved if network functionalities are being placed at radio access points. Many other advantages related to moving the functionalities across the network infrastructure are analyzed. In [27], some other examples can be found out, where Basta et al. drafted a statistical analysis of moving the charging functions in the cloud. Haneul et al. [38] developed an optimum function placement model in order to reduce the cost of packet transmission. The expected time for each function of each candidate location and the cost associated with the selected locations are together considered to calculate the total cost. The following benefits a network got from the location-based migration architecture: (a) In order to support dynamic network topology quick reconfiguration capability and improved resilience are required; (b) it has the potential to provide ultra-low latency and low end-to-end delay; (c) supporting a large number of elements. On the other side, a time critical state synchronization and network orchestration between the distributed mobile network

entities and central cloud are extremely essential. This process could include additional signaling overhead on the network [27]. In addition, with the concept of shifting functionalities out from the central cloud, few disadvantages may appear related to limited storage capabilities of cloud framework and minimal exploitation of processing as the overall management of the new network architecture.

4.3.3 Fundamental Overview of Network Architecture

A 5G network architecture is having multiple proposals depending on various applications [41–43], and defining a general architectural framework for 5G network is necessary to accommodate all types of application mechanisms. Network slicing architecture (also known as network leveling) is composed of two main blocks. One is purely dedicated for level implementation, and the other one is responsible for slice configuration and management [44]. The first block (Block A in Figure 4.13) has been designed with three levels of layers of slices, and they are service level, network function level, Resource & Functional Infrastructure level, respectively. Each one has been allocated with a dedicated task-based slice definition and deployment. The second block (Block B in Figure 4.13) is framed with a centralized architecture. Network Slice Controller (NSC) controls all the necessary functionalities and manages all the three levels in order to provide effective communication between multiple slices.

- **Service level**
  The service level acts as a common interface between various business entities (e.g., third party service providers and Multi-Vendor NOs (MVNO)). Common network layer service model has been provided by sharing the underlying physical network architecture. The service instances represented with individual services embeds the network features and characteristics in order to fulfil the requirement to create a suitable slice in the form of service level agreement (SLA) [41].

- **Network Function Level**
  The main job of the network function level layer is to form individual network slices depending upon the service license request's arrival from upper layers. The network functions that define the network behavior cumulatively make the network function level. Multiple interconnected network functions are developed on

a common physical network infrastructure depending upon the categories of network service requests arriving from end users. Network management operations are performed to check the full life cycle of a network service and de-allocate this whenever its demand gets expired. The reallocation of the same network service has been done to other users depending upon their requirements and needs. A service has been stopped when that is not being used for a long time [44]. Network operational complexity increases as single network functions are being shared by multiple slices but that also increases the efficiency of resources and is also effective in low-cost realizations. However, proposing a one-to-one usage in between services and slices will provide clarity to the network but will badly affect the efficiency of the network resources.

- **Resource and Functional Level**
  The RAN, transport network, and core network together make the infrastructure layers on which each slice of network has been multiplexed, and they also provide the necessary physical network resources to serve various network functions composing each network slice. A heterogeneous set of network elements like data centers (computation capacity resources and data storage), base stations (radio bandwidth resources), and different devices with network connectivity such as switches and routers (network routing resources) [45].

- **Network Slice Controller (Block B)**
  The fundamental job of NSC is to orchestrate. It intelligently manages and coherently communicates information between different network slices. In order to do system maintenance, it controls the Service Life Cycle Management Loop. It also manages different Domain Resources and Functions like RAN & Core Orchestration, Transport Orchestration, various infrastructure orchestration, etc. There are multiple benefits of this kind of network elements to provide on the go along with flexible and efficient slice configurations during their life cycle. Moreover, the NSC is responsible for several tasks to provide more effective coordination between the aforementioned layers [44]:

  - *End-to-End Service Management*: Allocating the network service instances to the appropriate network functions depending upon SLA requirements and satisfying the demands.

- *Virtual Resources Definition*: By virtualizing a network resource, flexible allocation of the same can be done to perform management operations to assign network resources.
- *Slice Life Cycle Management*: Dynamic reconfigurations of network slices to set the SLA requirement modifications across three layers can be done efficiently.

Each layer might be assigned with distinct complex functions to address different purposes. The NSC can be designed with multiple orchestration levels each to control the individual layers. In order to fulfil the requirements of network services all orchestration entities used to communicate and coordinate among each other by exchanging high level of information containing data regarding the slice creation and deployments inside the network model [46].

4.4 MAJOR CHALLENGES FOR NFV IN 5G

NFV brings major network challenges to deal with, and the community interested to bring it into action is trying their best. Challenges have been identified in the basic fundamental areas where the next-generation 5G networks are facing difficulties:

- *Portability and Interoperability*: Installing the virtual network appliances on the distinct but standardized hardware platform provided by various vendors using virtualization defines the portable nature of NFV. One of the fundamental challenges is to form a unified and supportive interface that precisely decouples the software instances from the underlying hardware, as per the demand of virtual machines and their corresponding hypervisors. Portability and interoperability are important to differentiate between data center vendors and virtual appliance vendors. Though both the systems were interrelated previously, NFV will provide the freedom of localization of services inside networks and required resources of the virtual elements without having further constraints.
- *Performance Trade-Off*: NFV conceptualizes things depending upon standard hardware platform avoiding any kind of proximity hardware as mentioned in *Network Function Virtualization – Introductory White Paper on Network Operator Perspectives on*

NFV priorities for 5G [47]. Here the major challenge is to keep the performance degradation of network elements as low as possible using layers of suitable network hypervisors and advanced software technologies. The effects of end-to-end latency, network throughput, and data processing overhead have to be minimized. The virtual appliances must have sufficient information about the backhaul resources to get the clarity in distributing them among the newly installed network function instances to serve the user demands.

- *Compatibility with Existing Platforms & Migration and Co-Existence of Legacy*: The newly defined virtual network functions must be well matched with the existing element system managements of the operators. That must co-exist with the previously used software and hardware platforms. To converge the IT orchestration to the new NFVO model, operations support system (OSS), business support system (BSS), and Network Management System should also be well compatible with respect to each other. The new NFV/SDN technology must allow the migration movement to go from today's proprietary physical network architecture-based solution to open standardized virtual network platforms. Clubbing the architectural concept of virtual network appliances with the classical physical network appliances in order to suffice the service demands has been objectified by NFV-/SDN-based network models. Virtual appliances must use the northbound interfaces to manage and control the internal and physical appliances implemented for the same functions.

- *MANO*: NFV requires constant management at the resource level and orchestration at the distribution level. The rapid incorporation of new network services has to be taken care of properly before allocating them on particular tasks. Northbound APIs and southbound APIs (Application Programming Interfaces) are well-defined standards and abstract specifications. The cost reduction and time-effective service deployment can be done using APIs. As far the NO's perspective is concerned, the new virtual appliances will be beneficial for them as well. SDN architecture re-enforces the technological aspects of NFV in system orchestration in control and data plane forwarding and managing the flows of data packets.

- *Automation*: The network functions can only be made suitable for automation to give high-end resilience of the service controlling functions.

- *Security & Resilience*: This is another challenge TSPs and NOs face while installing the NFV features. They have to take care of network security, resilience, and service availability so that they should not be hampered. The on-demand service creation feature of NFV will give proper resilience and availability to the network. Securing the hypervisor along with its configuration, the security of a network function can also be provided. The TSPs and NOs are looking for suitable tools to control and configure the hypervisor parameters. They are also aiming to get security-certified virtual appliances and hypervisors.

- *Network Stability*: The stability of the network has not been directly impacted while management or orchestrating of a large number of controllers and hypervisors are done. The stability criteria can only be taken under consideration when virtual network functions are being reloaded or reinstalled after a hardware or software failure inside the network architecture. Instability inside the network may occur due to improper placements of functions and erroneous mapping in between network services and commodity hardware. Synchronization in different time critical functions like congestion control, flow admission control, dynamic routing, resource allocation, etc. and cohesive reassignments of services are very important to maintain a stable network environment. It should be noted that *Network Functions Virtualization – Introductory White Paper* Issue 1 Page 12 of 16 [48] clearly mentioned about the network instability which includes compromising performance parameters and relaxing the optimized use of network resources. NFV will get further benefits with the inclusion of mechanisms capable of stability.

- *Simplicity*: Virtual network platforms will be simpler to manage, monitor, and operate than the existing networks. The probabilistic aim of NOs is to simplify the platforms of operations by virtue of which dynamic upgradation with maintaining the technological revolution will be very possible to implement. A significant focus is to reduce the unnecessary complexity of network architecture and bring more clarity and simplicity in realizations of all new network functionalities.

- *Service and Device Integration*: In NFV one of the key challenges is to accumulate multiple distinct network services onto existing high-volume servers or hypervisors of the industry. Telecommunication NOs and network service providers must be able to integrate servers, hypervisors, and virtual network appliances manufactured by different vendors easily without disrupting their flow functionalities. NFV provides this type of vendor independent platform. Third party support and maintenance of integrated services must be provided by the ecosystem; the internal conflicts between several parties can be removed. The system needs mechanisms to authenticate new NFV products that will arise in future network architectures. Dedicated service tools must be built to address these issues.

4.5 OPPORTUNITIES FOR NFV IN 5G

On the basis of the above-mentioned areas of challenges in Network Virtualization in 5G, researchers are trying to find suitable technologies that can provide appropriate solutions. The solutions under consideration to implement the discussed cases are explained below:

Network Slicing: The fundamental idea is to control different application environments in different vertical domains starting from broadband domain to critical industrial applications [41]. Requirement-based solutions are essential for custom design and a flexible network architecture. This defines the concept of network slicing in 5G [48]. The introduction of network slicing will make the network flexible in such a way that, instead of having a single core network with limited number of functionalities, multiple core networks can co-exist with a wide range of network functionalities. In 5G each core network could have a dedicated network slice working for it ensuring QoS accomplishments. Using network slicing each U/C-plane functionality may be used in each business use case. This further enables signaling overhead, scalability, and latency by truncating all unnecessary functionalities.

According to the 3GPP study mentioned in [49], a valid solution for simultaneous data handling in multiple verticals of 5G applications can be network slicing. Core networks dedicated for various applications are studied and evaluated in [50,51]. The architectural enhancements used to build dedicated core networks and monitoring the resource assignments and maintaining their association have been introduced by 3GPP. The idea of network slicing comes into reality due to network virtualization and softwarization. Unlike the 4G network, in 5G networks virtual slicing of network applications makes the cost of capital expenditure (CAPEX) and operational expenditure (OPEX) low, along with a considerable reduction in deployment cost, and it was previously high due to the existence of dedicated application-based core networks. Additional network features only require changes in software level, so no new physical infrastructures (e.g., routers, servers, switches, storages) are needed. This makes easy upgrading of current day networks. In [52] virtualization has been used as a high-level architectural requirement providing dedicated and logical networks with demand-specific functionalities. The impact of network slicing on various business domains sharing limited resources between multiple vendors and satisfying their demands of different applications at different levels are very high [53,54]. Integrating multiple service chains with a certain degree of freedom network services can be provided; in this case, NFV has a way out for deploying multiple interrelated VNFs and Service Function Chaining (SFCs) to suffice the same. However, SDN is also required to create data paths for the deployed VNFs and SFCs to meet its own QoS target.

Edge Computing: The raising demands and services are bringing the network functionalities closer to the edge [55]. This is also helpful to provide better QoS to the users by increasing the scalability of the network. To this aim, the Cloud/Centralized RAN (C-RAN) [56] is effectively coming under the area of importance in 5G network architecture. In 5G, C-RAN has been considered to have a central management unit with digital functionalities like Baseband Processing Unit (BBU) [57,58]. Mobile edge approach is bringing network scalability, reducing network complexity, and improving Quality of Network Services. The overall network functions are being transferred at the edge for faster communication [59].

NFV and softwarization are reconfiguring the network architectures by moving the network services or functions at reconfigurable positions. In case the service quality is degrading due to overload or excess service demands in a particular slice, then any neighboring slice with less traffic load can take the excess signaling requests and maintain the QoS of the network. This dynamic load sharing and load balancing in between multiple network slices can be done using SDN/NFV technology. Virtualization of low-level network functions at BSs have been addressed in [60]. In [60], major challenges are identified, and few of them have been elaborated as PHY layer, and compute-intensive baseband functions are installed on dedicated hardware or in some cases on general purpose hardware accelerators. The use cases of PHY layer virtualization have been explained in terms of acceleration technologies.

Decoupling: In order to bring flexibility inside network function deployment in 5G architecture re-designing of Control/Data Planes is required. From this perspective a disruptive concept of C/U-plane decoupling has been introduced with the concept of SDN and NFV.

Decoupling of Uplink and Downlink [61,62]

Decoupling of Control plane and User plane [63,64]

The solutions discussed above fall under the multi-connectivity philosophy of 5G, where multiple devices share network resources with multiple BSs [65]. Thus, softwarization and virtualization are essentially required to manage current network challenges in order to provide network functions availability of C/U plane keeping the user link/data link (UL/DL) or C/U association unchanged (see Table 4.1).

4.6 CONCLUSION AND FUTURE WORKS

The 5G technology which is expected to hit the global telecom market by the end of 2020 is within an intensive development phase. SDN/NFV technology plays a vital role in it. The need for network flexibility, smooth reconfigurability, and easy customization in 5G mobile networks is forcing the TSPs and NOs to exploit novel approaches for network infrastructure deployment and installation of new network services. The concept of network softwarization and virtualization are two key technology enablers to drive the future 5G systems and make it compatible with the latest technology developments. SDN & NFV make decoupling of the network functionalities from its underlying hardware

TABLE 4.1 The Relationship between Different Use Cases and Network Solutions of 5G Networks

Capabilities with 5G	Use Cases and Requirements in 5G	Solutions for 5G	Impact of SDN & NFV
Programmability & flexibility	Fast reconfiguration Resiliency Dynamic network topology Service aware QoS	Network slicing Service-based network functions Service-based network topology	NFV manages the functions and the SFCs of each slice NFV manages the function placement according to the QoS needs of the slices SDN manages the network topology (link/path configuration) of each slice SDN provides traffic isolation among slices
Scalability	End-to-end low latency Availability of the network Massive number of devices Easy integration and Interoperability with other technologies and protocols	Cloud-RAN Mobile edge Network functions closer to the UE Services closer to the UE Dedicated control signal management based on service requirements	NFV moves network functions closer to the edge SDN updates the topology to handle function relocation

(Continued)

TABLE 4.1 (*Continued*) The Relationship between Different Use Cases and Network Solutions of 5G Networks

Capabilities with 5G	Use Cases and Requirements in 5G	Solutions for 5G	Impact of SDN & NFV
Management of U/C-planes	Ultra-low latency Active QoS management Reliability UE-triggered signaling	Slice-based C-plane functions Slice-based U-plane functions C/U-plane decoupling (DUDe – Downlink and Uplink Decoupling) Multi-connectivity	NFV provides ad hoc C/U plane functions for each slice NFV handles function placement according to the DL/UL and C/U-plane configuration SDN updates the network topology according to the DL/UL and C/U-plane configuration

platform very easy for smooth network management and rapid upgradation. Despite additional network services and functionalities, there are many network-related issues that researchers are trying to solve using integrated SDN–NFV technology. The major issues have been explained here along with some important solutions. In this chapter we have considered the advancements of technology which have been added by 3GPP and 5G-PPP in order to standardize the 5G network architecture. The state-of-the-art survey along with comprehensive overview provided in recent research proposals, journals, and papers focusing on the omnidirectional impact of network softwarization and virtualization to improve overall 5G network functionalities. The effective analysis of SDN and NFV is having multiple directions like network scalability, security, efficacy, energy consumption, etc. The main motive of this chapter is to give an overall idea based on all the existing areas where problems are faced in NFV-/SDN-based 5G architecture implementation and how they are being mitigated. Though researches are continuing and new technologies are arising, there are still areas where proper inspection and more work are required. Moreover, this chapter will be helpful to understand the concepts of SDN and NFV from a very basic to advanced level. It will also help to have a clear picture of current networking challenges and opportunities and surely motivate them to pursue their research in this domain.

In future, the main aim will be developing new network protocols and architectures for the upcoming 5G network to eliminate various network-related issues. The Controller Placement Problems (CPP) in Distributed Networks, Network Scalability Management (NSM) in large numbers of controller deployment, new network services and component deployment cost optimization, seamless connectivity in high mobility, application of SDN/NFV in distributed systems, network security, etc. are the major areas that still require intensive research in order to standardize network parameters in all aspects. We will take care of these vital issues in our future research works. The task will also be like designing a new modified 5G network architecture with the concept of NFV integrated with software defined networking and simulating it to see whether the problems are mitigated or not. We aim for both software and hardware base implementations to validate the various network performances. Defining new algorithms to optimize the existing problems, improvement of overall network characteristics will be done. Due to the high number of service demands the density of users will be very high in the coming days.

Traffic analysis and network security in such a congested network is another issue, but the main focus will be based on how the overall network throughput can be improved with the help of NFV concept.

Bibliography

[1] D. Sinh, L.-V. Le, L.-P. Tung and B.-S. Paul Lin, "The Challenges of Applying SDN/NFV for 5G & IoT," *APWCS*, August 2017.

[2] S. K. Tayyaba and M. Ali Shah, "5G Cellular Network Integration with SDN: Challenges, Issues and Beyond," In *2017 International Conference on Communication, Computing and Digital Systems (C-CODE)*, 978-1-5090-4448-1/17/$31.00 ©2017 IEEE.

[3] D. Basu, R. Datta, U. Ghosh, and A.S. Rao, "Load and Latency Aware Cost Optimal Controller Placement in 5G Network Using sNFV," In *Proceedings of the 21st International Workshop on Mobile Computing Systems and Applications*, pp. 106–106, 2020.

[4] H.-H. Cho, C.-F. Lai, T.K. Shih, and Han-Chieh Chao, "Integration of SDR and SDN for 5G," *IEEE Access, Special Section on 5G Wireless Technologies: Perspective of the Next Generation Mobile Communications and Networking*, DOI: 10.1109/ACCESS.2014.2357435.

[5] U. Ghosh, P. Chatterjee, S. Shetty, C. Kamhoua, and L. Njilla, "Towards Secure Software-Defined Networking Integrated Cyber-Physical Systems: Attacks and Countermeasures," In *Cybersecurity and Privacy in Cyber Physical Systems*, CRC Press, ISBN: 9781138346673, 2019.

[6] U. Ghosh, P. Chatterjee, and S. Shetty, "A Security Framework for SDN-Enabled Smart Power Grids," In *IEEE 37th International Conference on Distributed Computing Systems Workshops (ICDCSW)*, Atlanta, USA, 2017.

[7] U. Ghosh, P. Chatterjee, D. Tosh, S. Shetty, K. Xiong, C. Kamhoua, "An SDN Based Framework for Guaranteeing Security and Performance in Information-Centric Cloud Networks," In *2017 IEEE 10th International Conference on Cloud Computing (CLOUD)*, 2017.

[8] E. S. Madhan, U. Ghosh, D. K. Tosh, K. Mandal, E. Murali and S. Ghosh, "An Improved Communications in Cyber Physical System Architecture, Protocols and Applications," In *2019 16th Annual IEEE International Conference on Sensing, Communication, and Networking (SECON)*, Boston, MA, USA, pp. 1–6, 2019.

[9] S. Abdelwahab, B. Hamdaoui, M. Guizani and T. Znati, "Network function virtualization in 5G," *IEEE Communication Magazine*, April 2016, vol. 54, no. 4, pp. 84–91.

[10] C. Liang, F. Richard Yu, and X. Zhang, "Information-Centric Network Function Virtualization over 5G Mobile Wireless Networks," In *IEEE Network*, May/June 2015, 0890-8044/15/$25.00 ©2015 IEEE.

[11] A. Al-Quzweeni, T.E.H. El-Gorashi, L. Nonde, and J.M.H. Elmirghani, "Energy Efficient Network Function Virtualization in 5G Networks," In *IEEE ICTON* 2015.

[12] M.K. Shin, S. Lee, S. Lee, and D. Kim, "A Way Forward for Accommodating NFV in 3GPP 5G System," *ICTC* 2017.

[13] T. Mahmoodi, V. Kulkarni, W. Kellerer, P. Mangan, F. Zeiger, S. Spirou, I. Askoxylakis, X. Vilajosana, H.J. Einsiedler, and J. Quittek, "VirtuWind: Virtual and programmable industrial network prototype deployed in operational wind park," *Transactions on Emerging Telecommunications Technologies*, June 2016, vol. 27, no. 9, pp. 1281–1288.

[14] M. Chiosi (AT&T), et al., "Network Functions Virtualization: An Introduction, Benefits, Enablers, Challenges & Call for Action," *White Paper*, Published on October 22-24, 2012 at the "SDN and OpenFlow World Congress", Darmstadt, Germany.

[15] GSMA, "5G Spectrum GSMA Public Policy Position," July 2019, https://www.gsma.com/spectrum/wp-content/uploads/2019/09/5G-Spectrum-Positions.pdf.

[16] ETSI Standards, "Multi-access Edge Computing (MEC)," https://www.etsi.org/technologies/multi-access-edge-computing.

[17] P. Marsch, et al. "5G radio access network architecture: Design guidelines and key considerations," *IEEE Communications Magazine*, vol. 54, no. 11 (2016), pp. 24–32.

[18] AT&T, "AT&T Reaches New Milestone in Unlocking RAN, Makes World's First eCPRI Call for Millimeter Wave," Technology Blog, September 9, 2019, https://about.att.com/innovationblog/2019/09/first_ecpri_call_for_millimeter_wave.html.

[19] P. Monti, et al. "A Flexible 5G RAN Architecture with Dynamic Baseband Split Distribution and Configurable Optical Transport," In *2017 19th International Conference on Transparent Optical Networks (ICTON)*, IEEE, 2017.

[20] H. Zhang, N. Liu, X. Chu, K. Long, A.H. Aghvami, and V.C. Leung, "Network slicing based 5G and future mobile networks: Mobility, resource management, and challenges." *IEEE Communications Magazine*, vol. 55, no. 8 (2017), pp. 138–145.

[21] X. Foukas, G. Patounas, A. Elmokashfi, and M.K. Marina, "Network Slicing in 5G: Survey and Challenges," *IEEE Communications Magazine*, vol. 55, no. 5 (2017), pp. 94–100.

[22] W. Roh, et al. "Millimeter-wave beamforming as an enabling technology for 5G cellular communications: Theoretical feasibility and prototype results," *IEEE Communications Magazine*, vol. 52, no. 2 (2014), pp. 106–113.

[23] M. Shafi, A.F. Molisch, P.J. Smith, T. Haustein, P. Zhu, P.D. Silva, F. Tufvesson, A. Benjebbour, and G. Wunder, "5g: A tutorial overview of standards, trials, challenges, deployment, and practice," *IEEE Journal on Selected Areas in Communications*, vol. 35, no. 6 (2017), pp. 1201–1221.

[24] Y.I. Choi, J.H. Kim, and N.I. Park, "Revolutionary direction for 5g mobile core network architecture," In *IEEE 2016 International Conference on Information and Communication Technology Convergence (ICTC)*, pp. 992–996, 2016.

[25] 3GPP, "5g: A Tutorial Overview of Standards, Trials, Challenges, Deployment and Practice," Technical Report TR38.913, 2016.

[26] V. Yazici, U.C. Kozat, and M.O. Sunay, "A new control plane for 5g network architecture with a case study on unified handoff, mobility, and routing management," *IEEE Communications Magazine*, vol. 52, no. 11 (2014), pp. 76–85.

[27] A. Basta, W. Kellerer, M. Hoffmann, K. Hoffmann, and E.-D. Schmidt, "A Virtual SDN-Enabled LTE EPC Architecture: A Case Study for S-/P-Gateways Functions in Future Networks and Services (SDN4FNS)," *2013 IEEE SDN*, pp. 1–7, Nov 2013.

[28] A. Bradai, K. Singh, T. Ahmed, and T. Rasheed, "Cellular software defined networking: a framework," *IEEE Communications Magazine*, vol. 53 (2015), pp. 36–43.

[29] P. Agyapong, M. Iwamura, D. Staehle, W. Kiess, and A. Benjebbour, "Design considerations for a 5G network architecture," *IEEE Communications Magazine*, vol. 52 (2014), pp. 65–75.

[30] P. Demestichas, A. Georgakopoulos, K. Tsagkaris, and S. Kotrotsos, "Intelligent 5G networks: Managing 5G wireless mobile broadband," *IEEE Vehicular Technology Magazine*, vol. 10 (2015), pp. 41–50.

[31] R. Trivisonno, R. Guerzoni, I. Vaishnavi, and D. Soldani, "Towards zero latency Software Defined 5G Networks," In *2015 IEEE International Conference on Communication Workshop (ICCW)*, pp. 2566–2571, June 2015.

[32] R. Guerzoni, R. Trivisonno, and D. Soldani, "SDN-based architecture and procedures for 5G networks," In *2014 1st International Conference on 5G for Ubiquitous Connectivity (5GU)*, pp. 209–214, Nov 2014.

[33] S. Sesia, I. Toufik, and M. Baker, *LTE: The UMTS Long Term Evolution.* Wiley Online Library, New Jersey, United States, 2009.

[34] B.-J. Kim and P. Henry, "Directions for future cellular mobile network architecture," *First Monday*, vol. 17, no. 12 (2012), doi:10.5210/fm.v17i12.4204.

[35] A. Bradai, K. Singh, T. Ahmed, and T. Rasheed, "Cellular software defined networking: A framework," *IEEE Communications Magazine*, vol. 53 (2015), pp. 36–43.

[36] K. Pentikousis, Y. Wang, and W. Hu, "Mobileflow: Toward software-defined mobile networks," *IEEE Communications Magazine*, vol. 51 (2013), pp. 44–53.

[37] M. Skulysh and O. Klimovych, "Approach to virtualization of Evolved Packet Core Network Functions," In *2015 13th International Conference on Experience of Designing and Application of CAD Systems in Microelectronics (CADSM)*," pp. 193–195, Feb 2015.

[38] H. Ko, G. Lee, I. Jang, and S. Pack, "Optimal middle box function placement in virtualized evolved packet core systems," In *2015 17th Asia-Pacific Network Operations and Management Symposium (APNOMS)*, pp. 511–514, Aug 2015.

[39] I. Giannoulakis, E. Kafetzakis, G. Xylouris, G. Gardikis, and A. Kourtis, "On the Applications of Efficient NFV Management towards 5G Networking," In *2014 1st International Conference on In 5G for Ubiquitous Connectivity (5GU)*, pp. 1–5, Nov 2014.

[40] F. Yousaf, J. Lessmann, P. Loureiro, and S. Schmid, "SoftEPC - Dynamic Instantiation of Mobile Core Network Entities for Efficient Resource Utilization," In *2013 IEEE International Conference on Communications (ICC)*, pp. 3602–3606, June 2013.

[41] NGMN Alliance, "Description of Network Slicing Concept," 2016.

[42] 5GPPP, "View on 5G Architecture," 2017.

[43] IEEE, "Network Slicing and 3GPP Service and Systems Aspects (SA) Standard - IEEE Software Defined Networks," sdn.ieee.org. Retrieved 2019-07-03.

[44] A., Ibrahim, T. Tarik, S. Konstantinos, K. Adlen, and F. Hannu, "Network slicing and softwarization: A survey on principles, enabling technologies, and solutions," *IEEE Communications Surveys & Tutorials*, vol. 20, no. 3 (2018), pp. 2429–2453. DOI: 10.1109/comst.2018.2815638. ISSN 1553-877X.

[45] H.-J. Einsiedler, A. Gavras, P. Sellstedt, R. Aguiar, R. Trivisonno, and D. Lavaux, "System Design for 5G Converged Networks," In *European Conference on Networks and Communications (EuCNC), 2015*, pp. 391–396, June 2015.

[46] J. Ordonez-Lucena, P. Ameigeiras, D. Lopez, J.J. Ramos-Munoz, J. Lorca, and J. Folgueira, "Network slicing for 5G with SDN/NFV: Concepts, architectures, and challenges," *IEEE Communications Magazine*, vol. 55, no. 5 (2017), pp. 80–87.

[47] J. Erfanian, B. Smith (Bell Canada), et al., "Network Functions Virtualisation - Network Operator Perspectives on NFV priorities for 5G," White Paper on NFV priorities for 5G, February 21st, 2017 at the NFV#17 Plenary meeting, Bilbao, Spain.

[48] R. El Hattachi (Editor – Bell Mobility), J. Erfanian (co-editor – AT&T), "5G White Paper" White Paper, February 2015.

[49] 3GPP, "Study on New Services and Markets Technology Enablers," TR 22.891, 3rd Generation Partnership Project (3GPP), Feb 2016.

[50] 3GPP, "Architecture Enhancements for Dedicated Core Networks; Stage 2," TR 23.707, 3rd Generation Partnership Project (3GPP), Feb 2016.

[51] 3GPP, "Study on Dedicated Core Network Enhancements," TR 23.711, 3rd Generation Partnership Project (3GPP), Feb 2016.

[52] 3GPP, "Study on Architecture for Next Generation System," TR 23.799, 3rd Generation Partnership Project (3GPP), Feb 2016.

[53] M. Jiang, M. Condoluci, and T. Mahmoodi, "Network Slicing Management and Prioritization in 5g Mobile Systems," In *22th European Wireless Conference*, pp. 1–6, May 2016.

[54] M. Jiang, M. Condoluci, and T. Mahmoodi, "Network Slicing in 5g: An Auction-Based Model," In *2017 IEEE International Conference on Communications (ICC)*, pp. 1–6, May 2017.

[55] M. Peng, Y. Sun, X. Li, Z. Mao, and C. Wang, "Recent advances in cloud radio access networks: System architectures, key techniques, and open issues," *IEEE Communications Surveys Tutorials*, vol. PP, no. 99 (2016), p. 1.

[56] A. Checko, H. L. Christiansen, Y. Yan, L. Scolari, G. Kardaras, M. S. Berger, and L. Dittmann, "Cloud RAN for mobile networks: A technology overview," *IEEE Communications Surveys & Tutorials*, vol. 17 (2015), pp. 405–426.

[57] G. Mountaser, M. L. Rosas, T. Mahmoodi, and M. Dohler, "On the Feasibility of MAC and PHY Split in Cloud Ran," In *2017 IEEE Wireless Communications and Networking Conference (WCNC)*, pp. 1–6, March 2017.

[58] G. Mountaser, M. Condoluci, T. Mahmoodi, M. Dohler, and I. Mings, "Cloud-Ran in Support of URLLC," In *2017 IEEEGlobecom Workshops (GC Wkshps)*, pp. 1–6, Dec 2017.

[59] E. Bastug, M. Bennis, and M. Debbah, "Living on the edge: The role of proactive caching in 5G wireless networks," *IEEE Communications Magazine*, vol. 52 (2014), pp. 82–89.

[60] D. Pompili, A. Hajisami, and T. X. Tran, "Elastic resource utilization framework for high capacity and energy efficiency in cloud ran," *IEEE Communications Magazine*, vol. 54 (2016), pp. 26–32.

[61] ETSI, "Network Functions Virtualization (NFV); Acceleration Technologies; Report on Acceleration Technologies & Use Cases." ETSI GS NFV-IFA 001, Dec 2015.

[62] F. Boccardi, J. Andrews, H. Elshaer, M. Dohler, S. Parkvall, P. Popovski, and S. Singh, "Why to decouple the uplink anddownlink in cellular networks and how to do it," *IEEE Communications Magazine*, vol. 54 (2016), pp. 110–117.

[63] A. Mohamed, O. Onireti, M. A. Imran, A. Imran, and R. Tafazolli, "Control-data separation architecture for cellular radio access networks: A survey and outlook," *IEEE Communications Surveys Tutorials*, vol. 18 (2016), pp. 446–465.

[64] Z. Corporation, "Comparison between CP solution C1 and C2," Tech. Rep. R2-132383, 3GPP TSG-RAN2, Aug 2013.

[65] L. Yan and X. Fang, "Reliability evaluation of 5g c/u-plane decoupled architecture for high-speed railway," *EURASIP Journal on Wireless Communications and Networking*, vol. 2014, no. 1 (2014), pp. 1–11.

CHAPTER 5

An Effective Deployment of SDN Controller in Smart City Renovation

Madhukrishna Priyadarsini and Padmalochan Bera

Indian Institute of Technology Bhubaneswar

CONTENTS

5.1 INTRODUCTION

A smart city is an urban area that uses automated data collection, providing necessary information to manage different assets and resources efficiently. The data collected about roads, vehicles, schools, community, etc., are analyzed to monitor and manage traffic and transportation systems, power plants, water supply networks, information systems, schools, libraries, hospitals, and other community services [1]. Smart city technology allows city officials to interact with the community and infrastructure, monitoring the activities and evolutionary tasks in the city.

Satellites are widely used in smart cities for the management of infrastructure using real-time remote sensing data from different regions [2]. This helps to avoid various unusual situations such as forest fires, floods, traffic deadlocks on roads, and environmental and infrastructural abnormalities [3]. Satellite control centers support in decision making and provide real-time updates to the citizens. Networking and communication technology play an important role in information gathering and dissemination. Communication using traditional TCP/IP network suffers from challenges like packet loss, delay, quality of service (QoS), low reliability, and scalability issues in a real-time environment. It is necessary to perform earth observations accurately and in time before an emergency arises. Software-defined networking (SDN) allows dynamic reconfiguration of the network according to the user's requirements. This, in turn, provides high bandwidth, reliability, scalability, QoS, less delay, and packet loss in the network, which are vital for smart city automation. The SDN platform achieves this by separation of control programs from the data plane. However, there exist certain network control function design challenges in SDN, such as traffic management followed by load balancing, reduction of energy consumption by the devices present inside the network, controller security, and finally, the design of efficient controller architecture. In this chapter, our focus is on presenting an SDN controller design in a smart city application that supports *function virtualization* and *energy efficiency.*

Now, we present a brief overview of SDN and satellite communication, which are the basis of this chapter.

5.1.1 SDN Overview

An SDN is a new paradigm that has been designed to enable more agile and cost-effective networks. SDN is emphasized on the fact that the basic architecture of the traditional network is so much complex and decentralized. In contrast, the present network needs a more flexible architecture with a simple troubleshoot option. SDN recommends the centralization of network intelligence by separating the forwarding process of network packets (data plane) from the routing process (control plane). Figure 5.1 shows the basic architecture of SDN, with three layers. The *infrastructure layer* or the data plane comprises of traffic/data forwarding network devices. The responsibilities of the data plane are mainly data forwarding, dropping/replicating of data, as well as monitoring local network information, and collecting flow statistics. The *control layer/control plane* is responsible for programming and managing the forwarding plane. The protocols, logic, and algorithms that are used to program the forwarding

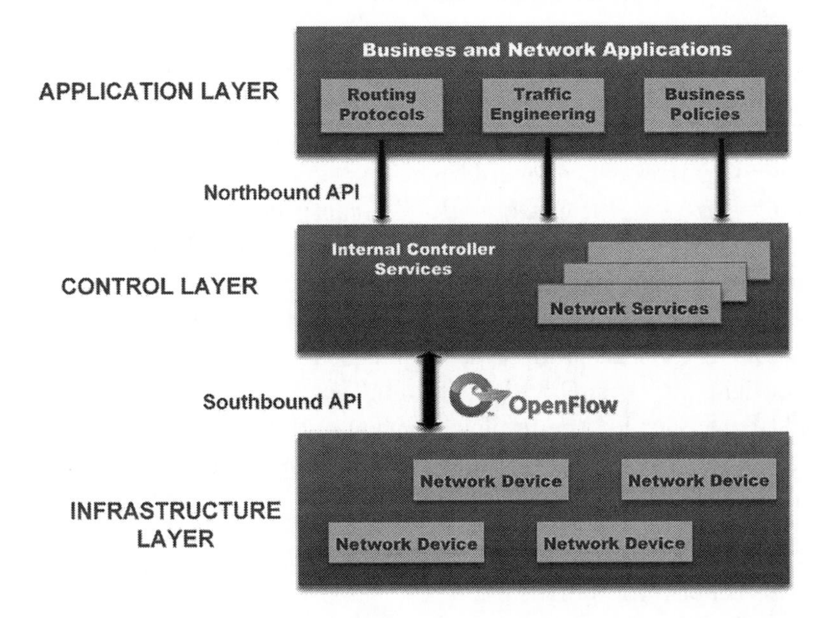

Figure 5.1 SDN architecture consists of three different layers, and are accessible through Open application programming interface (API).

plane are present in this layer. Many of these protocols and algorithms require global knowledge of the network. The control plane determines the generation of the forwarding rules and programming of the data plane. In summary, it makes use of the information provided by the forwarding plane and defines network operation and routing logic. It comprises one or more software controllers (which are logically centralized) that communicate with the forwarding network elements through the standardized interface, the southbound interface.

The application layer contains network applications that can introduce new network features, such as security and manageability, forwarding schemes, or assist the control layer in the network configuration. The application layer provides appropriate recommendations and feedback in the form of application policies/rules to the control layer. The interface between the application layer and the control layer is referred to as the northbound interface.

The OpenFlow architecture [4] enables the controller to determine the path for network packets across the network of switches (data plane). It manages switches from different vendors with different configurations and parameters using a unified OpenFlow protocol. An OpenFlow switch forwards the data packet of its hosts using the flow rules of its flow table. The flow table mainly consists of three fields: header, actions, and counter (expiration timer) [4]. A packet is processed with necessary actions if its header matches one or more rules in the flow table. If there is no match between the packet header and the flow table, then the corresponding packet header information is forwarded to the controller as a *Packet-IN* event. Subsequently, the controller generates flow rules for this *Packet-IN* request applying all running control functions. Finally, this flow rule is forwarded to the underlying switch that stores the rule in its flow table for further processing. There exist many open-source controllers that are widely used, such as NOX [5], POX, Beacon [6], Floodlight [7], OpenDayLight [8], and Ryu [9]. We used the OpenDayLight controller for the implementation and evaluation of our proposed framework. However, our proposed framework can also be used in integration with other controllers.

5.1.2 Satellite Communication

The communication between any two earth stations (ground stations (GS), user nodes) through a satellite is called as satellite communication *footprint*. The GS is designed to transmit and receive signals from satellites where the communication begins. The electromagnetic waves used for satellite communication carry information such as image, voice, audio, video, or any other data between ground and space and vice versa. The electromagnetic signals sent to the satellite from the GSs called uplink frequency, and the opposite is called downlink frequency. The satellites receive and re-transmit the signals back to earth where they are received by other earth stations in the coverage area of the satellite. The orientation of the satellite in its orbit depends on the Kepler's three laws [10].

The literature includes few directions towards implementations of SDN in satellites such as placement of controllers inside satellites [11] and in the GS [12]. One implementation research shows SDN in naval satellite communication (SDN–SAT) [13]. Along with these, some other directions of research in satellite communication using SDN exist [14–16], which are clearly explained in Section 5.2.

In this chapter, we present an SDN controller deployed design framework for traffic management and weather monitoring in a smart city. Our design helps in providing prior information to mobile devices in a different region for executing specific applications. When a mobile device is connected to the Internet, the satellite tracks its location and movement. It sends the device's movement information along with possible traffic and weather data within a range of 200 m radius to the GS. Our SDN controller deployed design framework runs in the GS, that processes the data, and sends the required information in advance to the mobile devices. The computation module of our framework includes mathematical models for the generation of valid information. We use constraint solving to verify the accuracy of the generated data and computed information. Our framework creates different virtual machines (VMs) inside the controller for the execution of various network functions (NFs). Multiple network packet schedulers intelligently allocate the traffic to appropriate VMs. The VM allocation method essentially reduces packet processing time and provides parallelism in executing the NFs. It also reduces the load on the controllers. Our architecture ensures energy-efficient routing using heuristic-based route selection. In smart cities, providing security to habitats and infrastructure is a major challenge. Our proposed framework can be used to implement adequate security for the citizens by accurately detecting malicious activities, natural calamities, damage to infrastructure, etc. using the verified satellite images. On detection, the necessary counter mechanism can be deployed towards resiliency against violation of security and privacy.

The performance of our SDN deployed design framework is evaluated in comparison with the TCP/IP satellite communication networks and state-of-art GPS systems. Our framework is also capable of processing other possible information and can be shared with the mobile devices present in the smart city region. We consider low earth orbit (LEO) satellites for communication in the smart city (approximately 50–60 km radius). The devices moving between more than one region can get the information depending on inter-satellite communication (with medium earth orbit (MEO) and geosynchronous equatorial orbit (GEO) satellites) [14].

The rest of the chapter is organized as follows. Section 5.2 presents the background literature on satellite communication using SDN. The proposed SDN deployed design framework for smart city communication is discussed in Section 5.3, with a detailed description of all the components. Section 5.4 presents NF virtualization for traffic dissemination in GS. The packet flow in the framework is mentioned in Section 5.5. The framework implementation with performance evaluation is presented in Section 5.6. This section also includes discussions about our proposed work, which contains extensibility, usability, and limitations. Finally, we conclude in Section 5.7 with future directions of research.

5.2 BACKGROUND STUDY

In this section, we present state-of-the-art satellite communication using the traditional TCP/IP network. Subsequently, we discuss related works on satellite communication using the SDN framework. In addition, we present works on the deployment of the SDN framework in real-life scenarios.

5.2.1 Satellite Communication Using Traditional Network

Satellites use TCP/IP as a communication protocol for space-based communication systems that need to access the broader terrestrial communication network. In paper [17], the author presents TCP/IP performance over satellite network links, identified problem areas for TCP/IP technologies, and made suggestions for optimizing TCP/IP over such links. Paper [18] explains all the available commercial network protocols for satellite communication along with their effects on QoS and routing. It also compares the performance of communication networks using TCP/IP and other protocols such as protocols for aeronautical applications and mobile Internet protocols. The paper [19] points the factors limiting TCP performance over satellite links and the TCP mechanisms affected by it. In addition, it proposes a solution for end users to access upstream satellite channels directly. The research work in [20] points out the significant problems in satellite links such as long propagation delay, high bit error rate, and bandwidth asymmetry. The authors propose an enhanced TCP flow control algorithm, which consists of a data loss recovery mechanism for adequate link utilization. Indian Space Research Organization (ISRO) released a document [21], which provides the signal and the data structure for standard positioning service (SPS) of the Indian Regional Navigational Satellite System (IRNSS). The document addresses the signal modulations, the frequency bands, the received power levels, the data structures, and their interpretations, algorithms, etc. ISRO released another document which provides the signal and the data structure for messaging service using IRNSS 1A spacecraft [22]. The document addresses the signal modulations, the frequency band, the received power levels, the data structures, their interpretations, etc. However, in each work GS suffers from information delay/loss due to dynamic requirement changes and high traffic load. Also, in due course, the computational complexity of the user devices becomes high.

5.2.2 Satellite Communication Using SDN Framework

SDN controllers solve traffic load, QoS, packet loss, link utilization, reliability, and scalability issues that are present in traditional TCP/IP networks. The satellite communication network with the SDN framework overcomes all the challenges in the TCP/IP network and can be deployed in any real-life scenario. Some of the works are mentioned in this section. In [23], all the benefits of satellite communication using SDN are highlighted, such as the flexibility and controllability of space dynamic routing algorithms, rapid deployment of flexible and fine-grained network management strategies, reduction of the system cost, and improvement of the collaboration between satellites and the compatibility of heterogeneous space systems. The paper [24] discusses the SDN and network virtualization in satellite network application system architecture. Here, the design principle of the satellite OpenFlow switch and controller, the realization methods of NF virtualization, are described, which provides the global dynamic load balancing, routing, and resource scheduling. The research work in [25] proposes a software-defined satellite architecture with high efficiency, fine-grained control, and flexibility. It also discusses some practical challenges regarding architecture deployment. In [26], the authors describe the benefits that SDN/network function virtualization (NFV) technologies can bring into satellite communications towards 5G. The paper [14] presents a space-ground integrated multi-layer satellite communication network, based on SDN and NFV. The network consists of an SDN-based three-layer satellite network which provides effective traffic scheduling and flexible and re-configurable satellite service delivery. In [12], SDN framework implementation inside the GS is focused, which provides efficient resource utilization and data management.

5.2.3 Application Deployment of SDN

The paper [13] implements the SDN framework in the naval network using satellite communication (SDN–SAT). In this work, each ship is an SDN switch with multiple satellite communication connections. A remote SDN controller handles the management and classification of flows between controllers and satellites. In paper [14], the authors consider a dynamic controller placement problem (DCPP) for LEO satellites, where the traffic demands change dynamically based on the user's geographical position and time zone. The paper [25] implements a terrestrial network using software-defined satellite networks and illustrates the fundamental aspects of integrated network application functions. In the next section, we present our proposed SDN deployed design framework for smart city renovation, which is implemented inside the GSs.

The literature study reveals that the application and deployment of SDN in smart city communication is in the premature stage. In addition, the satellite communication system using TCP/IP and GPS-based systems suffer from a performance bottleneck that limits the implementation of real-time network applications. Moreover, there are no works that exploit the advantages of NF virtualization for energy-efficient traffic management in applications.

5.3 PROPOSED SDN CONTROLLER DEPLOYED DESIGN FRAMEWORK

In this section, we present our proposed SDN deployed design framework for smart city communication. Figure 5.2 shows the overall architecture of our proposed framework and the flow of traffic through different modules.

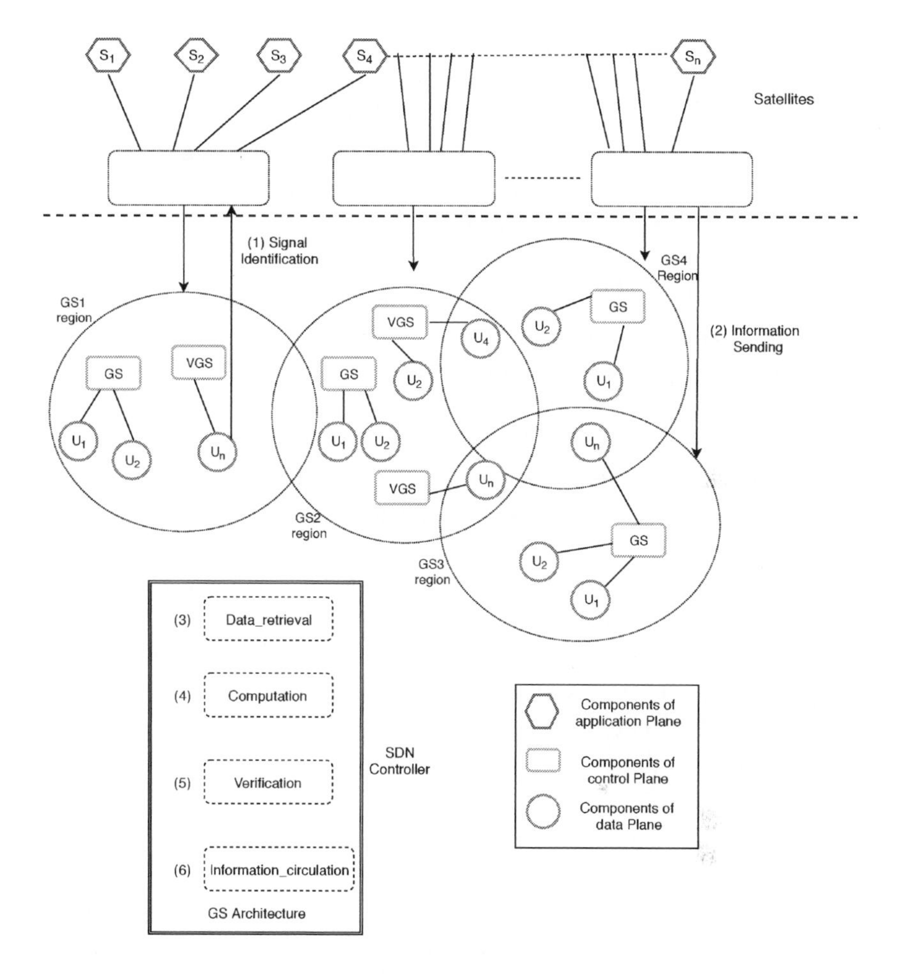

Figure 5.2 Proposed SDN deployed design framework for smart city communication.

Our design is targeted for implementing traffic management and weather forecasting applications in smart cities. It consists of three major entities, namely user nodes, satellites, and GSs. We deployed SDN controllers inside the GSs for the reduction of computation overhead, achieving accuracy in prediction, reliability, controlling packet drops, and reducing the delay. As the user is connected to the Internet, a signal is generated from its device. The satellites capture the signals and find its location (i.e., x and y coordinates). Then, the satellites record the user nodes' position and their movement [27]. Subsequently, it collects weather and road traffic-related data within a 200 m radius across the user's location and forwards the same to the GS.

The working procedure of the data collection phase is described as follows: After finding the user node's position, the satellite captures various images in the perimeter of the user node. Those images consist of traffic conditions (number of active user nodes present, road condition along the user node's path) and the weather conditions (cloud pattern). Here, we consider the LEO satellite constellation for tracking user position. We used four LEO satellites to ensure the correct detection of user location. The higher orbit satellites (MEO and GEO) are mainly used for wide range communication in the implementation of inter-smart city applications. The LEO satellites communicate with each other and with MEO and GEO satellites as required [15].

In general, the GS checks the correctness of the satellite's signal. In our framework, we used GS for computation and prediction of traffic and weather-related information for the active user nodes in its region and to ensure the correctness of the signal. It consists of four modules, namely *data retrieval, computation, constraint verification,* and *information circulation.* The *data retrieval* module retrieves the data from the images captured by the satellites. The *computation* module dynamically computes the weather and traffic condition around 200 m radius of the user node. The *constraint verification* module verifies the accuracy of the predicted information by the *computation* module. The *information circulation* module circulates the verified messages to all the user nodes within the said region. NF virtualization is used for implementing the controller function in GS.

Several VMs are created in each controller that retrieves data, computes tasks on it, and subsequently forwards the computed data to the verification module. After verification, the information is sent to the user nodes. We used the concept of the virtual GS (VGS) to provide reliability and responsiveness as well as to handle inter-GS communications effectively.

If the region of a GS is wide, then the prediction procedure is replicated in each VGS. Each VGS acts as one GS, and it communicates with the user nodes. The interaction between different GS is necessary as the user nodes move from one region to another. If one user node is present in the region of more than one GS, then the GS/VGS, which is present at the minimum distance from the user, sends the prediction information. The working procedure of each module of our proposed SDN deployed design framework is discussed in the following subsections.

The description of the notations used throughout the chapter is presented in Table 5.1.

TABLE 5.1 Notations Used in the Proposed SDN Deployed Design Framework

Notation	Description
U	User node
GS	Ground station
VGS	Virtual ground station
T	Time for which the information is predicted
D	Distance considered for node elimination
N	Line width
B	Buffer used to store node information
m_t	Message computation time
V	Maximum allowable difference of weather data between two frames for a single m_t

Here, we describe the functioning of all modules concerning a single user node. But, the procedures apply to all user nodes within the region of the GS.

5.3.1 Data Retrieval

The GS receives traffic and weather-related data that are collected by the satellites in the form of images. The traffic-related data are represented as user node's location (X, Y), movement of position through latitude and longitude $(o, ', '')$, road positions in the region, number of neighbor nodes, and road conditions (narrow or wide). The weather data are represented in terms of temperature (T), pressure (P), wind speed (S), and cloud pattern (C). These data are retrieved from the weather images.

To retrieve the required data from the images, we used the convolutional neural network's (CNN) [28] You Only Look Once (YOLO) framework.

It detects the objects and localizes them (finding exact coordinates) from the images using the concept of the bounding box. This detection is fast, as it processes approximately 45 frames per second [29], which improves the efficiency of the GS.

We consider dynamic calculation and prediction of traffic and weather data for a certain period (t).

So frames are sent continuously from the satellites to the GS. We consider each image as a frame, and the GS predicts the information present in m such frames. The retrieval module represents user nodes' position, neighbor nodes' position, and road positions using different matrices. The representation is given as follows:

User node position:

The data retrieval module processes m frames simultaneously and finds the location of the user nodes present in those images. It stores those points in a $1 * m$ matrix represented as

$$U^p = [(X_0, Y_0)\,(X_1, Y_1)\,(X_2, Y_2) \cdots (X_{m-1}, Y_{m-1})]_{1*m}$$

where m is the number of frames sent from the satellites. It can also be presented in the form of $<X_j,\ Y_j>$, where $j \in 0, m-1$.

Neighbor node position:

The GS receives all the neighbor node positions in a $1 * m$ matrix, where l is the number of neighbors in m frames. We represent the matrix as

$$\begin{aligned} NN = &((a00, b00)(a01, b01) \cdots (a0m-1, b0m-1)(a10, b10) \\ &(a11, b11) \cdots (a1m-1, b1m-1)(a(l-1)0, b(l-1)0) \\ &(a(l-1)1, b(l-1), 1) \cdots (a(l-1)(m-1), b(l-1)(m-1)))\, l * m \end{aligned}$$

It can also be represented as $<a_{ij},\ b_{ij}>$, $\forall i \in 0, l-1$ and $j \in 0,\ m-1$.

Road position detection:

The road positions are varying as U is moving. Let k be the no. of roads in m frames. For each road k, we consider two parallel points on the road to detect the line width (n) of the road. From the satellite image, the road points are retrieved as five-tuple $<M, N, P, Q, t>$, where $<(M, N), (P, Q)>$ is the coordinates of two parallel points on the road, which is received at time t. Here, we represent each five-tuple $<M, N, P, Q, t>$ as r, and the reference positions of the different roads are represented using matrix RB.

$$\begin{aligned} \text{RB} = &((r00)\, r01 \cdots r0m - 1r10\ r11 \cdots\ r1m - 1\ r\,(k-1)\,0 \\ &r\,(k-1)\,1 \cdots r\,(k-1)\,(m-1))\, k * m \end{aligned}$$

or in the form of $<r_{ij}>$ $\forall i \in 0, k-1$ and $j \in 0,\ m-1$.

Afterward, the module finds the average distance of each $<M, N>$, and $<P, Q>$ points to know the road's width and accommodation capacity of user nodes.

Then, it finds out the accuracy of the user position and its movement.

The accuracy of the user position is calculated using the timing information associated with each data packet (images) received from the satellites.

Here, it uses the GPS concept of constraint verification of data [27].

The weather-related data are retrieved from the weather data packets. The module retrieves temperature, pressure, wind speed, and cloud patterns received from the satellite images using CNN.

We represent weather data as $WR_i = <C\_i, T\_i, P\_i, S\_i>$.

The weather and traffic information over a specific period is forwarded to the *computation* module in GS for the necessary prediction.

If any user node falls in the overlapping region of two GS, the retrieved data are forwarded after consultation with the neighboring GS.

5.3.2 Computation

The computation module first finds the user direction according to its movement. Next, it eliminates out-of-path neighbor nodes and counts the nodes, which can increase traffic along the user node's path.

Afterward, it predicts traffic conditions for a certain distance (let d) from the current position of the user node. Similarly, it also computes weather data. The traffic and weather conditions are computed as follows:

User direction detection:

As a user node moves, we calculate its movement direction. Here, we consider the $U^p{}_{1*m}$ for the detection of user node direction. If $X_j > X_{j'}$, $\forall$, $j' < m$ then the user is moving in the east direction. Otherwise, its movement is in the west direction. For north and south direction, Y coordinate varies, and X coordinate remains constant. The movement in north direction satisfies the condition $Y_j < Y_{j'}$ $\forall j, j' < m$ and for movement in south the reverse is applied. Similarly, the conditions for the movements in north-east, north-west, south-east, and south-west direction, respectively, are as follows:

$$(Y_j < Y_{j'}) \text{ and } (X_j < X_{j'})$$

$$(Y_j < Y_{j'}) \text{ and } (X_j > X_{j'})$$

$$(Y_j > Y_{j'}) \text{ and } (X_j < X_{j'})$$

$$(Y_j < Y_{j'}) \text{ and } (X_j > X_{j'}),\ j, j' < m$$

Node consideration:

To predict the traffic condition accurately, the GS considers the nodes which are ahead of the user node and eliminates the nodes which are behind. Here, we consider $U^p{}_{1*m}$ and NN_{1*m} matrices. If the user is moving in the east direction, then the nodes that satisfy the condition $a_{ij} < X_j$ and $\forall j < m$ and $\forall i \in 0, l-1$ are considered. Similarly, for the other seven directions, the nodes behind the user nodes are eliminated from the traffic condition prediction method. The nodes which are ahead of the user node and within a distance of d are considered. We store these nodes in a buffer b for further processing. The following constraints show the conditions for storing the respective node information in the buffer.

$$a_{ij} > X_j \ \&\& \ (a_{ij} - X_j) <= d$$

$$b_{ij} > Y_j \ \&\& \ (b_{ij} - Y_j) <= d$$

$$((a_{ij}, b_{ij} > X_j Y_j) \ \&\& \ ((a_{ij}, b_{ij} - X_j Y_j) <= d \ \forall j \in 0, m-1 \text{ and } \forall i \in 0, l-1$$

After consideration of the required nodes, the matrix NN is represented as NC, which contains less number of nodes than NN.

Node count and comparison:

It counts the number of nodes present in the buffer of the user nodes along each possible path. Then, it compares the no. of nodes with the accommodation capacity (no. of nodes $> n$) of the road. If the above condition holds, it assesses more traffic in that area. Here, n is the road line width. Thus, road line width indicates the width of the single line along the road. The road's width and accommodation capacity of user nodes are known from the road position's detection phase. The information in the form of a message is forwarded to the constraint verification module for assessing the accuracy of the information.

For weather-related computation, the GS arranges the retrieved weather-related data from the retrieval module (temperature, pressure, wind speed, and cloud patterns) according to the movement of the user nodes. Subsequently, these information messages are forwarded to the constraint verification module. Here, we consider the validity of weather messages for the next few minutes. This is because the user node is moving, so it requires the weather information for a small period. In addition, the weather message for a long period will introduce computation complexity.

5.3.3 Verification of Design Constraints

This module verifies the accuracy of the computed information before sending it to the user nodes.

To verify the accuracy, we use the constraint verification method and consider three parameters, such as buffered data b_i for each user i, time of sending messages to the users, and location of the neighbor nodes.

During two consecutive time instance t and t', the sum of buffers b_i of all user node i for a particular road k should be equal. We represent it mathematically as

$$\sum b_i^{t,k} = \sum b_i^{t',k} \quad \forall (t, t') \in \text{time} \tag{5.1}$$

where l is the number of user nodes. So, for consecutive frames, the summation of buffered data for all user nodes should be equal. If m_t is the message generation time and c_t is the computation time, then

$$t + c(i) < m_t(i) < t + c_t(i) + c_t(i+1) \quad \forall 1 < i < n \tag{5.2}$$

where $c_t\ (i) < m/x$, m is the number of frames, and each frame takes x seconds for processing. Here, the above equation shows the message generation time is dynamic, and it lies between the first frame and the next consecutive frame computation time, i.e., whenever there is high probability traffic on the road, the message should be sent to the user nodes.

The accuracy of each neighbor nodes' position is verified using matrices U^p and NC. The nodes' position ahead of the user node is stored in an NC matrix with their coordinates. The nodes in the NC matrix are also available in the U^p matrix during its retrieval. We say if the coordinates match for the user nodes in both the matrices, then the prediction is accurate.

The weather-related data are verified, considering the derived temperature, pressure, wind speed, and cloud patterns. During the computation time period c_t, the values of temperature, pressure, and wind speed for different frames should be almost alike. If there is a large gap between values, then the derived values are not accurate. We represent it mathematically as $\text{WR}_i - \text{WR}_{i-1} <= V$, where V is the maximum allowable difference in temperature, pressure, wind speed, and cloud pattern values between two frames for a single m_t. But there is a possibility that $\text{WR}_i - \text{WR}_{i-1} <= V$ for different m_t.

After verifying the accuracy, this module arranges both traffic and weather-related information in a single message format and forwards the same to the *Information Circulation* module.

Algorithm 1 Proposed SDN Deployed Design Framework algorithm for single user node Scope: in GS for each user node

```
1: procedure DataRetrieval(l, m, k)
2:     for j= 0 to m-1 do
3:         U_j^p = (X_j, Y_j)
4:     end for
5:     for i= (0 to l-1) do
6:         for j=0 to m-1 do
7:             NN_ij = (a_ij, b_ij)
8:         end for
9:     end for
10:    for i= 0 to k-1 do
11:        for j=0 to m-1 do
12:            RB_ij = <(M_ij, N_ij), (P_ij, Q_ij)>
13:            n_ij = dist((M_ij, N_ij), (P_ij, Q_ij))
14:            n = avg(a_ij)
15:        end for
16:    end for
17:    for i= 0 to m-1 do
18:        WR_i = (C_i, T_i, P_i, S_i)
19:    end for
20: end procedure
21: procedure DataComputation(l, m, n, d)
22:    for i= 0 to l-1 do
23:        for j=0 to m-1 do
24:            if ((a_ij > X_j)&&(a_ij − X_j) ≤ d) ∨ ((b_ij > Y_j) && ((b_ij − Y_j) ≤ d) ∨
   (((a_ij, b_ij) > (X_j, Y_j)) && ((a_ij, b_ij) − (X_j, Y_j)) ≤ d) then
25:                b_i = (a_ij, b_ij)
26:            end if
27:        end for
28:    end for
29:    for i=0 to l-1 do
30:        if (∑ b_i > n) then
31:            msg ← ACCURACY(b_i, WR_i)
32:        end if
33:    end for
34: end procedure
35: procedure Accuracy(b, WR)
36:    for all do(t, t') ∈ time && c(t), k, V
37:        for i=0 to l-1 do
38:            if ((∑_{i=0}^{l-1} b_i^{t,k} = ∑_{i=0}^{l-1} b_i^{t',k}) && (t + c_t(i) < m_t(i) < t + c_t(i) + c_t(i+1)) &&
   (b_i ∈ NN_ij) && (WR_i − WR_{i−1} ≤ V)) then
39:                return 1
40:            else
41:                return 0
42:            end if
43:        end for
44:    end for
45: end procedure
46: procedure InformationCirculation(msg)
47:    if msg ≠ 0 then
48:        Send(msg)
49:    else
50:        Drop(msg)
51:    end if
52: end procedure
```

5.3.4 Information Circulation

This module sends the verified information to the user nodes. Within the region of a single GS, more than one node is connected to the Internet. If all the nodes move in the same direction, then the computation module computes for one user node, and the same information applies to others.

So, the circulation module forwards the same traffic and weather-related information to all other nodes using their IDs. This process improves the responsiveness of the GS. However, the computation is performed separately for each active node if the nodes change their direction.

We have deployed the SDN controller inside the GS. Thus, the processing time is efficient as compared to the TCP/IP network. We implemented message-passing based data sharing among the GSs for the nodes moving from one region to another. In real-life scenarios, if the region of the GS is wide, then VGSs can be implemented. In such a case, the procedures of GS are replicated in each of the stations. It effectively manages the performance of the network and reliability, and QoS is guaranteed. The overall process implemented in GS of our proposed SDN deployed design framework is presented in Algorithm 1.

In the next section, we describe the allocation of NFs into VMs of each controller.

5.4 NF VIRTUALIZATION FOR ENERGY-EFFICIENT TRAFFIC DISSEMINATION

A VM serves as an execution unit for running a finite set of NFs; this number is limited by a threshold value d. In other words, the parameter d indicates the maximum degree of parallelism supported by the virtualization platform. In a VM, an NF is associated with a unique tag (NF tag) that is used for referencing the function in the VM. Therefore, our architecture allows an NF to get executed in more than one VM simultaneously. A VM maintains an input queue (first-in, first-out) that keeps track of the packet processing requests received from the network packet scheduler (NPS). A packet processing request consists of a packet header and an ordered list of VMs the packet should navigate. To serve a packet, a VM takes the packet from the queue, reads the tag of the packet, matches this tag with its NF tag, and executes the function corresponding to the matched tag on the packet. On successful completion of the NF, the VM updates the state of the packet as "PROCESSED"

TABLE 5.2 Notations Used in NF Virtualization Architecture

Notation	Description
VM_i	Virtual machine
PS_i	Packet Scheduler
NF_i	Network function
$NF_{i,k}$	Tag of NF_i in VM_k
$Seq(NF_i)$	Sequence of NF
NFAT	Network function allocation table
VMAT	VM status table
$VM_i{}^q$	Current queue size in VM_i
$VM_{i,st}$	Current state of VM_i
T_{NFi}	Average completion time for NF_i

against the corresponding NF. Then, the VM sends an acknowledgment to the NPS (i.e., the corresponding packet scheduler (PS)) and subsequently sends the packet to the next VM in the ordered list. We realize the inter-VM communication is using message passing.

A VM can be loaded or updated with new NFs depending on the requirements. This configuration process is governed and synchronized by the respective PS. During this configuration period, the environment of the corresponding VM is saved, and the state is set as "BLOCKED." The information saved during this process is assigned NFs, the packet processing state of the VM, and the content of the queue. The VM becomes "ACTIVE" and resumes its execution once the configuration is over. Table 5.2 presents the notations used in virtualization architecture.

5.4.1 Network Packet Scheduler

A network packet scheduler keeps track of the information about the VMs. This information includes the state of each VM, NFs running in it along with their tags, and the processing statistics (average packet processing time, current queue size). A network packet scheduler has two tasks: NF-sequence determination and VM allocation.

1. NF-sequence determination: On receipt of a packet request from an OpenFlow switch, the NPS extracts the following fields from the packet header: source and destination IP addresses, source and destination ports, packet size, encryption type, and the required NFs. Then, it determines the ordered list of NFs ($Seq(NF_i)$) that is required to process the packet request. This list is created using filtering conditions based on the extracted fields. A packet should

execute all the NFs in the sequence for the successful execution of an application.

2. VM allocation: After determining the NF sequence for a packet request, the NPS allocates the packet to the respective VMs for its execution. It essentially outputs a mapping: $<\mathrm{VM}_k, \mathrm{NF}_{i,k}>$, for ith NF to kth VM. Our proposed VM allocation process is described in Algorithm 1. For the implementation of Algorithm 1, the NPS maintains two data structures, namely NF allocation table (NFAT) and VM status table (VMST).

    (a) NF allocation table (NFAT): It records the availability of NFs execution in the VMs. This table is indexed by NF number. One entry of NFAT is of the form: $< \mathrm{NF}_i, < \mathrm{VM}_j, \mathrm{VM}_j{}^q >, \ldots, < \mathrm{VM}_k, \mathrm{VM}_k{}^q >>$. The VM's queue size is increased after allocation of the packet to it and decreased after completion of the related NFs.

    (b) VM allocation table: This table stores the current state of a VM, a list of NFs available in it along with their tag number. The table is indexed by VM number. An entry in VMST is of the form: $< \mathrm{VM}_i, \mathrm{Vm}_{i,\mathrm{st}}, < 1, \mathrm{VM}_i, \mathrm{T}_{\mathrm{NF}i} >, \ldots, < k, \mathrm{VM}_k, T_{\mathrm{NF}i}>>$.

Algorithm 1 iterates through each entry of the NF sequence (generated in the previous step). For each ith entry NF_i, it adds a tuple $<\mathrm{VM}_k, \mathrm{NF}_{i,k}>$ to the ordered list. After updating the list by iterating through the NF sequence, the network packet scheduler sends the packet to the first VM in the sequence. Subsequently, the VM sends an acknowledgment to the corresponding PS after the successful execution of the corresponding NF in the VM, and the Packet scheduler updates the statistics of that VM.

5.4.2 Network Functions

An NF is an atomic network control operation or a task corresponding to a packet's processing requirements. Here, we consider three possible NFs: (i) routing control, (ii) energy efficiency, and (iii) load balancing. These NFs are integrated into a single NF named as energy-efficient load distribution. Energy-efficient load distribution provides efficient routing, minimum energy consumption, and load balancing according to the required changes. Our proposed NF is described in the following subsection:

Energy-Efficient Load Distribution: We propose a new direction of load balancing in the SDN controller, which drives both efficient routing and reduction of energy consumption by the devices. In our approach, when the load on the controllers increase, they find other lightly loaded controllers and their own heavily loaded switches through the *load collection phase*. This is realized using message passing between controllers and switches. Then, a set of suitable controller-switch pairs are selected for *load migration* from highly loaded controllers to the lightly loaded controllers. Afterward, an optimal route is selected based on hop count and energy consumption of the devices along the route. Our approach selects a route with the least hop count and minimum total energy consumption. Finally, load migration is performed between the source and target controllers. The heuristic used for route selection computes energy consumption by the network devices in each possible route. It then selects the best route, which provides less energy consumption by the devices as well as the link paths. This completes the energy-efficient load distribution process. One of the key features of our proposed load distribution mechanism is the integration of routing control, energy efficiency, and load balancing to improve the overall performance of the network.

5.5 MAPPING OF TRAFFIC CONTROL FLOW TO SDN VIRTUALIZATION

Here, we present a detailed description of traffic flows in the virtual controller architecture. The control flow consists of a sequence of four steps.

Step 1. A new packet is mapped to an appropriate packet scheduler

When any data packet arrives to the GS controller from a satellite, the tag bit associated to the packet determines the network packet scheduler to which the packet request is to be routed. This is realized through a hash function on the fields of the packet header.

$$f\text{:tag[packet]} => \text{PS}_i$$

Step 2. Packet Scheduler de-queues a packet from its queue and allocates appropriate VM

This step consists of the following tasks.

[2-a]: Extract tuple $<\text{PORT}_{\text{src}}, \text{PORT}_{\text{VM}}, \text{IP}_{\text{src}}, \text{IP}_{\text{VM}}>$ from packet header. Let this tuple be *H*. Create a sequence of NFs,

Seq(NF_i) based on H and available information about application requirements.

[2-b]: Initialize the ordered list $L = \varnothing$; Iterate through Seq(NF_i). For each NF_i; find an active virtual machine VM_k and corresponding tag $NF_{i,k}$. Add <VM_k, $NF_{i,k}$> to L. $NF_{i,k}$ signifies NF_i is running on VM_k with the tag $NF_{i,k}$.

[2-c]: Send the packet along with the sequence to the first VM in the list.

Step 3. Execution of NF in VM

On receipt of a packet, the corresponding VM removes an entry from the list, reads the tag, and runs corresponding NF on the packet. The VM sends an acknowledgment to PS after successful completion of the NF and forwards the packet to the next VM in the list for further processing.

Step 4. Iterate step 3 until all the NFs are executed on the packet.

Algorithm 2 presents the VM allocation procedure that is described in steps 2-b and 2-c of the control flow. Step 2-a explains the process of determining the port number and creation of the NF sequence for a data packet. The processing of the packet is completed in the computation model. After the creation of the required information, it is forwarded to the verification module. This ends the processing of particular traffic inside the controller. The controller sends the generated message to the user nodes if required in the information circulation module.

Algorithm 2 VM Allocation Process $Seq(NF_i)$

Input: $NFAT, VMST, Seq(NF)$
Output: L

1: **procedure** $VM\_Allocation()$
2: Initialize ordered list L to empty
3: **for each** $NF_i \in Seq(NF_i)$ **do**
4: $VM^q_{min} = \text{MAX(INT)}$
5: Initialize VM_{min} to empty
6: **for each** $< VM_j, VM^q_j >$ *in* NFAT[NF_i] **do**
7: **if** $VMST[VM_j].state = ACTIVE$ **then**
8: **if** $VM^q_j < VM^q_{min}$ **then**
9: $VM^q_{min} = VM^q_j$
10: Set VM_{min} to empty
11: Append VM_j to VM_{min}
12: **else if** $VM^q_j = VM^q_{min}$ **then**
13: Append VM_j to VM_{min}
14: **end if**
15: **end if**
16: **end for**
17: Append $< VM_j, NF_{tag} >$ to L
18: **end for**
19: Update queue entry for VM_j
20: Return L
21: **end procedure**

5.6 IMPLEMENTATION OF SDN DEPLOYED DESIGN FRAMEWORK IN SMART CITY

We evaluated our proposed SDN framework by analyzing its performance against state-of-the-art TCP/IP satellite communication protocols, and existing systems such as GPS and global differential GPS (GDGPS).

5.6.1 Experimental Setup

We used cars overhead with context (COWC) data set (Toronto, Canada) as satellite image data [30], and OpenDayLight controller [28] for implementation of GS functions.

We considered the CNN implementation with YOLO framework [29] from COWC data set images. YOLO framework first divides the input image into grids. Subsequently, image classification and localization are applied on each grid. Afterward, it predicts the bounding boxes and their corresponding class probabilities for objects. We used the trained neural network for accurate prediction of the data.

We used the OMNET++ [31] tool, specifically OS^3 and osg-earth framework for simulation of our design. The controller's (OpenDayLight) source code for the GS is integrated with the OsgEarthNet.ned file of the OMNET++ tool. The OS^3 and osg-earth framework together are used to visualize live weather data, high-resolution altitude data, and current satellite movement data with higher accuracy. For the simulation of real-time satellite communication, we considered seven satellites, 500 mobile nodes, seven GSs (controllers) over a perimeter of 60 km. Here, the mobile nodes join the network in different time instances, and it also can move from one GS region to another. Each GS region contains two VGSs, and the seven satellites send the signal to the seven GSs in various combinations. We executed our simulation for multiple iterations, with varying mobile nodes and satellite positions. For verification of computed information in the GS, we used an satisfiability modulo theory (SMT) based constraint solver, namely Z3 [32], that models the verification constraints in SMT-LIB (Library) language [33]. An instance of reducing verification constraints in SMT-LIB is shown in Table 5.3.

5.6.2 Performance Evaluation

We evaluated the performance of our proposed SDN deployed design framework for a smart city in terms of throughput, computation time, and information accuracy. Figure 5.3a shows the throughput (in Mbps)

TABLE 5.3 SMT-LIB Reduction of Query Specification

Constraints	SMT-LIB Model
Equation (5.1)	(define bf::(-> int int scalar(list)))
	(assert(bf(=> (timet) ∧ (userl) ∧ (roadk)))
	(= =(timet′) ∧ (userl) ∧ (roadk)))
	(check)
Equation (5.2)	(define c_t::(-> int real))
	(define m_t::(-> int real))
	(assert (m_t(=>(c_t t, i) OR (time t))
	(<(m_t t, i))(<c_t $t, i+1$) OR (c_t t, i) OR (time t)))
	(check)

with a varying number of mobile nodes under different GSs. It shows the comparison of throughput between our proposed framework, traditional TCP/IP satellite communication protocols, and state-of-the-art GPS systems. The figure shows that our framework provides higher throughput than traditional TCP/IP model and GPS [34] because of the multi-threaded NFV implementation [26,35] in the GS. Our framework provides less throughput than the GDGPS [36], because it uses the navigation technology of the next-generation GPS Operational Control Segment (OCX) and a complete array of real-time Global Navigation Satellite Systems (GNSS) state information.

Figure 5.3b presents the average message computation time (in ms) for different systems, including our proposed framework, with the varying number of user nodes. Our framework uses a single computed message for more than one node along the same path, which reduces the computation time. It uses OpenFlow for sending the data to the users that make message sending time negligible. In the case of traditional TCP/IP networks, the computation time is more due to delay in messages received from the satellites. The computation time of our framework and GDGPS is nearly equal because it also uses positioning accuracy for location computation.

Figure 5.3c presents the average energy consumption (in MJ) for different systems considering 200 packets/s as the traffic flow. Our proposed framework consumes less energy than the other systems as it uses a heuristic-based route selection algorithm, which reduces the energy consumption by the link paths. We also find that the average energy consumption for our framework lies within 700 MJ [37] (Threshold for real-life SDN application).

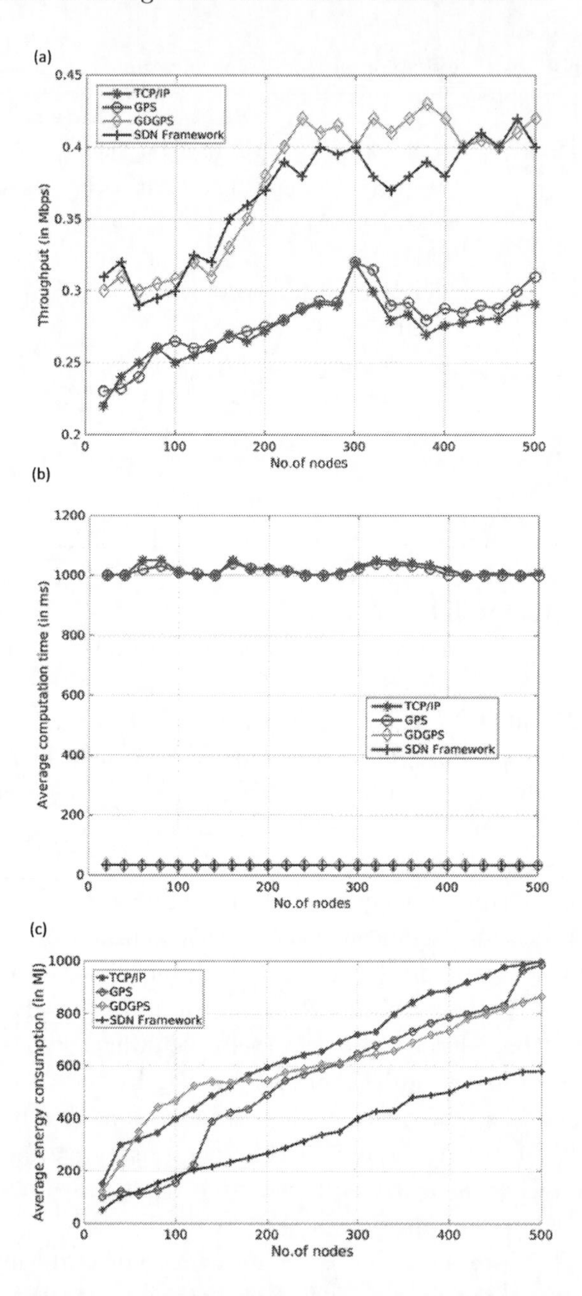

Figure 5.3 Performance comparison of proposed SDN deployed design framework and existing systems in terms of throughput, computation time, and energy consumption. (a) Throughput comparison of proposed SDN framework with state-of-the-art technologies. (b) Computation time comparison of various technologies along with proposed SDN framework. (c) Energy consumption comparison of four technologies.

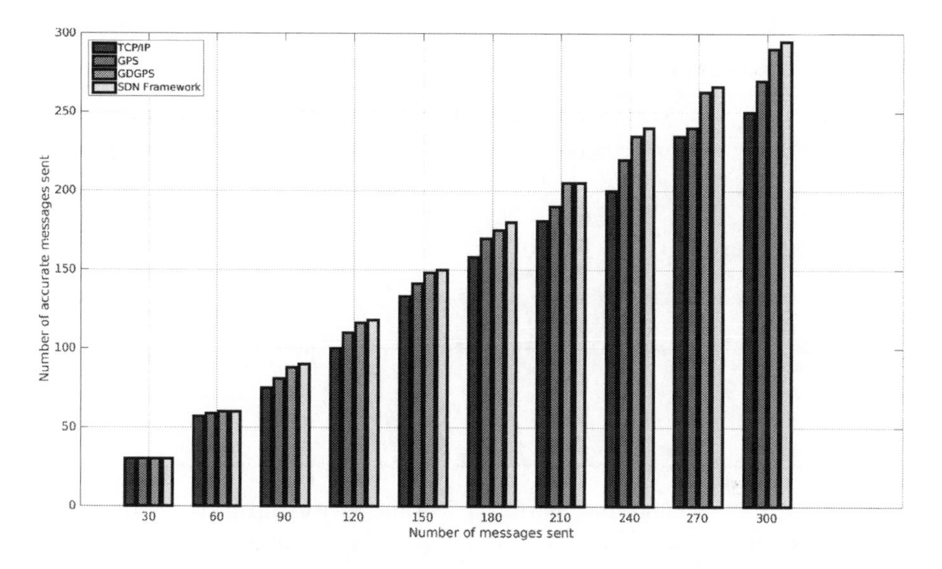

Figure 5.4 Accuracy comparison of the proposed SDN deployed design framework with state-of-the-art systems.

We also measured the accuracy of the sent messages to the user nodes by different systems and compared them with our proposed framework's information prediction accuracy. Figure 5.4 shows the comparison of accurate messages sent vs. the total number of messages sent in iterations by various systems. Here, we count the number of SAT messages generated by Z3 SMT solver in each iteration. As our framework uses Z3-based *Verification*, the accuracy of sent information is approximately 98%. In the case of other systems, due to the communication gap between the devices and the satellites, the system shows inaccurate information for 10%–15% of the cases.

Based on the extensive experimentation and the above discussions, we highlighted our overall observations in Table 5.4. The table reports overhead on user nodes, loss rate (packet drops), the average delay for three different state-of-the-art systems, and our proposed framework. Here M denotes the total number of messages sent from GS to the user nodes during the execution of the respective framework. The overhead on the user node is not present in our framework, as all the computations are implemented in the SDN controller deployed in GS. Also, the delay and loss rate are very less when compared with other systems. This comparison shows the efficacy of our proposed framework.

TABLE 5.4 Overhead, Loss Rate (Packet Drops), and Delay of TCP/IP, GPS, GDGPS, and SDN Framework

	Overhead	Delay (s)	Loss Rate (Packet Drops)
TCP/IP	12% of # M	15.371	0.25% of # M
GPS	11% of # M	13.67	0.15% of # M
GDGPS	10% of # M	0.04	0.05% of # M
SDN framework	NA	0.02	0.0132% of # M

The next subsection discusses the findings of our experimental results.

5.6.3 Discussion

Here, we discuss the extensibility, usability, and limitations of our proposed SDN deployed design framework using satellite communication in smart city renovation.

Extensibility: Our proposed framework is flexible to model heterogeneous networks consisting of OpenFlow controllers, switches, and protocols. In general, the existing OpenFlow network supports change in topologies, connectivity structures, and configuration parameters. Our framework adopts such scenarios without any changes in the design. On the other hand, if device configuration and communication parameters (i.e., link type, bandwidth, etc.) are changed, it is necessary to change the respective configuration parameters in our framework. Therefore, our framework is extensible for heterogeneous networks with minimal changes. In addition, our framework can be extended to detect security breaches in the smart city. First, it can be combined with other technologies such as visual surveillance to identify malicious users. Second, our framework can directly identify malicious activities, damage to infrastructure, etc. in the city using verified satellite images. In both the scenarios as mentioned earlier, the controller should be trained with well-known data sets. This describes the usability of our framework towards cyber-security resiliency for the citizens using application-level requirements. Our framework can be extensible for deployment in large-scale scenarios such as data centers, enterprise networks, and clouds.

Usability: The proposed SDN framework can be used in health care automation to detect the irregularity in the human body, i.e., body area network. It can be used in disaster management network to predict natural calamities and proactive action making. Also, it can be applicable for post-disaster scenarios, such as sending relief and evacuating people where detection of affected localities and human presence are needed. In addition, it can be effectively used in the smart transportation system that provides prior transport information to the users according to their locality of approach.

Limitations: The mathematical design and experimental results show that our proposed SDN framework is capable of effective dissemination of prior traffic and weather information to the users irrespective of the network type and size. Our framework requires a minimum of seven satellites to collect accurate information about any geographical location. The network connection plays a vital role in information send/receive by the users as the prediction is proceeding inside the GS. However, with a robust network connection and appropriate satellite communication infrastructure, our proposed design framework can be effectively deployed in the smart city network.

5.7 CONCLUSION

The SDN architecture promotes network innovation greatly due to its ability to separate the control plane and data plane. We design a new SDN deployed design framework for weather forecasting and traffic monitoring. Our design framework is implemented using the SDN controller deployed in the GSs. It predicts weather and traffic information in prior and disseminates the same to the mobile devices approaching the locality. It saves time and energy of the devices and improves robustness and efficiency in satellite operations. In addition, it provides security and privacy to the citizens of the smart city. The experimental results are shown to be adequate for large-scale satellite applications. It is also observed that our proposed framework is more effective in comparison to state-of-the-art GPS systems.

References

[1] Smart city [online]. Available: https://en.wikipedia.org/wiki/Smart_city.

[2] B. Kucherov and O. Privyl, "Increasing efficiency of getting results of satellite remote sensing for smart cities", *Smart Cities Symposium Prague*, 2017.

[3] I. Downey, Satellite solutions for cities of the future [online]. Available: http://iap.esa.int.

[4] OpenFlow Switch Specification, 1.4.0 ed.Open Networking Foundation, 2017.

[5] N. Gude, "Nox: Towards an operating system for networks", *SIGCOMM Computer Communication Review*, 38(3):105–110, 2008.

[6] D. Erickson, "The beacon openflow controller", The hot topics in SDN (HotSDN), 2015.

[7] Floodlight project [online]. Available: https://floodlight.atlassian.net/wiki/spaces/floodlightcontroller/pages/1343647/Floodlight+Projects.

[8] Opendaylight project [online]. Available: https://www.opendaylight.org/.

[9] M. Priyadarsini and P. Bera, "A new approach for performance enhancement in software-defined network", The Twenty Fifth International Conference on Computernetworks (CN), 2018.

[10] J. Keppler. Laws of planetary motion [online]. Available: https://en.wikipedia.org/wiki/Kepler%27s_laws_of_planetary_motion.

[11] J. Bao, "Opensan: A software-defined satellite network architecture", *IEEE SIG-COMM*, 2014.

[12] M. Schmidt, "Satellite operation improvement through efficient data combination in ground station networks", *PhD Thesis* submitted to University of Wurzburg, 2016.

[13] S. Nazari, "Software defined naval network for satellite communications (SDN-SAT)", Milcom, 2016.

[14] T. Li, "Using SDN and NFV to implement satellite communication networks", *International Conference on Networking and Network Applications*, 2016.

[15] A. Papa, "Dynamic SDN controller placement in a leo constellation satellite network", *IEEE GLOBECOM*, 2018.

[16] H. Song, "A research on the application of software-defined networking in satellite network architecture", *AIP Conference Proceedings*, 2017.

[17] R. Thompson, "Evaluating tcp/ip performance over satellite networks", *Master thesis* submitted to the University of Stellenbosch, 2004.

[18] J.H. Grinner W. D. Ivancic, "Satellite communications using commercial protocols", *The 8th International Communication Satellite Systems Conference*, 2000.

[19] Oueslati-Boulahia, S., Serhrouchni, A., Tohmé, S. *et al.*, "TCP over satellite links: Problems and solutions", *Telecommunication Systems*, 199–212, 2000.

[20] K.S. Raj, "Satellite-tcp: A flow control algorithm for satellite network", *Indian Journal of Science and Technology*, 8(17), 2015.

[21] Irnss sis icd for standard positioning service, IRNSS signal-in-space icd for SPS version 1.1, Published by ISRO, August, 2017.

[22] Signal-in-space icd for messaging service (irnss 1a), ISRO-IRNSS-ICD-MSG-1.0, Published by ISRO, June, 2018.

[23] Z. Tang, "Software defined satellite networks: Benefits and challenges", ComComAP, 2014.

[24] C. Wang, "Application of virtualization and software defined networking in satellite network", *International Conference on Cyber-Enabled Distributed Computing and Knowledge Discovery*, 2016.

[25] Y. Miao, "Software defined integrated satellite-terrestrial network: A survey", Springer Nature and CCIS, 2017.

[26] O. Sallent R. Ferrus, H. Kaumaras, "SDN/NFV-enabled satellite communications networks: Opportunities, scenarios and challenges", *Elsevier Physical Communication*, 18, 2016.

[27] R.B. Thompson, "Global positioning system: The mathematics of gps receiver". *The Mathematics Magazine*, 71(4), 1998.

[28] A comprehensive tutorial to learn convolutional neural networks from scratch [online]. Available: https://www.analyticsvidhya.com/blog/2018/12/guide-convolutional-neural-network-cnn/.

[29] A practical guide to object detection using the popular yolo framework [online]. Available: https://www.analyticsvidhya.com/blog/2018/12/practical-guide-object-detection-yolo-framewor-python/.

[30] Cars overhead with context [online]. Available: https://gdo152.llnl.gov/cowc/.

[31] Omnet++: Discrete event simulator [online]. Available: https://omnetpp.org/.

[32] N. Bjørner, L. de Moura, L. Nachmanson, and C. Wintersteiger, Programming z3 [online]. Available: https://theory.stanford.edu/~nikolaj/programmingz3.html, 2018.

[33] D.R. Cok. The smt-libv2 language and tools: A tutorial. GrammaTech Inc., version1.2.1, 2013.

[34] The global positioning system [online]. Available: https://www.gps.gov/systems/gps/.

[35] M. Priyadarsini and P. Bera, "A secure virtual controller for traffic management in SDN", *IEEE Letters of the Computer Society*, 2(3), 2019.

[36] Gdgps: The global differential gps system [online]. Available: http://www.gdgps.net/.

[37] R. Chai, H. Li, F. Meng, and Q. Chen., "Energy consumption optimization-based joint route selection and flow allocation algorithm for software-defined networking", *Journal of Information Sciences*, 60, 2017.

CHAPTER 6

Flying Ad Hoc Networks: Security, Authentication Protocols, and Future Directions

Aiswarya S. Nair and Sabu M. Thampi

Indian Institute of Information Technology and Management-Kerala (IIITM-K)

CONTENTS

UNMANNED AERIAL VEHICLES (UAVs), popularly known as drones, have created a revolution opening up a new world of flying "things" serving various industrial and academic sectors. Rendering attractive services on one hand, they pose serious security threats on the other hand. A fleet of drones coordinate and collaborate to form a Flying Ad hoc NETwork (FANET) or UAV Ad hoc NETwork (UAANET) controlled by a ground station. FANETs play promising roles especially in military services and disaster management; however, securing the drone network from cyberattacks is a vital and critical research area to be taken up. The limited resources and open access communication in the drone network unlock the attack loopholes making it vulnerable to innumerable cyber threats. Hence, designing a lightweight and efficient attack-resilient system is of superior importance and poses several trade-offs and challenges. Authentication is the primary step towards securing any ad hoc network which ensures the identity of the entity involved in the communication and gives the privilege of accessing the secret and sensitive network data only to the authentic users. In this chapter, we discuss about the applications, vulnerabilities, attacks, challenges, and security requirements in FANET, focusing on the existing authentication mechanisms and protocols along with the merits and demerits of the various techniques. Mutual authentication, user authentication protocols, and device and operator authentication are analyzed and the gaps in research are identified. This chapter also gives an insight into the latest trends in the field of authentication and the possibilities for developing lightweight and novel mechanisms for FANET authentication.

6.1 INTRODUCTION

Flying Ad hoc NETwork or FANET is a dynamic and self-organizing network of drones communicating and collaborating to accomplish a particular task. The term FANET was first introduced by Michael Müller and was made a separate area of research by Bekmezci [1]. A multi-drone system is highly efficient, reliable, and cost effective and provides a larger coverage reducing the failure rate and the overhead on an individual drone when compared with a single drone system. Such a system can deliver productive services, especially in applications like surveillance and disaster management where human interventions are not possible. Connecting the drones to form the Internet of Drones (IoD) will again help to increase security as well as the quality of service delivered. The complex/heavy processing such as machine learning or video processing of the data collected by the drones could be done at the cloud. The commonly used drones are Mavic, Phantom, and Inspire from Dà-Jiāng Innovations (DJI), the Augmented Reality (AR), Bebop from Parrot, and the Solo from 3D Robotics. Abraham Karem is known as the "dronefather" for his popular invention of the American Unmanned Aerial Vehicle (UAV) Predator, in the early 1990s, which is mainly used by the United States Air Force.

6.1.1 An Overview of FANET

The term ad hoc is a latin word with the meaning "for this," which implies for a special purpose, and an ad hoc network is a network in which the entities communicate with each other without relying on a central station. Drones are a common sight nowadays. Even during the Covid-19 pandemic, drones were used for monitoring the movement of vehicles and individuals to break up social gatherings, spraying disinfectants and for the delivery of medicines. An ad hoc network of drones is composed of several drones communicating and exchanging messages with each other with or without a central ground controller. FANET is a subgroup of Mobile Ad hoc NETwork (MANET) and a subset of Vehicular Ad hoc Network (VANET). The node mobility and dynamic topology changes are very high in FANETs and hence pose higher security challenges when compared with other ad hoc networks like MANET and VANET. The features are detailed in Table 6.1. Network architectures in general may be either infrastructure-based or ad hoc based. UAV networks are a combination of infrastructure-based (involves a control center) and ad hoc architectures (mobile nodes with no central control point), and the commonly deployed UAV network consists of a central ground station and UAV nodes that are highly dynamic changing their position/topology often. The popular topologies are star, mesh, and its variants. In star topology, also known as centralized architecture, all the

TABLE 6.1 Features of FANET

Parameter	Features
Propagation model	In the air with lower node density
Mobility	Very high compared to MANET and VANET
Change in topology	Highly dynamic in nature than MANET and VANET
LoS	LoS needed in most cases
Power consumption	Higher when compared to MANET and VANET
Security issues	Authentication, path planning, placement of UAV, UAV to UAV communication and collaboration, routing, integrity, confidentiality, availability of data, collision, hijack, and Global Positioning System (GPS) spoofing
System components	UAVs, ground station, satellite, mobile network (some cases), cloud (IoD)
Communication range	Longer range
Environmental changes	Largely affected by environmental changes
Topology	Centralized or star, Decentralized or mesh, multi-star, star-mesh
Cost	Single drone costs around 20$ to 250$.
Speed of a drone	50–70 mph
Controller operating frequency	2.400–2.483 GHz, 5.725–5.850 GHz
Antenna	Directional
Individual modules	Control and telemetry, antenna, sensors and actuators, processing units and memory.

drones communicate directly with the ground station maintaining line of sight (LoS), whereas UAV to UAV communication happens only in mesh topology or decentralized architecture, where the ground station maintains LoS with only a master drone, which in turn communicates with all the other drones like a mesh as shown in Figure 6.1. The variants include multi-star (ground connected to several UAVs in LoS, and each of these UAVs in turn is connected to many other UAVs in LoS), hierarchical mesh (single mesh connected to several meshes forming a multi-mesh), and star-mesh (ground to UAVs in LoS, and these UAVs

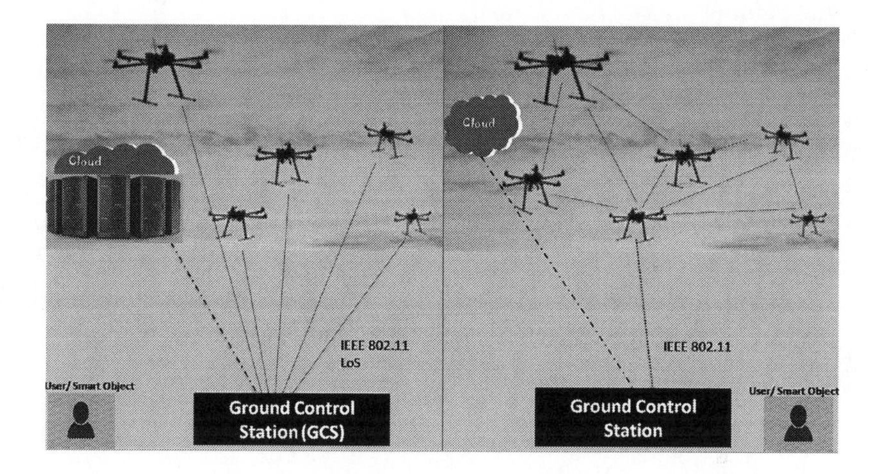

Figure 6.1 Centralized and decentralized FANET architecture.

form a mesh with other UAVs). Remote users may also try to connect with the drones and participate in the communication process.

Connecting FANET with the cloud improves the security of the system by storing the secret data in the cloud, masking the sensitive data from attackers. It also reduces the memory and processing requirements at the ground as the processing can be done at the cloud. UAVs can also directly communicate with the cloud eliminating the need of a ground controller. This opens up new arenas where the UAVs can venture into many services helping the society in many ways especially at the time of disasters. For instance, in a cloud- based UAV surveillance system, the burden of video processing is taken up by the cloud server rather than the UAV, increasing the security of the system at the same time minimizing the drone processing cost. Cloud connectivity to FANET will also help in giving cognitive power to the drones by running the artificial intelligence (AI) based algorithms in the cloud. Cognitive based algorithms are computationally expensive to be implemented on resource constraint drones, and hence computing them on cloud makes the system fast and efficient. To summarize, the advantages of cloud connected FANET are availability of more storage, unlimited processing power, possibility to run machine learning algorithms, ability to perform difficult task, and deployment of IoT devices on a UAV.

The data from the UAV can be communicated using Bluetooth, WiFi, or cellular network depending on the altitude, distance, and application. As the number of drones in a network increases, it becomes difficult for the ground control station (GCS) to maintain the quality of service delivered, and hence the UAVs can be given cellular connectivity to increase the coverage and data rate, reduce the transmission delay, and provide secure communication. Cellular connected UAVs are aerial mobile users controlled by a ground base station (BS). Inter-cell interference and managing handover issues are the main challenges in such a network. UAVs can also act as aerial BSs and can provide broadband internet services at the time of disasters and also to expand the internet coverage. In addition, they can assist the terrestrial networks like VANETs for better transmission of data and for escalating the network security [2].

6.1.1.1 5G- and Blockchain-Enabled FANET

Fifth generation of wireless network technology or 5G is the latest wireless technology that can provide high-speed data at an affordable cost. This high data rate can support fast internet applications and enormous connectivity like Internet of Things (IoT), fog computing, autonomous vehicles, cellular connected VANETs and FANETs, blockchain (BC) services, and many more. 5G supports data rate higher than 10 Gbps and operates in the frequency band of 28–60 GHz offering 99.99% reliability. The major advantages of 5G are its high data rate, lower latency, and enormous connectivity. It can be used as a communication link for UAV networks reducing the network latency. 5G enabled FANET supports beyond LoS services, extension of coverage area, expansion of the application services, better connectivity between the drones and between drones and GCS, cloud data processing, less device memory requirements, and even provides network service recovery in disaster areas. In spite of these merits, there are many security issues to be considered, mainly data security, integrity, and privacy, especially when more and more heterogeneous devices get connected to the network.

BC has emerged as one of the prominent solutions for developing a secure 5G enabled IoT system. BC is a decentralized database technology developed in 2008 by Satoshi Nakamoto used for cryptographically protected data storage and tracking connected devices. Santoshi used the terms "blocks" and "chains" together, making a chain of blocks for data organization in a decentralized manner spread across all the participating nodes, and the first block is termed as genesis. It was originally used for the management of financial records where each block represents a transaction. The blocks are connected by the hash of the previous block, and each block consists of information such as block hash, hash of the previous block, timestamp, nonce, number of transactions, and merkle root of all the transactions. Since the blocks are connected by the hash of the previous block, whenever a transaction happens, the BC is updated on the consensus of all the participants in the network, and if any unauthorized data modifications in a block occur, it will affect the hash of the block as well as the blocks succeeding it. This ensures security and integrity of the data in the BC. There are three types of BC – private, public, and community. Ethereum and bit-coin are public BCs where there is no central controlling authority and is 100% decentralized. Whereas private BCs are controlled by a single authority and need permission. Gem OS BC comes under this category. Community BCs are partially decentralized and need permission; examples are Hyperledger and Ethereum. BC can empower the 5G services in many ways. It helps to overcome the data security challenges in 5G, eliminates the need for a central controlling station, and ensures data availability with high efficiency and reduced failure rate.

Integrating 5G and BC into UAV network builds an epoch of a highly secure data communicated at a rapid rate. There are a few papers incorporating these three concepts and are mentioned below. Mehta et al. [3] propose BC as one of the security technique that is possible in the 5G era for safeguarding the data and the devices from network attacks. The authors have described the security issues, attacks, authentication, integrity, reliability, energy efficiency, and latency in 5G networks. The concepts of 5G, BC, and UAV are integrated in the works of Mehta et al. and Nguyen et al. [3] and [4]. Nguyen et al. [4] survey the applications of BC for 5G and the possibilities of inclusion of BC into smart health, smart city, smart grid, UAVs, and smart transportation. A BC framework using Hyperledger Fabric for a UAV swarm is explored in [5] and discusses the security aspects and how the network security could be enhanced with BC technology. A technique using BC for detecting an attacked UAV from a UAV swarm is proposed in [6]. The security in a 5G-enabled FANET is a highly relevant topic, and an upcoming research area and BC is one of the potent state-of-the-art countermeasures.

6.1.2 UAV Subsystems

The basic component of FANET is the UAV, and this subsection discusses the main subsystems in a UAV. They are sensors, actuators, communication unit, processing unit, control unit, and energy unit as shown in Figure 6.2. The sensor unit includes multiple sensors such as cameras,

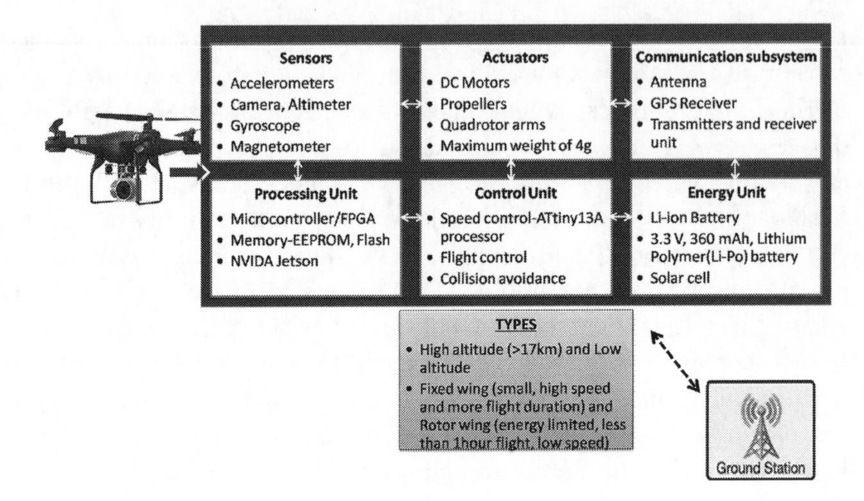

Figure 6.2 UAV subsystems.

accelerometers for monitoring the acceleration and vibration, gyroscopes for maintaining a reference direction, magnetometers for finding the changes in the magnetic field at a particular location, and altimeters for measuring the altitude. The combination of accelerometer, gyroscope, and magnetometer is termed as the inertial measurement unit which reports drone's orientation, angular rate, and force. The actuator subsystem includes motors and propellers; the processing unit processes the data from the sensors and aids in the decision-making process, and it contains microcontrollers such as ARMCortex, field programmable gate arrays or FPGAs, memory units of flash and electrically erasable programmable read-only memory (EEPROM), and random access memory (RAM) and Nvidia Jetson for image processing.

The control unit takes decisions for controlling the movements of the UAV and sends commands to the actuators and has a speed control processor (ATtiny13A) and altitude control and collision avoidance mechanisms. The data from the ground is communicated using the WiFi (IEEE 802.11b) in the frequency band of 2.4–5 GHz. The communication subsystem mainly deals with the transmission and reception of data between the UAV and the ground and the satellite and UAV. This unit contains antennas or antenna arrays, amplifiers, GPS receivers, and other transmitters and receiving units. The energy unit supplies the power for the various subsystems. The commonly used batteries are lithium ion or lithium-polymer with 3.3 V, 36 mAh ratings; certain UAVs even have solar cells for powering the circuits.

PapparaziUAV is a drone-specific open-source project for designing the drone hardware (with sensor interfacings and all the subsystems discussed above) and the software (written in C code) contains an autopilot system with flight controls (for example, adjusting the motor, localization) and ground station links. It also contains the ground controller software with simulators and graphical user interface (GUI) compatible with Linux and Mac OS X.

UAVs are of many types; according to the size, they are nano (<250 g and needs no permit), micro (<2 kg), small (<25 kg), medium (<150 kg), and large (>150 kg); according to altitude, they are high altitude (>17 km) and low altitude and they cannot be flown above 400 feet, and according to the structure they are fixed wing and rotor wing. Drones other than nano have to be registered with the central authority of the corresponding country; in India, it is done by the Directorate General of Civil Aviation, Government of India. The selection of the UAV depends upon the application for which they are deployed.

6.1.3 Applications

As in the past, UAVs are no longer limited for military applications. They are now used in a spectrum of applications including disaster management, agriculture, search and rescue, surveillance and critical area services like borders, power plants, and many more. Drones can be used for road traffic monitoring, traffic analysis, illegal vehicle monitoring, easy passage of emergency vehicles, accident monitoring beyond working hours, and also for monitoring metro/railway lines for faulty rail lines or cracks, and animals or people in the track. They are also used by companies like Amazon, Nokia, Dubai Ministry, Precision Hawk, Google, Walmart, and Alibaba and Courier service companies including UPS, DHL, and Facebook. Amazon Prime Air from Amazon is expected to begin its goods delivery services by 2020. In 2016, Facebook's unmanned plane Aquila completed its first test flight as an alternative Internet delivery platform for remote parts of the world. Germany's express delivery company, Deutsche Post (DHL), is planning to deliver medicines to remote locations using drones. DRDO Rustom is a medium altitude UAV being developed by the Defence Research and Development Organization (DRDO) for the Indian Army, Indian Navy, and the Indian Air Force. NASA has estimated around 700,000 drones flying all round the city by 2020 and operates a research center for Unmanned aircraft system Traffic Management (UTM). UTM manages the traffic at low altitudes eliminating the need for human operators and ensures that only authenticated drones occupy the airspace.

UAVs are also part of other services such as UAV-assisted VANET, UAVs as relay nodes for IoT devices, UAVs as aerial BSs, cellular connected UAVs for network service recovery, cloud connected UAVs and UAV swarm as shown in Figure 6.3. UAVs can act as a link between two VANETs, and they can pass the information from these networks to the cloud, collect data from IoT devices, and store in cloud acting as relays. They can also provide internet services in disaster affected areas and can even act as BSs for other ad hoc networks.

Consider a simple FANET decentralized system with several drones for traffic monitoring and traffic management in a busy city. Each drone is positioned near the traffic lights separated by two kilometers on ground. Drones are equipped with several sensors including cameras for taking high-resolution pictures and videos of the traffic, and the data from the drones can be used for intelligent traffic management. They can even detect emergency vehicles in the traffic and control the traffic light system saving lives of people. A drone can communicate with other drones in its vicinity and pass important messages about the traffic and

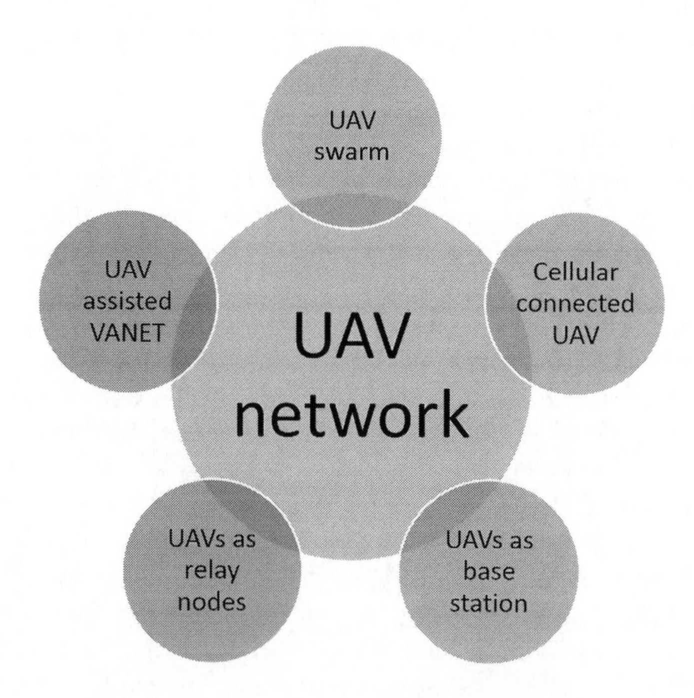

Figure 6.3 Applications of UAV network in various domains.

intimate a remote user about the traffic congestions and less crowded routes. Similarly in cases of road accidents, drones can intimate the ground controller, and rescue operations could be carried out more proficiently and in a rapid way saving life and time. In case of any illegal activities like a terrorist attack, drones can play vital roles in creating alert well in advance.

Another application of UAV swarm is discussed in [7], where a UAV swarm acts as a relay and collects data from the IoT devices and transmits it to the nearby server. BC is used as the data acquisition system for storing the data from IoT devices. UAVs can be employed as relays in places where human intervention is difficult and UAVs can deal the situation in an effortless way, saving the battery life of the IoT devices. A secret key is shared between the swarm and the IoT devices in the initial stage. Pi-hash bloom filter and digital signature are used by the UAVs for authenticating the IoT devices, and the data is encrypted before forwarding to the server. The authors have simulated the scenario using four IoT devices, three UAVs, one mobile edge computing server, and bluetooth for communication. Pi-hash bloom filter was simulated using MATLAB® and Python.

Table 6.2 showcases the important surveys in the area of FANET. Drones can change the world providing better amenities and

TABLE 6.2 A Few Important Surveys in the Area of FANET

Author	Year	Short Description
Bekmezci et al. [1]	2013	Reviews the design issues, applications, and research issues in FANETs
Motlag et al. [8]	2016	Discusses the UAV architecture for IoT services
Sedjelmaci et al. [9]	2018	Surveys the cybersecurity attacks and solutions in UAV networks
Lagkas et al. [10]	2018	Discusses UAV, 5G, IoT, security and privacy in FANET
Bithas et al. [11]	2019	Surveys the integration of machine learning into UAV communications
Mozaffari et al. [2]	2019	Reviews the challenges, tools, research problems, and cellular connected UAVs
Shakhatreh et al. [12]	2019	Discusses the applications and challenges in UAV

sophistications for a better living. But securing the drone network is an important area mostly left out as irrelevant. Securing the drone network in turn provides security to the infrastructure, including buildings, vehicles, and human beings.

There are several vulnerabilities that lead to attacks in the UAV network and are discussed in Section 6.2. The security requirements in FANETs are detailed in Section 6.3. And the primary step towards network security is authentication, which checks the genuineness of the entities in the network and provides access to the system resources only after verifying the identity of the entities and ensuring that they are legitimate and trustworthy. A survey of the existing authentication schemes with comparison and gaps are discussed in Section 6.4. Finally, in Section 6.5 the recent trends and emerging areas in FANET authentication with future directions are discussed.

6.2 VULNERABILITIES AND ATTACKS

Drones, which form the integral part of FANET, are characterized by its compact size, limited battery and storage, low power consumption, and minimum processing capabilities. There is always a trade-off between these lightweight features and the network efficiency, performance, and most significantly, the network security. Safeguarding the network from attackers and simultaneously coordinating the communications in the network is a challenging task. The first major attack on a UAV system was using the software SkyGrabber by the Iranian terrorists in December 2009, who were found to be recording the video feeds from the drones. The unencrypted video feed was the identified vulnerability that led to the attack. Another malware attack was in September 2011 on the US BS control and command network threatening the national security. A physical attack on a US drone RQ-170 by Iran was reported in December 2011 due to vulnerabilities of the GPS.

Network vulnerabilities are the weak points, limitations, or flaws existing in a network through which an attacker invades into a system and demolishes it. The network discrepancies or vulnerabilities make it easy for the attackers to delete, modify, block/jam the data in the network that may lead to drone damage, unsafe landing, drone collisions, collision with buildings, and even loss of life. The major vulnerabilities identified in a UAV network are the wireless connectivity, physical availability of the drones in the air, dynamic topol-

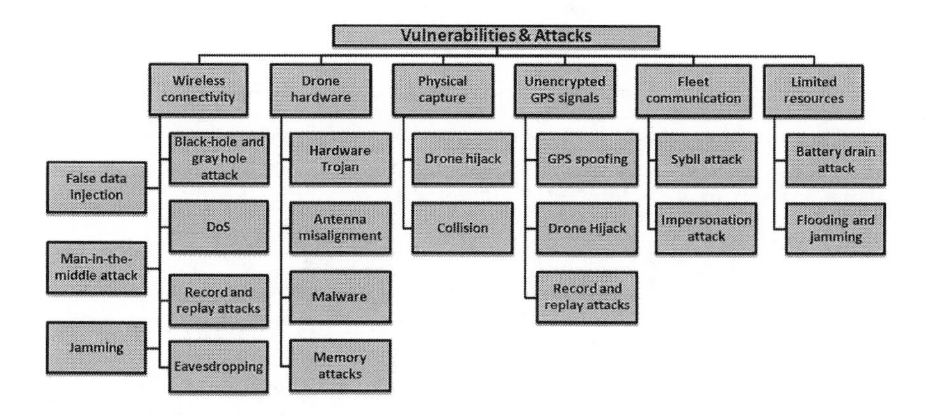

Figure 6.4 Vulnerabilities and attacks in FANET.

ogy, fleet communication, unencrypted GPS data, limited resources, and the drone hardware vulnerabilities. Each of these loopholes paves way for several attacks that need to be mitigated. The vulnerabilities and the attacks in FANETs are detailed in Figure 6.4. The definition of the attacks and the corresponding countermeasures are given in Table 6.3.

Due to the increased deployment of drones in today's world and the possibilities of occurrence of cyberattacks in these networks, security in FANETs is a significant and developing research area. Security techniques implemented should be attack-resilient, efficient, and lightweight in terms of memory, communication, and computation overhead, suiting the resource constraint drone network.

6.3 SECURITY REQUIREMENTS IN FANET

Strong security solutions to combat the attacks need to be developed in the area of FANET security. The security requirements include authentication of the data communicated between the drones or between the ground and drone, which includes confidentiality and integrity of the data, authentication of the entities in the network which could be a remote user or an IoT device or a drone, trust evaluation of the entities participating in the communication, intrusion detection and prevention to detect the abnormalities in the data pattern and flight changes, secured routing to provide a collision free service delivery, and hardware security of the drone and the ground (as shown in Figure 6.5).

TABLE 6.3 Attacks, Definition, and Countermeasures

Attacks	Definition	Countermeasure
Denial of service	The network resources (data collected by the drone, memory access) are made unavailable to the legitimate users	Authentication
Sybil attack	A node uses multiple fake identities	Authentication
Impersonation	Sending messages in the network pretending to be an authentic sender	Authentication
ADS-B attack	Automatic dependent surveillance - broadcast: it is a message broadcast by the UAV to the ground about its location. Attacker can generate fake ADS-B data to destroy the communication link.	Authentication
Drone hijack	Drones are misled to a wrong destination	Hardware security, authentication, and secured routing
GPS spoofing	The drone gets the wrong GPS data from an attacker who sends stronger GPS signals resulting in drone hijack	Authentication and trust
Record and replay	Recording the signals from the valid drone and later replaying it with certain modifications	Authentication
Battery drain	Energy drain of UAV by flooding the network with data causing failure of UAV	Intrusion detection
Malware	Damage causing software that may affect the ground station or the UAV system	Intrusion detection
Insider attack	An authenticated person may use the secret keys and misuse the data either intentionally or by accident	Intrusion detection

(Continued)

TABLE 6.3 (*Continued*) Attacks, Definition, and Countermeasures

Attacks	Definition	Countermeasure
Man-in-the-middle	Attacker in the communication channel eavesdrops the messages either from drone or ground and modifies it	Intrusion detection and authentication
False data injection	Inserting wrong data into the message communicated between the drones or with the ground	Intrusion detection and trust
Jamming	The signals are blocked and the drones lose connection with the ground station	Intrusion detection
Black-hole	All the data packets communicated between the UAVs or with the ground are lost	Intrusion detection, Trust
Gray-hole attack	Only few of the data packets communicated between the UAVs or with the ground are lost	Intrusion detection and trust
Eavesdropping	Secretly capturing data packets communicated between the UAVs or with the ground	Intrusion detection
Collision	Colliding with buildings, humans, vehicles, and other UAVs	Secured routing
Routing attack	Changing the flight path leading to hijack	Secured routing
Hardware Trojan	Modification of the hardware circuit or integrated circuit (IC) at the drone or ground control during the manufacturing process	Hardware security
Antenna misalignment	Since directional antennas are used for UAV to UAV communication, the data will be lost if the antenna is aligned in the wrong direction	Hardware security
Memory attack	Accessing the memory for secret data or keys, and flooding the drone memory with irrelevant data	Hardware security

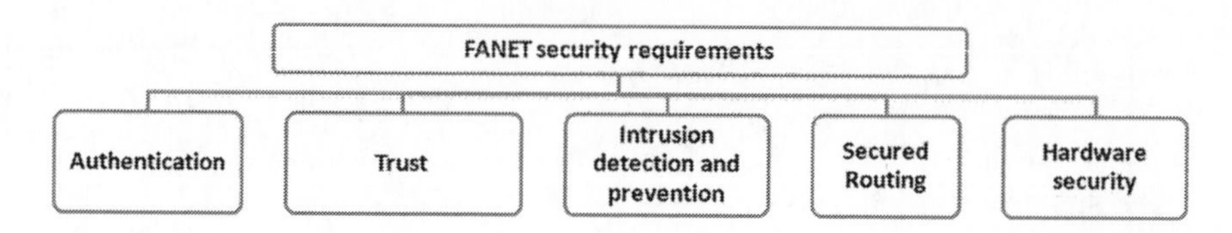

Figure 6.5 FANET security solutions.

Authentication: Authentication is the process of confirming the identity of the entities in a network; it can be either the drone or ground control or a remote user in the network. Message authentication is different from user authentication. Message authentication is the process of checking the integrity of the data and verifying its source. User authentication is the process of verifying the identity of the remote user and giving access to the system resources after verifying the identity. Smart objects such as smart phone or a Bluetooth device may try to connect with the UAV; connection can only be established if the entities are already registered with the trusted center. Mutual authentication is the process by which two entities mutually verify their identities and exchange secret data among them. Hardware authentication is confirming the identity of the physical device.

Trust: Evaluating the reliability of the drones in the network based on the history of their communications is termed as trust. It is an expectation or confidence that the entity is trustworthy and not an enemy computed based on a prediction of its next level of communication status. Authentication and trust are closely related, and the former is the primary step towards network security, and trust is established between the authenticated entities based on their communications and interactions. Trust can only be evaluated between authenticated devices. Assessing the trustworthiness of the entities also helps in preventing insider attacks. Many works on trust-based routing have been proposed by researchers recently. The commonly used trust computation methods are fuzzy model, Bayesian trust model, game theory model, probability theory, petri-net, Markov chain model, and BC. Some of the recent works on UAV trust models and trust management are [13–15] as follows:

Intrusion Detection and Prevention: It is the process of monitoring a network detecting the attackers who intrude into the network and modify the data. Intrusion can happen from an outsider or an insider who has passed the authentication process and has complete rights to access the system resources. Host-based intrusion detection identifies the abnormalities in the system logs, whereas network-based intrusion detection checks the network traffic for abnormalities. Network intrusion detection methods are again of two types: signature based and anomaly based. Signature-based method maintains a database of the previous attacks and alerts when a match occurs, whereas anomaly-based technique looks for changes in the normal data pattern and any changes in the pattern are categorized as an intrusion. UAV intrusion detection system (IDS) analyzes the data flow in the network and can be placed in the GCS or UAV. Existing UAV IDS methods are behavior rule based, game theoretic approach, bio-inspired methods and wavelet based IDS. A survey of IDS in vehicular networks is presented in [16].

Secured Routing: It is the process of routing the data or transmission of data packets among the drones in a UAV network and can be either single or multi-path routing. Single-path mechanism identifies a single route towards the destination and multi-path routing finds many parallel paths for the data transmission. The main challenges involved in this step are the high mobility and dynamicity of UAV networks. The existing FANET routing protocols are static routing, proactive routing, reactive routing, hybrid routing, geographic routing, and hierarchical routing. Recent surveys on the routing protocols in FANETs are presented in [17] and [18] that discuss in detail about each of the routing protocols and their advantages and limitations. The simulators that can be used are MATLAB, NS2, QUALNET, OPNET, NS3, and OMNET++.

Hardware Security: Securing the hardware at the drone and the ground from malwares and hardware trojans and providing system-resiliency by protecting the secret data even after physical capture are the measures in hardware security.

Authentication is the stepping stone towards FANET security. In today's world, drones have become a part of human life for several critical applications as discussed in Section 6.1.3, and hence authentication of the entities and development of secure authentication protocols is an urgent need. Also due to the alarming increase of attackers and unauthorized users trying to sneak the network secret data, authentication is a highly critical area to be addressed and investigated. Even though there are many surveys related to authentication ([19,20]), most of them concentrate on the domains of IoT, Internet of Vehicles (IoV), and Wireless Sensor Networks (WSN) and a very few on IoD [21]. In this chapter we discuss in detail about user, device, and mutual authentication protocols in FANETs, proposing a new taxonomy of the countermeasures and finally address the research issues and the possibilities of developing novel authentication methods for the upcoming technological era.

6.4 AUTHENTICATION MECHANISMS IN FANET

Authentication provides a mechanism to filter out unauthorized users from valid users and provide network privacy and security. Can an intruder eavesdrop the data communicated in a UAV network and use it for personal purposes? Can a new drone participate in the services of an already established UAV network? Can a smart device access the video data collected by a surveillance drone in a network? The answer is, *'yes, only if they are authenticated and registered in the network'*.

The real-time data from the drones can be accessed only by an authenticated entity. Consider a disaster affected area as an example scenario where the rescue team has to find the people stranded in houses and buildings during a devastating flood condition. By authenticating the rescue team member, the data can be accessed and several lives could be saved. In the event of any suspicious activity in a drone monitoring area, the data from the drones can be accessed through the GCS by a person who passes the authentication process.

Authentication is followed by a secret key exchange between the entities and can be explained in simple terms considering an example as in Figure 6.6. The elements in the architecture are the GCS, UAVs, and a remote user. The GCS is assumed to be a trusted entity. UAV_i and UAV_j are two UAVs in the system. The foremost step in authentication and key exchange is the setup phase in which the drones and the users have to register with GCS, and it assigns a unique identity for the UAVs for the sharing of confidential and secret data among the entities in the network. The user and drone registration processes are slightly different. For user registration, the user selects an ID and password and sends a message to the GCS. The GCS calculates a unique ID for the user and passes the secret data to the user. The user stores the data in memory. In the UAV registration phase, the UAV selects a random number and sends a request to the GCS. The GCS computes the ID and the secret credentials for the UAV and sends it back, which gets stored in UAV memory. The registration phase is followed by the authentication phase and an update phase. The implementation of the protocol differs according to the application and the scenario considered. A simple scenario is

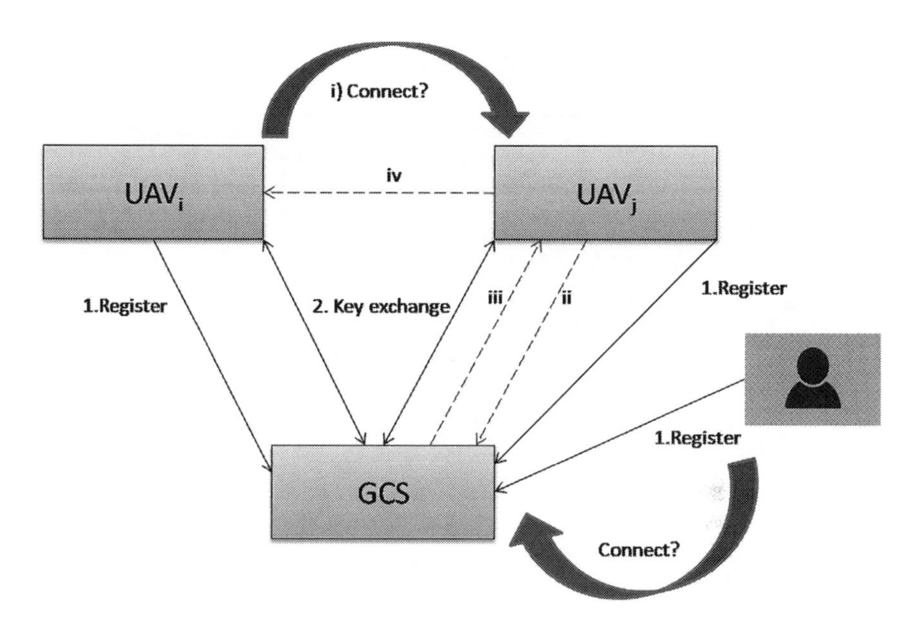

Figure 6.6 Authentication process.

considered in Figure 6.6. When UAV$_i$ wants to communicate with UAV$_j$, it sends a request to UAV$_j$. UAV$_j$ in turn notifies the GCS about the UAV$_i$'s desire and checks whether the UAV$_i$ is authenticated. The GCS sends the identity of UAV$_i$ to UAV$_j$. After proving to be authentic, they start exchanging secret keys and data. However, this is not a fixed standard for the protocol, and different variants of this protocol have been proposed by the researchers using different architectures, scenarios, new parameters, fuzzy logic, biometrics, smart credentials, etc.

There are different levels of authentication in an ad hoc network: mutual authentication between two entities, user authentication, device authentication, and operator authentication. Mutual authentication between the devices is more important in a decentralized network architecture where the UAVs communicate the significant messages to the master UAV and the master, in turn, communicates with the ground station. It could be between UAV and UAV or UAV and GCS. Mutual authentication in FANET is a less explored area, and novel techniques need to be developed in the near future. User-level authentication mechanisms prove that the users in the network are rightful and genuine and they are ready for secret key exchange. Device-level authentication checks for the physical integrity of the devices added in the network. Operator authentication is verifying the integrity of the drone operator.

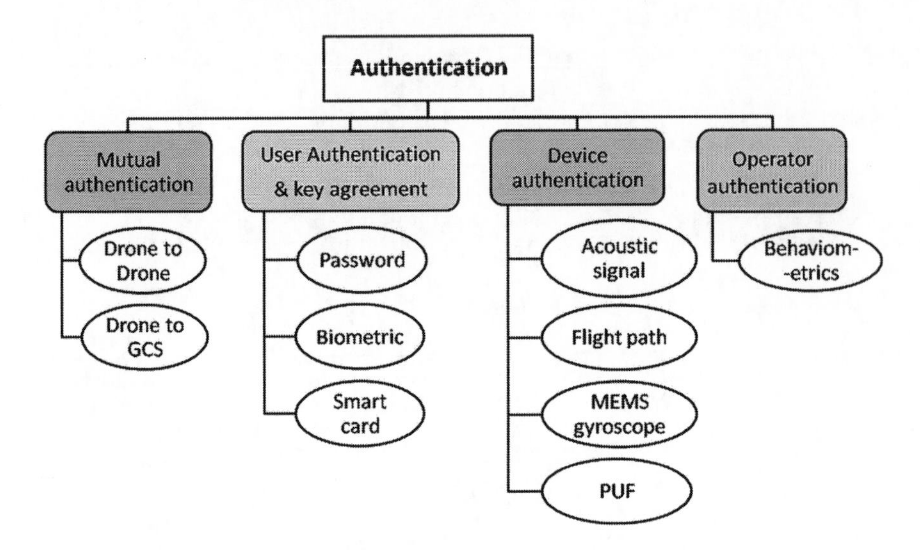

Figure 6.7 Proposed taxonomy of the authentication schemes.

A taxonomy of the authentication techniques is presented in Figure 6.7. Several lightweight authentication techniques have been proposed in this area and are compared in Tables 6.4–6.6.

6.4.1 Mutual Authentication

Mutual authentication is a process by which the participating entities in a network mutually verify their identities and authenticate each other for the transmission of secret keys between them to establish a secure communication channel. It can be between UAV and GCS or between UAV and UAV.

A mutual authentication scheme between drone and end device is described in [22], where both the entities can mutually verify their identities using their signatures. The system consists of a group of drones, a set of end devices, and a remote management centre (RMC) to manage and generate the private keys for the end devices and the drones based on their identities. A member of the end device's group has to verify its identity with the drone and gets its broadcast key from the drone for communicating with other end devices. It is a hierarchical scheme with mainly three steps, namely system setup, register, and secure communication phase. System setup is run by the RMC to generate the master key, and all the network entities need to first register with the RMC to generate their private keys. Registration is done using signcryption based

TABLE 6.4 Mutual Authentication Methods

Author	Methodology	Limitations	Tools
He et al. [22] 2017	Mutual authentication using identity-based signcryption	More computations needed	NS3
Won et al. [23] 2017	Mutual authentication between smart object and drone	Computationally expensive	TinyECC
Chen et al. [24] 2018	Mutual authentication between network connected UAVs and the ground station	Incurs high cost and no security analysis is performed	TPM emulator
Semal et al. [25] 2018	Certificateless group authentication	Only for static groups, and dynamic addition of UAVs is not considered.	Scyther tool, RaspberryPi
Li et al. [26] 2019	Mutual healing group key distribution using BC	Not effective for a large drone network, more computation	OPNETModeler 14.5, BC: Hyperledger Fabric 2.0 -open-source BC development platform
Chen et al. [27] 2020	ECC-based mutual authentication	High computational and communication costs.	Study with Burrows–Abadi–Needham (BAN) logic proof

TABLE 6.5 User Authentication Techniques

Author	Methodology	Communication Cost	Computation Time	Storage	Security Analysis	Limitations	Tools Used
Wazid et al. [28] 2018	Cryptographic one-way hash function, bit-wise XOR operation and biometric fuzzy extractor	1696 bits	0.027 s	$480 + (m + 1)\ln(p)$	Automated Validation of Internet Security Protocols and Applications (AVISPA)	Throughput is less and slightly higher packet loss rate	NS-2 with 50 drones
Jangirala et al. [29] 2019	TCALAS: Three factor-cryptographic hash function and biometric fuzzy extractor	2536 bits	0.026 sec	Not analyzed	Real-or-random (ROR) model and AVISPA	Slightly higher computation cost	Simulation study not done

TABLE 6.6 Drone Authentication Techniques

Author	Methodology	Limitations	Tools Used
Benzarti et al. [35] 2018	ID-based signcryption	Mutual authentication and efficiency are not considered	RFID
Son et al. [36] 2018	Offset of MEMS gyroscope used as the device fingerprint	Suitable for small network	MEMS gyroscope
Ramesh et al. [37] 2019	Acoustic fingerprinting using the difference in the motor noise-machine learning (ML)	Not suitable for large number of drones, and manufacturing defects in propellers are not considered.	Arduino Uno, Blue Yeti Pro microphone
Karimibiuki et al. [38] 2019	Flight path validation using K-NN, SVM, LR	More computations	ArduPilot

on the identity of the end device and the drone. Drones are responsible to authenticate the end devices if they want to enter into the network. The system is resistant to denial of service attacks. Game strategy is used for analyzing the security of the protocol. Simulation is done using NS3, and performance is analyzed by means of average message loss rate and average message delay. As the packet size increases, the average delay increases but the loss rate is unchanged. As the number of entities in the network increases, the loss rate increases. The real identity of the devices is masked using a pseudonym or a temporary identity.

Won et al. [23] propose a certificateless mutual authentication scheme between a smart object and a drone. Three different cases are considered; the first case is the communication between a smart object and a drone, the second case consists of a drone communicating data with many smart objects, and the third case includes the case where multiple smart objects communicate their data to a drone. For this, they have postulated three protocols – a Certificate-Less Signcryption Tag Key Encapsulation Mechanism (eCLSC-TKEM) for one-to-one communication, a Certificate-Less Multi-Recipient Encryption Scheme (CL-MRES) for one-to-many, and a Certificate-Less Data Aggregation (CLDA) protocol for many-to-one communication. The first scheme makes use of a partial private key that expires after a specific time period, CLDA makes use of elgamal homomorphic encryption and an optimized batch verification technique, and CL-MRES is a hybrid encryption scheme. But this method is computationally expensive.

Chen et al. [24] use asymmetric bi-linear pairing for mutual authentication between network connected UAVs and the ground station. Authenticating the trusted platform module (TPM) using platform identity authentication in a cellular connected UAV context incurs high cost, and no security analysis is performed.

A certificateless-group authenticated key agreement (CL-GAKA) scheme for secure UAV-UAV communication is discussed in [25]. The protocol consists of two stages – initialization phase and group key agreement phase. The initialization phase involves the generation of user's partial private key and public key by the server. This scheme is possible only for static groups, and a dynamic addition of UAVs is not considered. Scyther tool is used for the security analysis of the protocol. This method offers several advantages such as mutual key agreement, key escrow elimination, joint key control, key freshness, known-key security, entity revocation, non-repudiation, conditional privacy, known-key security, and forward secrecy. The test bed consists of RaspberryPi and a wireless router.

Recently, BC has also received major attention for aiding the authentication process by storing the membership certificates and recording the group keys, thus maintaining a trusted database for UAV Ad hoc NETwork (UAANET) operations as in [26] in a dynamic way. Since the nodes in FANET are highly dynamic, the nodes may miss some of the broadcasted information and will have to leave the network. The data may include the group keys and messages from the GCS. Self-healing mechanism checks the previous group keys and tries to recover the lost key without contacting the GCS, whereas the mutual healing mechanism retrieves the lost key by communicating with the neighboring nodes. The authors propose a private BC created by the GCS as a database for saving all the issued group keys and managing the UAV joining and exiting the network. In case of lost data, the UAV can directly get it from the private database. Unlike public BC where anyone can join the BC network, in a private BC, only permitted members can enter into the network. The UAVs have to update the BC periodically and have to recover the lost blocks from its neighbors. The UAVs joining the network are authenticated by verifying its signature using ECDSA-secp256k1. Mutual healing protocol has three main steps: a broadcast request to all neighboring nodes, a unicast message to the requested node, and message confirmation. The requested node sorts the received blocks according to the length of the blocks and selects the longest block; this method is termed as "longest lost chain." This mutual healing process also achieves mutual authentication between the requested UAV and its neighbor. The method is resistant to replay attacks but incurs more overhead.

An elliptic curve cryptography (ECC) based method for mutual authentication is proposed in [27] consisting of a trusted authority, UAV manufacturer, drone operator, and the GCS. Mutual authentication between a drone operator or player and the UAV manufacturer, then between the operator and the GCS followed by mutual authentication between the drone operator and the UAV, and finally between the ground station and UAV are considered in this paper. The system is resistant to denial of service (DoS) attacks and spoofing attacks. Security proof using Burrows–Abadi–Needham (BAN) logic is presented. This method, however, incurs high computational and communication costs.

6.4.2 User Authentication and Key Agreement Protocols

A remote user at a distant place may intend to communicate with the drones in a UAV network. Only an authentic user is given the privilege to do so. User authentication is done by using either passwords or smart cards or personal biometrics. After validating the user, secret keys can be exchanged between the drone and the user for future communications. Several works are done in the domain of IoT and WSNs, but only very few works are in the specific domain of FANET. Two-factor schemes make use of two user credentials, whereas three-factor schemes use three user credentials for verifying the identity of the remote user who has requested services from the drone network. The user has to first register with the GCS and then proceed for data transmission after a successful key agreement procedure with the ground station.

Two-factor schemes in [30] and [31] use smart cards and passwords for user authentication in IoT environments. However, there are several attacks that make them insecure, such as password-guessing, stolen smart card, impersonation, man-in-the-middle attack, and insider attacks. In [32,33] and [34] a three-factor scheme using smart card, user password, and biometrics is employed for user authentication in IoT scenario. The method proposed in [32] is susceptible to DoS attacks and [33] and [34] do not provide a security analysis.

The works in [28] and [29] are based on three-factor user authentication in an IoD environment. In [28], there are seven steps in the key agreement protocol which also includes the secure communication and key establishment between two communicating drones. It uses cryptographic hash functions and biometric fuzzy extractor, and the method is resilient to replay attacks, man-in-the-middle attacks, drone capture attack, secret leakage attack, and password update attack. It uses Dolev-Yao (DY) threat model, and security verification is done using Automated Validation of Internet Security Protocols and Applications (AVISPA) tool. Simulations in NS2 are done using 50 drones, three users, and one gateway node. The computational cost is 0.027 seconds, communication cost is 1696 bits and storage cost is $480 + (m + 1)\ln(p)$, where p is a very large prime number and m is the degree of symmetric bivariate polynomial.

A lightweight three-factor user authentication protocol is proposed in [29] using a combination of cryptographic hash function, bit-wise XOR operation, and fuzzy extractor method. The drone to drone key management as in [28] is not considered here. The communication cost is

1536 bits, computational cost is 0.026 seconds, and storage overhead is not analyzed. Security analysis is performed using Real-or-Random model for validating the security of the session keys. Formal security verification using AVISPA tool has also been done.

6.4.3 Drone Authentication

Consider a military border surveillance scenario with five drones in the network. If a sixth drone enters into the network, how to identify if it is real or fake? A real drone may also be replaced by a spy drone or malware drone that resembles the original drone by a terrorist. The fake drone will be able to record the messages in the network and later replay it or process it to retrieve the secret keys. Device authentication or, specifically, drone authentication becomes a necessity in such cases to confirm the identity of the drone participating in the network.

A drone authentication and tracking scheme using Radio-frequency identification (RFID)-based signcryption are discussed in [35]. The system consists of six entities, namely BS, BS controller, civilian cloud, database, identity server, and routers. Each drone is equipped with an RFID tag for registering into a network. An RFID tag is installed in every drone in the network with the RFID reader at the BS. The drone should be in the vicinity or range of the BS. When a drone enters into the range of a BS, the drone RFID is read by the BS, and it is forwarded to the BS controller which in turn requests the cloud for a temporary identity which expires after a certain time. Private keys are generated using the drone identities and is used for signature generation. Drone to drone and drone to multi-drone algorithms are discussed. Security proofs and performance analysis are not presented. Mutual authentication and communication between the drones are not considered, and efficiency of the scheme is not evaluated in this method.

Fingerprinting of the drones is done in [36] using the special features of the gyroscope sensor on drones. Micro-electro mechanical systems (MEMS) gyroscopes are used for measuring the orientation and rotation of the drones, and the output of each sensor is distinct from the outputs of other sensors for identical inputs. This difference occurs due to the variations in the manufacturing processes. Hence this feature can be used as an identifier or a fingerprint of an authentic drone. But the results narrow down as the number of drones in a system increases and is more suitable for small networks.

The differences in the noise characteristics of the drones (due to manufacturing defects of the drone motors) are used for identifying the drone and authenticating it in [37]. This acoustic drone fingerprinting aims to eliminate the drone impersonation attack. It is a two-factor authentication scheme where digital signature is the first factor of authentication and acoustic fingerprint is the second factor. The electromagnetic and mechanical noise features of the valid drone motors are extracted and trained using Support Vector Machines (SVM) classifier with radial basis function as the kernel. The acoustic signals from the motors are captured using microphone and preprocessed to remove the noise values and normalize the data. The features are extracted and trained and further used as a database for authentication. It can be used to predict the authenticity of a drone by recording and analyzing the motor sound.

Another work uses K-Nearest Neighbor (KNN), SVM, and Logistic Regression (LR) machine learning methods to predict and validate the drone flight path from the flight traces [38]. Drone authentication is done by checking its flight path. If authentication fails, it is a drone hijack. The models are trained using both real and false data and can be used for predicting the new drone paths as authentic or not. In KNN model, Euclidean distance function is used for finding the wrong data. Simulations are carried out using ArduPilot simulator. SVM detects the changes from the original data, whereas LR finds a relation between the features. The experiments prove that K-NN is the best classifier for validating the flight path, but the process is time consuming.

A real-time behavior-based UAV identification is proposed in [39]. Here the authors have discussed about a UAV identification mechanism by predicting the real-time UAV path and spotting the illegal users who try to manipulate the flight path. Real-time data from the drones are collected, mainly the location data and sensor data, to study the behavior of the drones, and a model is created that can predict the trajectory of the drone in future and verify the flying route, thereby authenticating the UAV. The real-time sensor data considered are longitude, latitude, and speed along with the drone characteristics like weight and maximum speed. These data are learnt using a Kalman filter online Bayesian learning method, and the authentication is known as Gaussian-Processes based Authentication (GPA). The data processing is done by a server and stored in a database management system. UAV identification is done using a serial number and a QR code; the operator also has to input their identity credentials upon which a license will be issued by the server. However, this work is confined to only a single UAV system.

Physical Unclonable Functions (PUFs) and TPMs could be implemented on the drones in future for generating device specific keys and authenticating the device hardware respectively.

6.4.4 Operator Authentication

A drone operator's behavioral biometrics is used for the flight data authentication in [40]. The system consists of a UAV, a UAV operator, and a transmitter. The mechanism is based on the idea that the pattern of operating the drone joystick by each operator is unique. The control signals analyzed are roll, pitch, and yaw that are extracted by the drone processing system and sent to the ground controller. The command pattern of the flight operator can be studied and used for identifying the malicious operator from the authorized ones. Random forest classifier is used to classify the operator behavior, but this method needs more computation time and memory.

The existing FANET authentication protocols are studied and compared in detail and the following observations were made. The existing authentication protocols are based on public key cryptosystems like elliptic curves and digital signatures that suffer from many drawbacks as mentioned in Tables 6.4–6.6. The computational expense of the existing schemes is higher for resource constraint devices. The mechanisms demand for larger memory and more processing time. Reducing the key size to minimize the communication overhead is another concern. The presented techniques are based on the hardness of solving the discrete logarithm problem and the integer factorization problem which can be solved very quickly and efficiently by a quantum computer and hence are vulnerable to quantum attacks. There are only a few techniques for device authentication. User authentication protocols are confined to passwords, smart cards, and biometrics. Only very few FANET authentication techniques make use of the state-of-the-art technologies like BC. Consequently, there is necessity of more secure and lightweight authentication protocols for FANETs.

6.4.5 Tools and Techniques

UAV simulators are used for analyzing the performance and estimating the values of the different parameters in the particular designed algorithm. Particularly, the different drone authentication mechanisms can be tested, analyzed, and validated using these simulators specially designed for UAANET. The popular UAV simulators are SUAAVE, Simbeeotic, X-Plane, AVENS, RAVEN, UAVSim, and D-MUNS. Table 6.7 outlines some of the important features of these seven simulators. The open-source simulators are AVENS and UAVSim. Other network simulators such as AVISPA, NS-2, OMNET++, and MATLAB are also widely used for network testing and analysis.

6.5 RECENT TRENDS AND FUTURE DIRECTIONS

Recently, there have been many works relating authentication with emerging concepts like artificial intelligence and chaos and quantum theory. Authentication using AI is a recent makeover in the area of authentication. Also, there has been a shift in the research towards hardware-based authentication protocols. FPGA implementation of authentication protocols for better performance is also an upcoming research area. Another trend is the chaos-based and quantum-based authentication protocol. Quantum cryptography is an upcoming field which will soon replace the public key cryptosystems and emerge as alternate field in the area of mutual authentication. The emerging areas in authentication are shown in Figure 6.8.

6.5.1 AI-Enabled Authentication

In the case of a large UAV system, machine intelligence plays a significant role in learning the behavior of the nodes and predicting the future activities. AI is giving human-like learning and decision-making skills to the devices for better performance and making the system less vulnerable to attacks. AI can be used for the prediction of threats in a large-scale network where the BS (in a cellular connected UAV) or the gateway does not have the complete information about all the UAVs in the network. Cellular connected UAVs are aerial mobile users and can support longer range services, beyond LoS services, lower delay, internet connectivity, and wider range of real-time applications.

The BS may not be able to store the records of all the UAVs, and hence they make use of AI to predict the vulnerable nodes based on the vulnerabilities and attack history. In [41] and [42] deep learning based intelligence is used for predicting the future network scenarios and identifying the attack vulnerable UAVs and classify them as

TABLE 6.7 UAV Simulators

Simulator	Description	Open Source
SUAAVE	Sponsored by Engineering and Physical Sciences Research Council (EPSRC) UAV swarm coordination and control are the main functions	–
Simbeeotic	O/S simulator designed by Harvard University Virtual 3D simulation of UAVnet communication Uses 211 JBullet, a Java-based physics library for modeling of swarms	–
X-Plane	Created by Laminar Research Runs on Windows, 257 Linux, and Android and can communicate with User Datagram Protocol (UDP) or Transmission Control Protocol (TCP)-based networks	–
AVENS	Merges X-Plane, 262, and the OMNeT++ integrated with LARISSA Networking protocols can be developed	✓
RAVEN	Indoor test bed developed by researchers at MIT. It can manage a swarm network of ten UAVs and is used for prototyping the control and coordination algorithms.	–
UAVSim	An OMNeT++ based test bed designed at The University of Toledo for the simulation of UAV-network communication Contains separate modules for different attacks and UAV models UAVSim in combination with GNSSim is to design and simulate GPS-related attacks such as GPS jamming and spoofing on UAVs.	✓
D-MUNS	Distributed Multiple UAVNet Simulator Includes UAV Simulator Controller, Network Simulator Controller, and a Monitor.	–

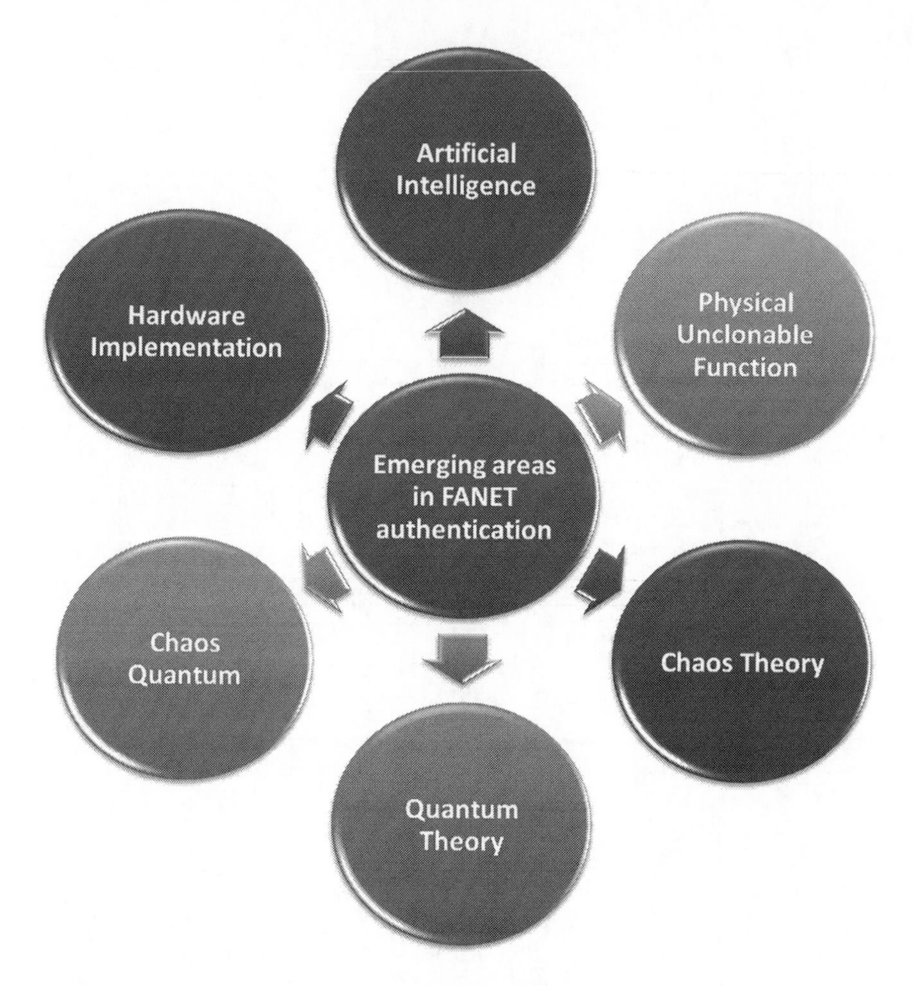

Figure 6.8 Emerging areas in authentication.

non-authentic. In [39], a deep reinforcement learning algorithm is implemented in the BS of a cellular connected UAV system to learn the previous behavior of the UAVs and predict its future behavior. This in turn helps to authenticate the UAVs by identifying the non-vulnerable UAVs for its future communications. They have discussed three scenarios for UAV-based delivery system, UAV-based real-time multimedia streaming, and UAV-based intelligent transportation system. Ferdowsi et al. [42] use deep long short-term memory (LSTM) in a large IoT system to create a watermark in the data transmitted by an authentic IoT device. The technique extracts features from the IoT signals and generates a watermark that is inserted into the original one, acting as the authenticator. It also helps to predict the vulnerable IoT devices in the network and disconnect them from future communication/data transfer in the network. IoT signal features extracted are spectral flatness, skewness, and central moments.

6.5.1.1 Physical Attribute-Based Authentication

Physical layer authentication means device-level authentication using time varying characteristics of the device and the communication channel by means of measurable and computable quantities such as received signal strength, carrier frequency shifts, and frequency response. Lately, AI has became popular in the area of physical layer authentication. Fang [43] discusses the need for intelligent authentication in 5G networks and beyond using physical layer attributes like received signal strength, impulse response, and frequency response. Multiple multi-dimensional attributes from the received signal could be collected and analyzed for self-organization and automatic authentication. They propose three main steps for intelligent authentication – collection of attributes, analysis using machine learning, and finally, performance evaluation. Different machine learning variants are also discussed but case studies are not presented.

A machine learning assisted physical attribute-based sensor authentication is described in [44]. A unique pseudo-random binary sequence is generated for every sensor in a large-scale IoT network between the sensor and the gateway, using its physical layer attributes. The features collected are converted into binary data by a new SVM- based quantization technique which eliminates the boundary values and the non-linear boundary is sent to the sensor. The sensor and the gateway estimates the pseudo-random sequence, and if the access time or frequency of the sensor is not matching with the unique number at the gateway, then the sensor node will be denied access. SVM improves the security of the system by classifying the non-linear boundary values. Physical attribute-based authentication methods are also discussed in [45] and [46].

6.5.2 PUF-Based Authentication

Physical unclonable functions (PUFs) are ICs that have special fingerprints or identities that are unique created due to the manufacturing process variations. The responses generated by different PUFs are highly random and not predictable for the same PUF input (generally termed as a "challenge"); however the integrity of the response is maintained for a particular PUF. These challenge-response pairs (CRPs) are device dependent and can be used as the device fingerprint for the identification of the device equipped with the PUF. CRPs are analogous to the secret keys in cryptography. Since they are unique and unpredictable, they can be used for the generation of secret keys in cryptographic protocols. PUFs can be designed as a separate circuit onto the devices as an application-specific IC (ASIC) or made a part of system-on-chip. PUFs can also be implemented on FPGAs. They are of two main types: strong PUF and weak PUF. Strong PUFs are characterized by large number of CRPs and weak PUFs have small set of CRPs. Strong PUFs are used for authentication, whereas weak PUFs are used for secret key generation. A few examples of PUFs include ring oscillator PUFs (most compatible to be implemented on FPGA) and are based on frequency variations, arbiter PUFs that are based on propagation delay, and SRAM PUFs that are based on the differences in the initial values during start up. Different variants of PUFs have been discussed by the researchers all over the world. PUFs can be used for device fingerprinting, which eliminates the key storage and memory attack issues. Nowadays, PUFs are extensively used in resource constraint devices and IoT for the authentication of the nodes in the network as a substitute for cryptographic solutions as they are heavier. PUF-based authentication is depicted in Figure 6.9 in a simple way.

The device with PUF has to initially register with the GCS, shown in step 1. The GCS records the CRP database in its memory. To initiate communication, the drone sends a request to the GCS, shown in step 2. The GCS checks the authenticity of the drone by sending a random challenge and waits for a response from the drone, shown as step 3. The drone on receiving the challenge gives it as input to the PUF and obtains the response, which is sent back to the GCS as shown in step 4. GCS verifies the response with the CRP database and clears the device in authentication process. Once authenticated, they can start data transfer as shown in step 5.

In [47], a PUF for a cloud-edge system is proposed, where the edge devices are involved in the partial computation process and the cloud does the complex computations. Since the edge systems are prone to spoofing attacks, mutual authentication between the nodes and the gateway is necessary to prevent unauthentic interactions. The CRPs of the PUFs of the nodes are initially enrolled in a trusted server to create a database.

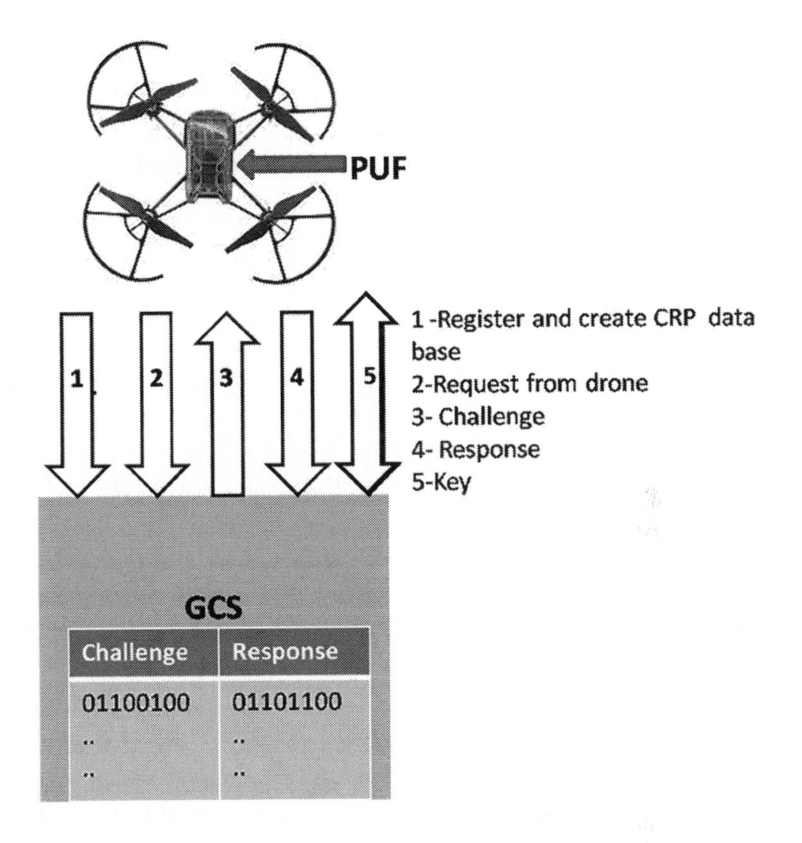

Figure 6.9 PUF authentication.

The authentication protocol involves sending of a challenge from this database and verifying the response from the device. Strong PUFs are used by both the nodes and the gateway. The authentication protocols described are Physical Hardware-Enabled Mutual Authentication Protocol (PHEMAP) and salted PHEMAP in [48] and [47]; they use PUF chains. But these schemes are vulnerable to impersonation attacks.

An IoT authentication scheme using the Elliptic Curve Diffie–Hellman (ECDH) key exchange protocol is described in [49], with the keys derived from a Static Random-Access Memory (SRAM) PUF. The IoT devices register their CRPs with the trusted server, and it generates the PUF-based private key and elliptic curve-based public key. The device certificate is stored in the cloud. The architecture is implemented on an FPGA using four systems, namely control unit, PUF unit, ECC scalar multiplication unit, and pseudo-random generator unit. A fuzzy extractor is used along with the SRAM PUF to remove the noise and rebuild the keys; however, security verification is not performed.

An optimized configurable ring oscillator based PUF for device authentication to prevent theft attacks is presented in [50]. The PUF is used to generate a signature for the chips, and they are encoded to generate a system-level signature for the IoT device. A PUF-based key agreement scheme is discussed in [51].

In [52], two levels of authentication – cryptographic and non-cryptographic methods – are discussed. In the first level, PUF along with the channel-specific characteristics are used for authentication, and in the second level, physical layer attributes like signal-to-noise ratio are considered for authentication. The CRPs are updated for every communication cycle.

A lightweight user authentication protocol integrating PUFs is proposed in [53]. Several features such as PUF, bit-wise XOR, fuzzy extractor function, and one-way hash function are used to make the system lightweight. Physical security of the user smart card and IoT device are ensured by the use of PUF. NS3 is the simulation tool used and AVISPA for security analysis. ROR-based formal security verification is also done.

A smart health care system with PUF-based device authentication to transmit the patient's health-related sensor readings to the cloud is proposed in [54]. The sensor equipments and the edge devices are equipped with a PUF module that ensures the authenticity and physical security of the devices. The end devices are first authenticated before receiving the critical data from it. PUFs are also used for the generation of secret keys as per the necessity, and this eliminates the need of key storage in server memory. The PUF unit is a Hybrid Oscillator Arbiter PUF that is implemented on an FPGA board. The total system power consumption also is less.

Mutual authentication between an IoT device and a server and between two IoT devices is discussed in [55]. Each IoT device is equipped with a PUF, and the CRP is enrolled in the server along with the ID using time-based one-time password (TOTP) algorithm. When an IoT device wants to communicate with the server, the server reads the CRP stored in its database and checks the ID for authentication. Security analysis is done using Mao and Boyd logic and is found to be secure against several attacks and is computationally efficient.

An IoT device authentication method using PUF-aided ECC for device enrollment, digital signal generation, and authentication is described in [56]. Private keys are generated using PUFs and authentication is done using zero knowledge proof protocol. PUF-based ElGamal Encryption algorithm and PUF-based digital signature are proposed and tested on Xilinx Artix 7 FPGA. Fuzzy extraction engine is also used to remove the noises from the PUF output.

Authentication of an IoT node with a server is addressed in [57] using several key concepts such as Identity-Based Encryption (IBE), PUFs, and keyed hash function. The IoT node contains the PUF module, and the server contains a keyed hash function to verify the responses from the PUFs. PUFs generate the secret keys for the IBE providing better security as it is built on hard problems. The system is simulated in a video surveillance environment.

6.5.3 Hardware Implementation of Authentication Protocols

Hardware implementation helps to evaluate the performance of a proposed algorithm by implementing the design on Advanced RISC Machine (ARM) controllers and FPGA. This helps in better evaluation of parameters like computation cost, storage requirements, and the suitability in a resource constraint device. FPGA implementation does the assessment at the gate level concerning the number of hardware components like multiplexers, lookup tables, and gates that are required for the design. In general, the implementations are done using RaspberryPi and Xilinx Artix FPGA boards.There are a few papers that discuss this type of implementation. An FPGA implementation of a user authentication and key agreement protocol based on ECC is discussed in [58]. It is a three-factor mutual authentication protocol between the user and the smart device for accessing the medical records of patients in a medical WSN. Hardware implementation of the protocol was using the Altera QuartusII FPGA board.

An RFID-based mutual authentication protocol has been proposed in [59] with FPGA and ASIC implementations. It is a ultra lightweight protocol using bit-wise logical operations namely, bit-wise XOR, and pseudoi-Kasami codes. The computational costs are minimum and can be easily implemented on an RFID tag.

6.5.4 Chaos-Based Mutual Authentication

Chaotic behavior denotes the irregular activities of a system, and chaos theory is a special branch of mathematics that deals with non-linear system behavior. A small change in the input conditions will produce large variations in the output. Chaos is different from randomness. Chaotic behavior can be determined, whereas randomness is not deterministic. The features of chaotic functions such as random-like behavior, sensitivity, and ergodicity makes it an ideal choice for cryptographic function generation. Key generation using chaos functions brings down the computational costs and key management issues. It is computationally efficient when compared to the traditional cryptographic techniques like ECC and Rivest–Shamir–Adleman algorithm (RSA). The key exchange algorithm using chaos is similar to Diffie–Hellman key exchange algorithm. A few research works in the area of chaos-based authentication are given below.

A chaotic hash function based group key agreement protocol was proposed in [60]. PUF is used for device identification in [61], and better security is assured using chaos theory. Chaos-based Differential Chaotic Shift Keying modulation scheme is used for the transmission of CRPs and it ensures low power and simplicity of the system. Authentication using Chebyshev Chaotic Maps is used in [62] to achieve lightweight authentication in an IoT system. It is a hierarchical system consisting of a bottom layer of sensors and upper layer of data centers. The devices mutually authenticate using semigroup property of the Chebyshev polynomials, and the system is resistant to tracing attack, sensor insider attack, replay attack, impersonation attack, and DoS attack.

Sahoo et al. [63] describes a user authentication scheme in a system with three entities – network manager, application provider, and a user. The main stages in the authentication process are registration phase, login phase, authentication phase, and password change phase, and the main functions used are cryptographic hash function and chaotic map function. Security proof is done with an ROR model. In [64], a system is described in which a trusted server uses Chebyshev chaotic maps for the generation of the secret key for authenticating the identity of the users in the network. Semigroup property of the extended Chebyshev chaotic map removes the key management issues and is used their work for authentication and exchange of session key between the two users in the system. A three-factor user authentication and key agreement protocol for an industrial IoT background is discussed in [65] using chaotic maps. Smart card, biometrics, and password are used for the identity authentication. Fuzzy extractors are used for removing the noise and better results. NS2 simulators are used for the analysis of the protocol.

A similar three-factor user authentication protocol is described in [66] for a smart health care system.

6.5.5 Quantum Authentication

Quantum computers are based on the principles of quantum mechanics which uses the undefined quantum states of an object (such as the spin of an electron or polarization of a photon) to produce a quantum bit (qubit). They are based on Shor's algorithm for integer factorization and hence can solve complex mathematical problems very fast, making them superior to the traditional computers. When quantum computers become a reality, the existing cryptographic techniques will become obsolete and insecure. Thus, it is trivial to develop quantum-resistant cryptographic algorithms also known as post-quantum cryptography (PQC) for the future quantum world.

There are mainly four types of quantum-resistant cryptographic methods – code-based, hash-based, multi-variate, and lattice based. The lattice-based cryptosystems (LBC) offer better trade-off between security and performance and hence are suitable for resource constraint applications like IoT. LBC relies on the hard problems of Learning With Errors (LWE), Shortest Integer Solution (SIS), Shortest Vector Problem (SVP), Closest Vector Problem, and their ring variants [67]. There are different variants of lattice-based systems, one of which is NTRUEncrypt, which uses polynomial convolution and is characterized by short keys, low memory requirements, and high efficiency and is better suited for IoT applications [68]. The other LBC algorithms that have been implemented on microcontrollers are cryptosystems based on Ring-LWE problem, signature scheme BLISS, NewHope and FrodoKEM (implemented on FPGA), Falcon, Saber, and Kyber; out of which NewHope is reported as the best in performance and memory requirements.

A lattice-based authentication protocol based on SIS for VANETs is developed in [69]. Quantum key distribution (QKD) using Bennett and Brassard protocol (termed as BB84 protocol) was proposed for VANET identity verification in [70]. A mutual user authentication scheme based on lattice is presented in [71]. A lattice-based group authentication is proposed in [72]. ECC-based digital signature and lattice PQC with LWE are used for user authentication in a smart health care system in [73].

Thus, there is a need to develop quantum-resilient, lightweight authentication mechanisms for FANETs that can resist the attacks in the upcoming powerful age of quantum computers.

6.5.6 Quantum Chaos Authentication

Systems using a combination of non-linear dynamics and quantum mechanics is an upcoming research area. An introduction to quantum chaos is given in [74] and [75]. It can also be used in the field of image encryption. Hematpour et al. [76] have incorporated the theories of quantum mechanics and chaos and proposed a digital signature based on quantum whose security is enhanced with the help of chaotic maps. It is secure against forging and repudiation. Researchers claim that integrating both these theories enhances the security of the system to a very high level.

6.6 CONCLUSION

Flying things always create interest and curiosity in engineering minds. But the security of the devices is a question to be pondered upon. In this chapter, we have discussed about the FANET features, architecture, UAV subsystems, applications, vulnerabilities, attacks, security solutions, and the authentication techniques in FANET. This chapter mainly focuses on the authentication protocols at user level, device level, and entity level. The authentication can be done between drone and drone, drone and ground, drone and smart object, or between drone and user through the ground station. The remote users can also become part of the network using passwords, biometrics, or smart card and exchange secret keys and confidential messages. The device-level authentication methods are very few and hence more novel techniques can be developed in this area in future. The chapter also introduces some new areas that can be used in the area of FANET authentication, such as PUFs, AI, chaos theory, quantum cryptography and quantum chaos theory, and hardware implementation of the authentication protocols. To summarize, there is a need for novel, lightweight post-quantum authentication techniques of superior performance that can overcome the challenges in FANETs.

Bibliography

[1] Ilker Bekmezci, Ozgur Koray Sahingoz, and Şamil Temel. Flying ad-hoc networks (fanets): A survey. *Ad Hoc Networks*, 11(3):1254–1270, 2013.

[2] Mohammad Mozaffari, Walid Saad, Mehdi Bennis, Young-Han Nam, and Mérouane Debbah. A tutorial on uavs for wireless networks: Applications, challenges, and open problems. *IEEE Communications Surveys & Tutorials*, 21(3):2334–2360, 2019.

[3] Parimal Mehta, Rajesh Gupta and Sudeep Tanwar. Blockchain envisioned UAV networks: Challenges, solutions, and comparisons. *Computer Communications*, 151:518–538, 2020.

[4] Dinh C. Nguyen, Pubudu N. Pathirana, Ming Ding, and Aruna Seneviratne. Blockchain for 5g and beyond networks: A state of the art survey. *arXiv preprint arXiv:1912.05062*, 2019.

[5] Isaac J. Jensen, Daisy F. Selvaraj, and Prakash Ranganathan. Blockchain technology for networked swarms of unmanned aerial vehicles (uavs). In *2019 IEEE 20th International Symposium on A World of Wireless, Mobile and Multimedia Networks(WoWMoM)*, pages 1–7. IEEE, 2019.

[6] Iván García-Magariño, Raquel Lacuesta, Muttukrishnan Rajarajan, and Jaime Lloret. Security in networks of unmanned aerial vehicles for surveillance with an agent-based approach inspired by the principles of blockchain. *Ad Hoc Networks*, 86:72–82, 2019.

[7] Anik Islam and Soo Young Shin. Bus: A blockchain-enabled data acquisition scheme with the assistance of uav swarm in internet of things. *IEEE Access*, 7:103231–103249, 2019.

[8] Naser Hossein Motlagh, Tarik Taleb, and Osama Arouk. Low-altitude unmanned aerial vehicles-based internet of things services: Comprehensive survey and future perspectives. *IEEE Internet of Things Journal*, 3(6):899–922, 2016.

[9] Hichem Sedjelmaci and Sidi M. Senouci. Cyber security methods for aerial vehicle networks: taxonomy, challenges and solution. *The Journal of Supercomputing*, 74(10):4928–4944, 2018.

[10] Thomas Lagkas, Vasileios Argyriou, Stamatia Bibi, and Panagiotis Sarigiannidis. Uav iot framework views and challenges: Towards protecting drones as "things". *Sensors*, 18(11):4015, 2018.

[11] Petros S. Bithas, Emmanouel T. Michailidis, Nikolaos Nomikos, Demosthenes Vouyioukas, and Athanasios G. Kanatas. A survey

on machine-learning techniques for uav-based communications. *Sensors*, 19(23):5170, 2019.

[12] Hazim Shakhatreh, Ahmad H. Sawalmeh, Ala Al-Fuqaha, Zuochao Dou, Eyad Almaita, Issa Khalil, Noor S. Othman, Abdallah Khreishah, and Mohsen Guizani. Unmanned aerial vehicles (uavs): A survey on civil applications and key research challenges. *IEEE Access*, 7:48572–48634, 2019.

[13] Ezedin Barka, Chaker A. Kerrache, Hadjer Benkraouda, Khaled Shuaib, Farhan Ahmad, and Fatih Kurugollu. Towards a trusted unmanned aerial system using blockchain for the protection of critical infrastructure. *Transactions on Emerging Telecommunications Technologies*, 1–15, doi:10.1002/ett.3706, issn: 2161-3915, 2019.

[14] Kuldeep Singh and Anil K. Verma. A fuzzy-based trust model for flying ad hoc networks (fanets). *International Journal of Communication Systems*, 31(6):e3517, 2018.

[15] Ezedin Barka, Chaker A. Kerrache, Nasreddine Lagraa, Abderrahmane Lakas, Carlos T. Calafate, and Juan-Carlos Cano. Union: A trust model distinguishing intentional and unintentional misbehavior in inter-uav communication. *Journal of Advanced Transportation*, 1–12, 2018.

[16] George Loukas, Eirini Karapistoli, Emmanouil Panaousis, Panagiotis Sarigiannidis, Anatolij Bezemskij, and Tuan Vuong. A taxonomy and survey of cyber-physical intrusion detection approaches for vehicles. *Ad Hoc Networks*, 84:124–147, 2019.

[17] Atul Malhotra and Sanmeet Kaur. A comprehensive review on recent advancements in routing protocols for flying ad hoc networks. *Transactions on Emerging Telecommunications Technologies*, e3688:1–32, 2019.

[18] Muhammad Y. Arafat and Sangman Moh. Routing protocols for unmanned aerial vehicle networks: A survey. *IEEE Access*, 7:99694–99720, 2019.

[19] Mohamed A. Ferrag, Leandros A. Maglaras, Helge Janicke, Jianmin Jiang, and Lei Shu. Authentication protocols for internet of things: A comprehensive survey. *Security and Communication Networks*, 2017:1–41, 2017.

[20] Mohammed El-hajj, Ahmad Fadlallah, Maroun Chamoun, and Ahmed Serhrouchni. A survey of internet of things (iot) authentication schemes. *Sensors*, 19(5):1141, 2019.

[21] Mohammad Wazid, Ashok K. Das, and Jong-Hyouk Lee. Authentication protocols for the internet of drones: taxonomy, analysis and future directions. *Journal of Ambient Intelligence and Humanized Computing*, 1–10, doi:10.1007/s12652-018-1006-x, 2018.

[22] Shuangyu He, Qianhong Wu, Jingwen Liu, Wei Hu, Bo Qin, and Ya-Nan Li. Secure communications in unmanned aerial vehicle network. In *International Conference on Information Security Practice and Experience*, pages 601–620. Springer, 2017.

[23] Jongho Won, Seung-Hyun Seo, and Elisa Bertino. Certificateless cryptographic protocols for efficient drone-based smart city applications. *IEEE Access*, 5:3721–3749, 2017.

[24] Liquan Chen, Sijie Qian, Ming Lim, and Shihui Wang. An enhanced direct anonymous attestation scheme with mutual authentication for network-connected uav communication systems. *China Communications*, 15(5):61–76, 2018.

[25] Benjamin Semal, Konstantinos Markantonakis, and Raja N. Akram. A certificateless group authenticated key agreement protocol for secure communication in untrusted uav networks. In *2018 IEEE/AIAA 37th Digital Avionics Systems Conference (DASC)*, pages 1–8. IEEE, 2018.

[26] Xinghua Li, Yunwei Wang, Pandi Vijayakumar, Debiao He, Neeraj Kumar, and Jianfeng Ma. Blockchain-based mutual-healing group key distribution scheme in unmanned aerial vehicles adhoc network. *IEEE Transactions on Vehicular Technology*, 68(11): 11309–11322, 2019.

[27] Chin-Ling Chen, Yong-Yuan Deng, Wei Weng, Chi-Hua Chen, Yi-Jui Chiu, and Chih-Ming Wu. A traceable and privacy-preserving authentication for uav communication control system. *Electronics*, 9(1):62, 2020.

[28] Mohammad Wazid, Ashok K. Das, Neeraj Kumar, Athanasios V. Vasilakos, and Joel J.P.C. Rodrigues. Design and analysis of secure lightweight remote user authentication and key agreement scheme

in internet of drones deployment. *IEEE Internet of Things Journal*, 6(2):3572–3584, 2018.

[29] Srinivas Jangirala, Ashok K. Das, Neeraj Kumar, and Joel Rodrigues. Tcalas: Temporal credential-based anonymous lightweight authentication scheme for internet of drones environment. *IEEE Transactions on Vehicular Technology*, 68:6903–6916, 2019.

[30] Mohammad S. Farash, Muhamed Turkanović, Saru Kumari, and Marko Hölbl. An efficient user authentication and key agreement scheme for heterogeneous wireless sensor network tailored for the internet of things environment. *Ad Hoc Networks*, 36:152–176, 2016.

[31] Wei-Liang Tai, Ya-Fen Chang, and Wei-Han Li. An iot notion–based authentication and key agreement scheme ensuring user anonymity for heterogeneous ad hoc wireless sensor networks. *Journal of Information Security and Applications*, 34:133–141, 2017.

[32] Ruhul Amin, S.K. Hafizul Islam, G.P. Biswas, Muhammad K. Khan, Lu Leng, and Neeraj Kumar. Design of an anonymity-preserving three-factor authenticated key exchange protocol for wireless sensor networks. *Computer Networks*, 101:42–62, 2016.

[33] Sravani Challa, Mohammad Wazid, Ashok K. Das, Neeraj Kumar, Alavalapati G. Reddy, Eun-Jun Yoon, and Kee-Young Yoo. Secure signature-based authenticated key establishment scheme for future iot applications. *IEEE Access*, 5:3028–3043, 2017.

[34] Mohammad Wazid, Ashok K. Das, Vanga Odelu, Neeraj Kumar, Mauro Conti, and Minho Jo. Design of secure user authenticated key management protocol for generic iot networks. *IEEE Internet of Things Journal*, 5(1):269–282, 2017.

[35] Sana Benzarti, Bayrem Triki, and Ouajdi Korbaa. Privacy preservation and drone authentication using id-based signcryption. In *SoMeT*, pages 226–239, 2018.

[36] Yunmok Son, Juhwan Noh, Jaeyeong Choi, and Yongdae Kim. Gyrosfinger: Fingerprinting drones for location tracking based on the outputs of mems gyroscopes. *ACM Transactions on Privacy and Security (TOPS)*, 21(2):1–25, 2018.

[37] Soundarya Ramesh, Thomas Pathier, and Jun Han. Sounduav: Towards delivery drone authentication via acoustic noise fingerprinting. In *Proceedings of the 5th Workshop on Micro Aerial Vehicle Networks, Systems, and Applications*, pages 27–32. ACM, 2019.

[38] Mehdi Karimibiuki, Michal Aibin, Yuyu Lai, Raziq Khan, Ryan Norfield, and Aaron Hunter. Drones' face off: Authentication by machine learning in autonomous iot systems. In *IEEE 10th Annual Ubiquitous Computing, Electronics and Mobile Communication Conference (UEMCON)*, pages 0329–0333, October 2019.

[39] Changjun Jiang, Yu Fang, Peihai Zhao, and John Panneerselvam. Intelligent uav identity authentication and safety supervision based on behavior modeling and prediction. *IEEE Transactions on Industrial Informatics*, 3203:1–12, 2020.

[40] Abdulhadi Shoufan. Continuous authentication of uav flight command data using behaviometrics. In *2017 IFIP/IEEE International Conference on Very Large Scale Integration (VLSI-SoC)*, pages 1–6. IEEE, 2017.

[41] Ursula Challita, Aidin Ferdowsi, Mingzhe Chen, and Walid Saad. Machine learning for wireless connectivity and security of cellular-connected uavs. *IEEE Wireless Communications*, 26(1):28–35, 2019.

[42] Aidin Ferdowsi and Walid Saad. Deep learning for signal authentication and security in massive internet-of-things systems. *IEEE Transactions on Communications*, 67(2):1371–1387, 2018.

[43] He Fang, Xianbin Wang, and Stefano Tomasin. Machine learning for intelligent authentication in 5g and beyond wireless networks. *IEEE Wireless Communications*, 26(5):55–61, 2019.

[44] He Fang, Angie Qi, and Xianbin Wang. Fast authentication and progressive authorization in large-scale iot: How to leverage ai for security enhancement? *arXiv preprint arXiv:1907.12092*, 2019.

[45] Linda Senigagliesi, Marco Baldi, and Ennio Gambi. Performance of statistical and machine learning techniques for physical layer authentication. *arXiv preprint arXiv:2001.06238*, 2020.

[46] He Fang, Xianbin Wang, and Lajos Hanzo. Adaptive trust management for soft authentication and progressive authorization relying on physical layer attributes. *IEEE Transactions on Communications*, 68(4):2607–2620, 2020.

[47] Mario Barbareschi, Alessandra De Benedictis, Erasmo La Montagna, Antonino Mazzeo, and Nicola Mazzocca. A puf-based mutual authentication scheme for cloud-edges iot systems. *Future Generation Computer Systems*, 101:246–261, 2019.

[48] Mario Barbareschi, Alessandra De Benedictis, and Nicola Mazzocca. A puf-based hardware mutual authentication protocol. *Journal of Parallel and Distributed Computing*, 119:107–120, 2018.

[49] Haji Akhundov, Erik van der Sluis, Said Hamdioui, and Mottaqiallah Taouil. Public-key based authentication architecture for iot devices using puf. *arXiv preprint arXiv:2002.01277*, 2020.

[50] Zhao Huang and Quan Wang. A puf-based unified identity verification framework for secure iot hardware via device authentication. *World Wide Web*, pages 1–32, 2019.

[51] An Braeken. Puf based authentication protocol for iot. *Symmetry*, 10(8):352, 2018.

[52] Hassan N. Noura, Reem Melki, and Ali Chehab. Secure and lightweight mutual multi-factor authentication for iot communication systems. In *2019 IEEE 90th Vehicular Technology Conference (VTC2019-Fall)*, pages 1–7. IEEE, 2019.

[53] Soumya Banerjee, Vanga Odelu, Ashok K. Das, Samiran Chattopadhyay, Joel J.P.C. Rodrigues, and Youngho Park. Physically secure lightweight anonymous user authentication protocol for internet of things using physically unclonable functions. *IEEE Access*, 7:85627–85644, 2019.

[54] Venkata P. Yanambaka, Saraju P. Mohanty, Elias Kougianos, and Deepak Puthal. Pmsec: Physical unclonable function-based robust and lightweight authentication in the internet of medical things. *IEEE Transactions on Consumer Electronics*, 65(3):388–397, 2019.

[55] Muhammad N. Aman, Kee Chaing Chua, and Biplab Sikdar. Mutual authentication in iot systems using physical unclonable functions. *IEEE Internet of Things Journal*, 4(5):1327–1340, 2017.

[56] John R. Wallrabenstein. Practical and secure iot device authentication using physical unclonable functions. In *2016 IEEE 4th International Conference on Future Internet of Things and Cloud (Fi-Cloud)*, pages 99–106. IEEE, 2016.

[57] Urbi Chatterjee, Vidya Govindan, Rajat Sadhukhan, Debdeep Mukhopadhyay, Rajat S. Chakraborty, Debashis Mahata, and Mukesh M. Prabhu. Puf+ ibe: Blending physically unclonable functions with identity based encryption for authentication and key exchange in iots. *IACR Cryptology ePrint Archive*, 2017:422, 2017.

[58] Venkatasamy Sureshkumar, Ruhul Amin, V.R. Vijaykumar, and S. Raja Sekar. Robust secure communication protocol for smart healthcare system with fpga implementation. *Future Generation Computer Systems*, 100:938–951, 2019.

[59] Umar Mujahid, M.Najam-ul Islam, and Madiha Khalid. Efficient hardware implementation of kmap: An ultralightweight mutual authentication protocol. *Journal of Circuits, Systems and Computers*, 27(02):1850033, 2018.

[60] Xianfeng Guo and Jiashu Zhang. Secure group key agreement protocol based on chaotic hash. *Information Sciences*, 180(20): 4069–4074, 2010.

[61] Wang Hong, Li Jianhua, Lai Chengzhe, and Wang Zhe. A provably secure aggregate authentication scheme for unmanned aerial vehicle cluster networks. *Peer-to-Peer Networking and Applications*, 13(1):53–63, 2020.

[62] Aida Akbarzadeh, Majid Bayat, Behnam Zahednejad, Ali Payandeh, and Mohammad R. Aref. A lightweight hierarchical authentication scheme for internet of things. *Journal of Ambient Intelligence and Humanized Computing*, 10(7):2607–2619, 2019.

[63] Shreeya S. Sahoo and Sujata Mohanty. Chaotic map based privacy preservation user authentication scheme for wbans. In *TENCON 2019-2019 IEEE Region 10 Conference (TENCON)*, pages 1037–1042. IEEE, 2019.

[64] Yousheng Zhou, Junfeng Zhou, Feng Wang, and Feng Guo. An efficient chaotic map-based authentication scheme with

mutual anonymity. *Applied Computational Intelligence and Soft Computing,* 2016:422, 2016.

[65] Jangirala Srinivas, Ashok K. Das, Mohammad Wazid, and Neeraj Kumar. Anonymous lightweight chaotic map-based authenticated key agreement protocol for industrial internet of things. *IEEE Transactions on Dependable and Secure Computing*, 97:185–196, 2019.

[66] Sandip Roy, Santanu Chatterjee, Ashok K. Das, Samiran Chattopadhyay, Saru Kumari, and Minho Jo. Chaotic map-based anonymous user authentication scheme with user biometrics and fuzzy extractor for crowdsourcing internet of things. *IEEE Internet of Things Journal*, 5(4):2884–2895, 2017.

[67] Chi Cheng, Rongxing Lu, Albrecht Petzoldt, and Tsuyoshi Takagi. Securing the internet of things in a quantum world. *IEEE Communications Magazine*, 55(2):116–120, 2017.

[68] Rajat Chaudhary, Gagangeet S. Aujla, Neeraj Kumar, and Sherali Zeadally. Lattice based public key cryptosystem for internet of things environment: Challenges and solutions. *IEEE Internet of Things Journal*, 6(3):4897–4909, 2018.

[69] Sankar Mukherjee, Daya S. Gupta, and G.P. Biswas. An efficient and batch verifiable conditional privacy-preserving authentication scheme for vanets using lattice. *Computing*, 101(12):1763–1788, 2019.

[70] Zhiya Chen, Kunlin Zhou, and Qin Liao. Quantum identity authentication scheme of vehicular ad-hoc networks. *International Journal of Theoretical Physics*, 58(1):40–57, 2019.

[71] Sung-Wook Park and Im-Yeong Lee. Mutual authentication scheme based on lattice for nfc-pcm payment service environment. *International Journal of Distributed Sensor Networks*, 12(7):9471539, 2016.

[72] Jheng-Jia Huang, Yi-Fan Tseng, Qi-Liang Yang, and Chun-I Fan. A lattice-based group authentication scheme. *Applied Sciences*, 8(6):987, 2018.

[73] Amit Dua, Rajat Chaudhary, Gagangeet S. Aujla, Anish Jindal, Neeraj Kumar, and Joel J.P.C. Rodrigues. Lease: Lattice and ecc-based authentication and integrity verification scheme in e-healthcare. In *2018 IEEE Global Communications Conference (GLOBECOM)*, pages 1–6. IEEE, 2018.

[74] Mason A. Porter. An introduction to quantum chaos. *arXiv preprint nlin/0107039*, 2001.

[75] Denis Ullmo and Steven Tomsovic. Introduction to quantum chaos. *Encyclopedia of Life Support Systems (EOLSS), Oxford, UK*, https://www.lptms.upsud.fr/membres/ullmo/Articles/eolss-ullmo-tomsovic.pdf, 2014.

[76] Nafiseh Hematpour, Sodeif Ahadpour, and Sohrab Behnia. Digital signature: Quantum chaos approach and bell states. In *Chaotic Modeling and Simulation International Conference*, pages 85–93. Springer, 2018.

CHAPTER 7

Investigating Traffic of Smart Speakers and IoT Devices: Security Issues and Privacy Threats

Davide Caputo, Luca Verderame, and Alessio Merlo
University of Genoa

Luca Caviglione
National Research Council of Italy

CONTENTS

7.1 INTRODUCTION

Smart speakers and voice-based virtual assistants are important building blocks of modern smart homes. For instance, they can be used to retrieve information, interact with other devices, and command a wide range of Internet of Things (IoT) nodes. Moreover, they can be used as hubs for managing IoT deployments or implementing device automation services, e.g., to perform routines in smart lighting or provide remote connectivity for domestic appliances. According to Canalys[1], there are over 200 million of smart speakers installed in private properties (with the wide acceptation of private habitative units and small office home office settings), and the trend is expected to culminate in 2030 when the number will exceed 500 million of units.

In general, smart speakers and voice-based virtual assistant take advantage of cloud-based architectures: vocal commands of the user are sampled and sent through the Internet to be processed. As a result, the smart speaker or the appliance running the virtual assistant receives a textual representation as well as optional, companion multimedia data. Then, it executes the command or route it to a proper hub, e.g., to communicate via Zigbee or Bluetooth links with IoT nodes. To enforce privacy and security, the prime mechanism is the encryption of traffic (see, e.g., reference [28] and references therein). However, features of the flows such as, the throughput, the size of protocol data units or the `(address, port)` tuples, can leak important information about the habits of the users [9] or the number and the type of IoT nodes [4,22]. As a consequence, an attacker can collect traffic from the local IEEE 802.11 wireless loop or between the home gateway and the Internet and then try to guess the state of the IoT nodes, the number of devices deployed, and the state of sensors and actuators. With such a knowledge, the malicious entity can launch a wide array of offensive campaigns, such as profile users, plan attacks to the physical space, or perform social engineering campaigns [4,22].

Despite the underlying technology or the complexity of deployment, there is an increasing interest in investigating risks arising for the statistical analysis of the traffic exchanged by a smart speaker and the cloud. For instance, in [4] authors showcase how passive network analysis can be used to identify devices and correlate some user activities, e.g., traffic flows produced by switches and health monitors can leak the sleep cycle of a user. In [13], the traffic produced by state transitions of home devices (i.e., a thermostat and a carbon dioxide detector) can be used to infer if a user is present in the home. Such idea is further refined in [1], where passive measurements are used to develop models of the daily routine of individuals (e.g., leaving/arriving home). Concerning works aiming at identifying devices, possibly by adopting efficient machine learning or statistical tools, in [22] several machine learning techniques are used to identify IoT devices by exploiting "poor" information like the length of packets produced during normal operations. Additionally, in [3] the risks of HTTP-based communications are discussed, both from the perspective of inferring data about the devices (e.g., the state or the intensity of a light source) and performing session-highjacking attacks. In addition, sensitive data contained in IoT nodes and smart speakers can be relevant for forensic investigations [18], and traffic patterns can be manipulated by malware to exfiltrate data, for instance through information hiding schemes [14] or covert channels [11,19].

In this vein, the chapter discusses risks of machine-learning-capable techniques to develop black-box models for automatically classifying traffic and implement privacy leaking attacks. Differently from previous works [1,3,13,22], we concentrate on understanding whether it is possible to recognize the presence of a user when no queries are performed. In fact, when a request is sent towards the Internet, the produced traffic volumes or the appearance of specific network addresses trivially leak the presence of a human operator in the house. Nevertheless, attention will be devoted in proposing ideas to mitigate such kind of threats by acting at the traffic level. In fact, the design of suitable mitigation techniques is often neglected (see, e.g., [3] for a notable exception) or addressed at an Application Programming Interface (API)-permission level [2], which is definitely out of the scope of the chapter.

Summing up, the contributions of this chapter are: (a) review the architectural blueprint used by smart speakers and voice-based virtual assistant and elaborate an affective model to conduct privacy leaking attacks; (b) evaluate the effectiveness of using machine learning techniques for black-box modeling of traffic; and (c) present some design rules to mitigate identification attacks.

The remainder of the chapter is structured as follows. Section 7.2 discusses the general architecture used by smart speakers to control IoT devices, introduces the threat model and the machine learning mechanisms that can be exploited by the attacker. Section 7.3 deals with the test bed used to collect data, while Section 7.4 presents numerical results. Section 7.5 proposes some countermeasures, and Section 7.6 concludes the chapter and showcases some possible future directions.

7.2 SMART SPEAKERS: ARCHITECTURE AND THREAT MODEL

As hinted, smart speakers and voice-based virtual assistants are a core foundation for smart homes. In essence, they provide a user interface allowing humans to issue requests or commands in a natural manner, i.e., by simply talking. Such devices can also be used as hubs for other IoT nodes and network appliances or to perform tasks like playing music and video, purchase items, and make recommendations. Besides, smart speakers and virtual assistants can provide a variety of information including directions and weather forecasts.

As today, most popular smart speakers implementing the aforementioned features are Google Home[2], HomePod[3], and Amazon Echo[4], whereas virtual assistants are Amazon Alexa[5], Apple Siri[6], and Google Assistant[7]. Literature still lacks a unified terminology for this class of devices and services. In fact, smart speakers and virtual assistants are identified as intelligent personal assistants, virtual personal assistants, home digital voice assistants, voice-enabled speakers, smart speakers, and voice-based virtual assistant, just to mention the most popular names. Therefore, in the following, we only use the terms smart speakers or Intelligent Virtual Assistant (IVA) interchangeably, except when doubts arise.

Even if each smart speaker is characterized by specific design choices and some setups are implemented via a complex interplay of technologies and services, the core architectural blueprint is quite standard and depicted in Figure 7.1. The overall set of components is often defined as the *ecosystem* so as to emphasize the end-to-end pipeline at the basis of such services, i.e., hardware or software entities allowing the interaction of end users, computing and communication services, and software

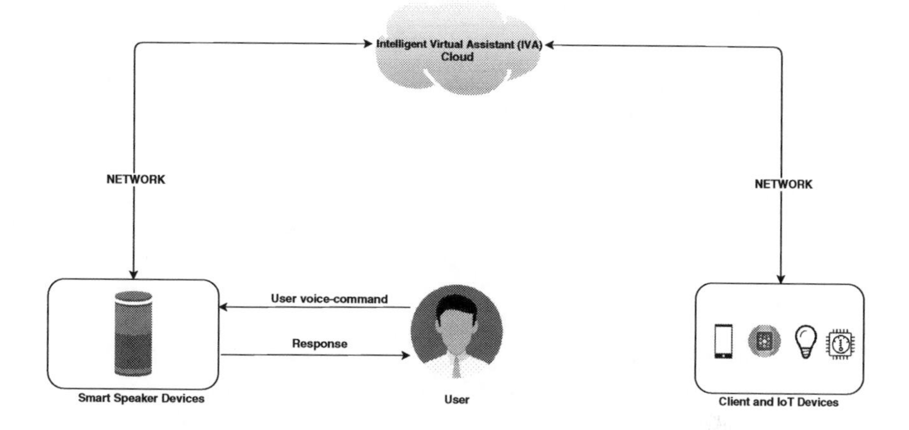

Figure 7.1 General system architecture used by smart speakers to control nodes in smart home scenarios.

running in IoT nodes. Even if each vendor usually implements its own blueprint, the typical one is composed of four major components:

- *Smart Speaker* or *IVA*: It is in charge of collecting vocal commands, sample them, and transmit the data trough the Internet to a back-end. Upon receiving a response, the smart speaker or the software IVA agent can provide a feedback to the user or directly interact with other devices. For instance, the smart speaker could start the playback of a music stream received from a Content Delivery Network (CDN) or send through a Zigbee link a command to a smart lightbulb. In some cases, it can also act as a sort of "router," thus delivering commands to the suitable hub. To avoid security and privacy threats, communications are encrypted via the Secure Socket Layer (SSL) [9,16].

- *Client* and *IoT Devices*: They are the targets of commands of the ecosystem. Typical nodes deployed in a smart home are sensors, actuators, Bluetooth/Zigbee bridges, wireless speakers, or IoT-capable appliances. As previously said, some entities belonging to this class can be colocated within the smart speaker.

- *IVA Cloud*: It is the backend in charge of processing data and delivering back text/binary representations of commands to be executed, including additional contents like multimedia streams, geographical information, or JavaScript Object Notation (JSON) files containing a composite variety of information. With the advent of open ecosystems promoting the interaction among services provided by multiple vendors, the borders of the IVA cloud are blurring [9,11,16]. For instance, vocal stimuli could be processed in a datacenter and sent back to the IVA while contents can be delivered by a third-part CDN, and some IoT nodes could establish a direct point-to-point connection with the computing infrastructure of their manufacturer.

- *Network*: It connects the smart speaker or the IVA with IoT nodes as well as the Internet. Typical deployments use a single local (wireless) network, which is connected via a router/gateway to the Internet. However, in most complex scenarios, different networks could be present, e.g., a local access cabled network for some IoT nodes and hubs and multiple wireless loops to connect smart devices and grant access to the user via a smartphone. Concerning protocols used to exchange data between the IVA and the cloud, the Transmission Control Protocol (TCP) is the main choice, with the multipath variant to optimize performances and reduce delays [9]. A notable exception is the Google ecosystem. In fact, it exploits QUIC [6], a protocol originally engineered to improve performance issues of HTTP/2 and based on transport streams multiplexed over User Datagram Protocol (UDP). We point out that the presence of QUIC can represent a signature to ease the identification of the ecosystem (e.g., Apple HomeKit vs. Google). However, this requires to understand its behaviors, which can be highly influenced by the underlying network conditions (see, e.g., [8] for a sensitivity/performance analysis of the SPDY counterpart in different wireless settings).

Concerning the typical usage scenario, smart speakers rely upon a microphone to sense commands, which are processed by a vocal interpreter running locally. In fact, only wake-up commands are executed within the device, while others are transmitted remotely to the cloud. Each IVA is activated via its own phrase or keyword, and the most popular are "Ok Google," "Alexa," and "Hey Siri" for the case of Google Assistant, Amazon Echo/Alexa, and Apple/HomeKit ecosystem, respectively. As it will be detailed later on, a relevant fragility is due to the continuous data exchange from the IVA and the cloud. Even if several frameworks could be considered "secure" both from the architectural and technological viewpoints, still they are prone to a variety of privacy-breaking attacks targeting a composite set of features observable within the encrypted traffic flows [2,4,9,28].

7.2.1 Threat Model

We aim at investigating the class of attacks targeting the encrypted traffic in a black-box manner, i.e., without trying to decipher the payload of protocol data units. Literature abounds of works dealing with techniques against SSL flows, for instance, [12] provides an extensive survey on Man-in-the-Middle (MitM) attacks for SSL/Transport Layer Security (TLS) conversations as well as techniques to highjack or spoof different protocol entities and nodes (e.g., Border Gateway Protocol (BGP) routes, Address Resolution Protocol (ARP)/reverse ARP (RARP) caches, and access points). Moreover, [21] reports an MitM attack expressly crafted for the Alexa IVA. Specifically, it targets "skills" which are extensions introduced to integrate third-part devices and services in the Amazon ecosystem. An attacker can redirect the voice input of the victim to a malicious node, thus highjacking the conversation. However, such attacks are definitely outside the scope of this chapter. Rather, we consider an adversary wanting to profile the user, for instance, for reconnaissance purposes or to plan a physical attack. To this aim, the adversary can exploit the traffic to infer "behavioral" information, e.g., when the victim is not at home. Figure 7.2 depicts the reference threat model.

In more detail, we assume the presence of an adversary (denoted as *malicious user* in the figure) that can only perform a passive attack, i.e., he/she can observe and acquire the traffic produced by the victim but cannot alter or manipulate it. To this aim, the adversary should access the home router. However, this is not a tight constraint as he/she can abuse the IEEE 802.11 wireless loop to gather information to be sent to the IVA (see, e.g., [10] for an analysis of threats that can be done by moving throughout the attack surface). We also assume that the adversary is not able to use the contents of the packets to launch the attack: in other words, he/she is not able to attack the TLS/SSL or VPN schemes usually deployed. Therefore, by inspecting the traffic produced by the smart speaker, the adversary can only rely on statistics and metadata of conversations. As an example, the attacker inspects (or computes by performing suitable operations) values like the throughput,

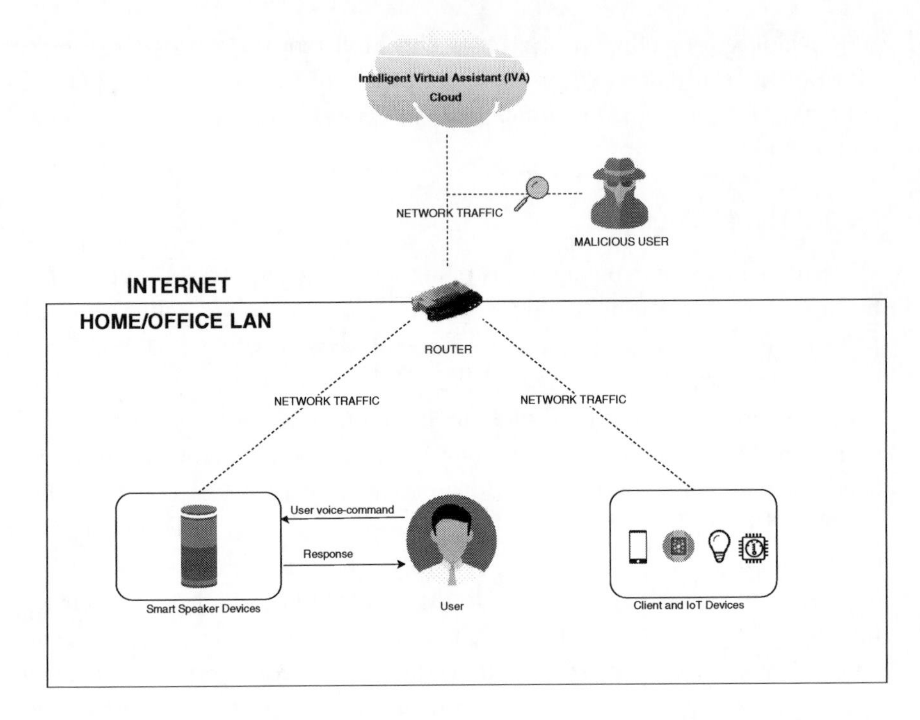

Figure 7.2 Reference threat model targeting the encrypted traffic for privacy-breaching attacks.

the size of protocol data units, the IP address, the number of different endpoints, flags within the headers of the packets, or the behavior of the congestion control of the TCP. Finally, as we usually done in similar works, we assume that the attacker is able to isolate and recognize traffic that comes from different IoT devices [5,20,25].

Even if the deployment of encryption schemes is not sufficient to prevent the leakage of important information about the habits of users [9] and the number or the type of IoT nodes deployed [1,3,4], this was a suitable countermeasure to mitigate a wide variety of threats. Alas, the advent of computational-efficient statistical tools brings into a feasibility zone a new wave of attacks. As a prototypal example, the work in [27] demonstrated how to leak the language of the talker by inspecting the bit rate of voice over IP (VoIP) conversations. In essence, authors used a sort of "signatures" produced by the variable bit rate codec to feed various classifiers, such as the k-Nearest Neighbors (kNN), Hidden Markov Models, and Gaussian Mixture Models, and a computational-efficient variant of the χ^2 classifiers, to identify the language with different performances (e.g., they can discriminate between English and Hungarian and from Brazilian Portuguese and English with a 66.5% and 86% of accuracy, respectively). We then review the most suitable tools that the adversary can use to extract information obtained from the gathered traffic and then scardinate the privacy of smart speakers and part of the IoT subsystem.

7.2.2 Machine Learning Techniques for Attacking the IoT Ecosystem

Nowadays, gathering and analyzing traffic is a core technique used during the reconnaissance phase of an attack [26], e.g., to enumerate devices or to fingerprint hosts for searching known vulnerabilities. In this work we consider the attacker wanting to infer high-level information, for instance to launch social engineering campaigns or plan physical attacks. Literature showcases different machine learning approaches and their adoption to solve networking duties is becoming a *de-facto* standard, see, [23] for a recent survey on the use of deep learning for different traffic classification problems. However, in the perspective of endowing an attacker with the suitable tools to gather information on the state of the smart speaker or the IVA, we shortlisted the following most promising algorithms.

7.2.2.1 k-Nearest Neighbors (kNN)

kNN is a supervised learning algorithm for solving classification and regression problems [17]. The kNN algorithm assumes that similar things exist in close proximity. Accordingly, kNN finds the distances between a query and all the data with a known label and selects the entries (i.e., k) closest to the query. Then it votes for the most frequent label (in the case of classification) or averages the labels (in the case of regression).

7.2.2.2 Decision Tree (DT)

Decision Tree (DT) is a family of non-parametric supervised learning methods suitable for classification and regression problems [17]. The DT builds classification or regression models in the form of a tree structure. To this aim, it breaks down the data into smaller subsets while developing an associated DT. The process is iterated by further splitting the dataset, and the final result is a tree with decision nodes and leaf nodes.

7.2.2.3 Adaptive Boosting – AdaBoost (AB)

AB exploits the idea of creating a highly accurate prediction rule by combining many relatively weak and inaccurate rules [17]. AB can be used in conjunction with many other types of learning algorithms to improve performances. In this case, the output of the other learning algorithms (defined as *weak learners*) is combined into a weighted sum that represents the final output of the boosted classifier.

7.2.2.4 Random Forest

Random forest is a model composed of many DTs [7,17]. It relies upon the following key steps: random sampling of training data points when building trees, and definition of random subsets of feature generated when splitting nodes. Specifically, when training, each tree in a random forest learns from a random sample of the data points. The samples are drawn with a replacement technique, known as bootstrapping [17], which means that some samples will be used multiple times in a single tree. The idea is that by training each tree on different samples, the entire forest will have lower variance but not at the cost of increasing the bias. We point out that, when using the random forest, only a subset of all the features are considered for splitting each node in the DT.

7.2.2.5 Support Vector Machine (SVM)

Support Vector Machine (SVM) is a group of supervised learning models for solving both regression or classification problems [17]. SVM builds non-probabilistic binary classifiers whose purpose is the search for the optimal hyperplane of separation between the two possible classes within the feature space. This type of classifiers is used more when the input data is not directly separable as it allows to map the initial data into a higher dimension space where it is possible to find a separation hyperplane.

7.2.2.6 Neural Networks

Neural networks are a family of algorithms, loosely modeled after the human brain, that are designed to recognize patterns. Neural networks are organized in a series of layers, where the input vector enters at the left side of the network, which is then projected to a hidden layer. Each unit in the hidden layer is a weighted sum of the values in the first layer. This layer then projects to an output layer, which is where the desired answer generates.

7.2.2.7 K-Fold Cross-Validation

K-fold cross-validation avoids the risk of missing important patterns or trends in the dataset. To this aim, the data is randomly partitioned into k equal-sized subsamples. Each single subsample is retained as the validation data for testing the model, and the remaining $k-1$ subsamples are used as training data. The process is then repeated k times, with each of the k subsamples used exactly once as the validation data. The k results from the folds are averaged to produce a single estimation. An interesting property of this method is its ability of limiting overfitting phenomena.

7.3 EXPERIMENTAL TEST BED

To prove the effectiveness of privacy threats of smart speakers leveraging machine learning techniques, we developed an experimental test bed. Due to the lack of public datasets containing networks traffic of smart speakers, we have also developed an automated framework for generating and collecting the relevant network traffic.

Concerning the device under investigation, we used a Google Home Mini[8] since it is one of the most popular smart appliance. Our version is equipped with an IEEE 802.11 L2 interface, an internal microphone to sense commands and the surrounding environment, and a loudspeaker for audio playback, and LEDs for visual feedbacks. The configuration of the device must be done via a companion application[9]. To this aim, we provided the Service Set IDentifier (SSID) and the password of our test network, which allowed the smart speaker to communicate remotely with the cloud running Google services and to exchange data with other devices connected to the same network (e.g., smart tv, smart light bulbs, etc.). We did not performe other tweaks as to reproduce an average installation usually accounting for the device deployed by the user in an out-of-the-box flavor.

Since we are focusing on privacy leakages related to the behavior of the microphone when disabled or when sensing various situations, i.e, the presence of humans or a quiet condition, we performed three different measurement campaigns, each one lasting 3 days. In particular, for the first round of tests, the microphone of the smart speaker was manually set off so as to investigate the traffic exchanged between the device and the remote cloud datacenter. Then, for the second round, the microphone was manually set on and the device put in a quiet condition, i.e., the microphone did not receive any stimuli from the surrounding environment, which was completely without noise or voices. For the last round of tests, we set the microphone on and we simulated the presence of humans speaking each others or background noise. We underline that human talkers will not issue the "Ok Google" phrase or will not inadvertently activate the smart speaker. In the following, we denote the different tests as `mic_off` for the case when the microphone is disabled, `mic_on` and `mic_on_noise` for tests with the microphone active and the smart speaker placed in a silent or noisy environment, respectively. To the aim of having proper audio patterns, we selected videos from YouTube in order to stimulate the smart speaker with a wide variety of talkers and settings (e.g., female and male speakers of different ages).

To capture data, we prepared a standard computer to act as the IEEE 802.11 access point, and we deployed ad hoc scripts for running `tshark`[10], i.e., the command line interface provided by the Wireshark tool. To process the dataset and perform computations, we used a computer with an Intel Core i7-3770 processor, with 16 GB of RAM running the Ubuntu 16.04 long-term support (LTS) operating system.

To implement the machine learning algorithms presented in Section 7.2.2, we used the scikit-learn[11] library. In essence, it is an open-source library developed in Python that contains the implementation of the most popular machine learning algorithms. However, after some preliminary trials, we shortlisted the considered techniques. Specifically, in the perspective of investigating the feasibility of performing an attack via off-the-shelf methods to scardinate the privacy of encrypted traffic generated by IVAs, we discarded SVMs and neural networks. Even if both techniques have several pros, i.e., there is a vast literature on black-box modeling leading to a variety of tools and insights, they require a relevant amount of time to complete the training phase (e.g., more than 8 days on an i7 CPU). Another important consideration from our preliminary trials is the double-faceted nature of the random forest technique. In fact, it does not only avoid overfitting problems, but it also gives qualitative indicators to assess the importance of features observable in the traffic. Such an aspect can be relevant when engineering countermeasures, as discussed in Section 7.5.

7.3.1 Data Handling

As said, we only collected traffic without performing any operation aimed at breaking the encryption scheme. In other words, we consider a worst-case scenario where the attacker is not able to perform deep packet inspection or more sophisticated actions (e.g., pinning of SSL certificates). Instead, the threat model we investigate deals with a malicious entity wanting to infer the smart speaker state by only using statistical information observable within the encrypted network traffic. To this aim, the attacker can extract/compute indicators by using two different "grouping" schemes, as depicted in Figure 7.3. In more detail, we computed the desired metrics by considering a suitable amount of packets obtained according to the windowing mechanisms considered as follows:

- time spans of length Δt (see Figure 7.3a);
- bursts of a fixed length of N (see Figure 7.3b).

We point out that the size of the windows affects the amount of information to be processed by the machine learning algorithm. In fact, even if the dataset still remains unchanged, the number of windows is directly proportional to the volume of information offered to the statistical tool

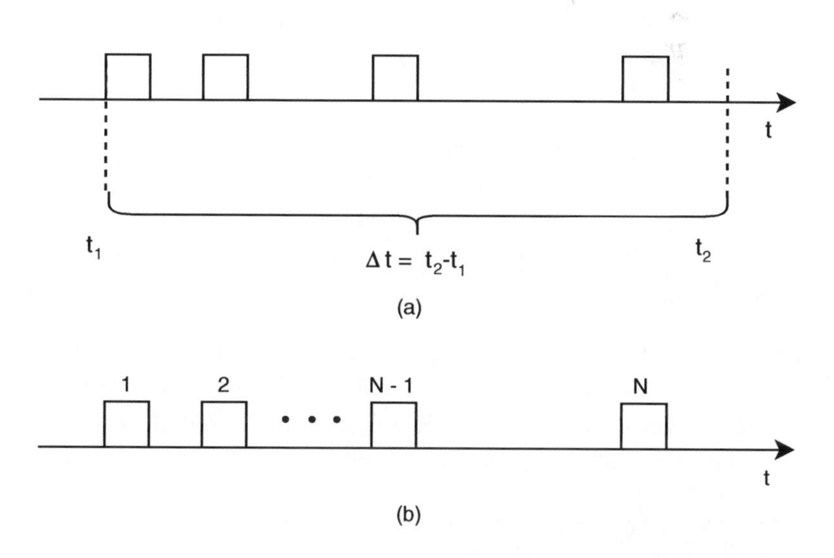

Figure 7.3 Different policies for grouping packets used for the computation of statistical information. (a) Packets grouped in a window of Δt seconds. (b) Packets grouped in a window of N data units.

(i.e., for each window a statistical indicator is computed). Concerning the statistical indicators that an attacker can obtain from the traffic exchanged between the IVA and the cloud, we consider

- Number of TCP, UDP, and Internet Control Message Protocol (ICMP) packets: allow to quantify the composition of the traffic in terms of observed protocols. For instance, UDP datagrams indicate the presence of signaling carried by the QUIC protocol, whereas TCP segments can represent the exchange of additional data such as multimedia material.

- Number of different IP addresses and TCP/UDP ports: the presence of different endpoints could be used to spot interaction between the smart speaker and the IVA cloud, including actions requiring to contact third-part entities or providers, IoT nodes, private datacenters, or CDN facilities.

- *per*-window Inter packet time (IPT) or packet count: allow to consider how traffic distribute within the two windows used to group packets described in Figure 7.3. Aggressiveness of the source could be used to reveal user activity or stimuli triggered by a vocal input.

- Average value and standard deviation of the TCP window: describe the behavior of the flow in terms of burstiness and bandwidth usage. Such information could lead to indications about how the IVA and its cloud exchange data.

- Average value and standard deviation of the IPT: similar to the previous case, they can be used to complete information inferred from the packet rate. For instance, the IPT could be used to recognize whether a flow is generated by an application with some real-time constraints.

- Average value and standard deviation of the packet length: hint at the type of the application layer, for instance, small packets can suggest the presence of voice-based activities requiring a low (bounded) packetization delay.

- Average value and standard deviation of the Time To Live (TTL): can be used to mark flow belonging to different portions of the network and possibly indicating that the smart speaker has been activated for a task also requiring the interaction with additional providers or actuators (e.g., IoT nodes).

We point out that many indicators are intrinsically "privacy leaking" as they allow a malicious observer to infer some information about the smart home hosting the device [4]. For instance, counting different conversations and the number of protocol data units in a time frame could reveal the presence of specific IoT nodes or the type of the requested operation, e.g., retrieving a summary of the news. At the same time, considering such values could impact on the performance of the classification framework owing to the exploitation of interactions among the different architectural components, which are difficult to forecast.

7.4 NUMERICAL RESULTS

In this section, we showcase numerical results obtained in our trials. First, we provide an overview of the collected dataset, then we present the performances of machine learning algorithms used to leak privacy of users with particular attention on the time needed for the training phase.

7.4.1 Dataset Overview

As presented in Section 7.3, the dataset has been generated in a 9-day long measurement campaign composed of three trials of 3 days with different conditions of the microphone of the smart speaker. Specifically, for the `mic_off` case, we collected 203,596 packets for a total size of 69 Mbytes. Instead, when the microphone is active, we collected 216,456 packets in the `mic_on` scenario and 282,656 packets `mic_on_noise` one, for a total size of 74 and 173 Mbytes, respectively.

Figure 7.4 depicts the average values characterizing the dataset in each scenario. It is worth noting that the average packet length and the average size of the TCP window for the `mic_off` and `mic_on` cases are very similar. Instead, for the `mic_on_noise` case, the average packet length doubles, whereas the average TCP window size halves.

Figure 7.5 depicts the correlation matrices computed for each test composing the dataset. In general, there are no strong relationships among the data. The only exception is when considering the packet length and the TTL when the microphone is sensing noise. In fact, as shown in Figure 7.5c, a positive correlation is present.

7.4.2 Classifying the State of the Smart Speaker

We now show the results obtained when trying to classify different states of the smart speaker to conduct a privacy leaking attack.

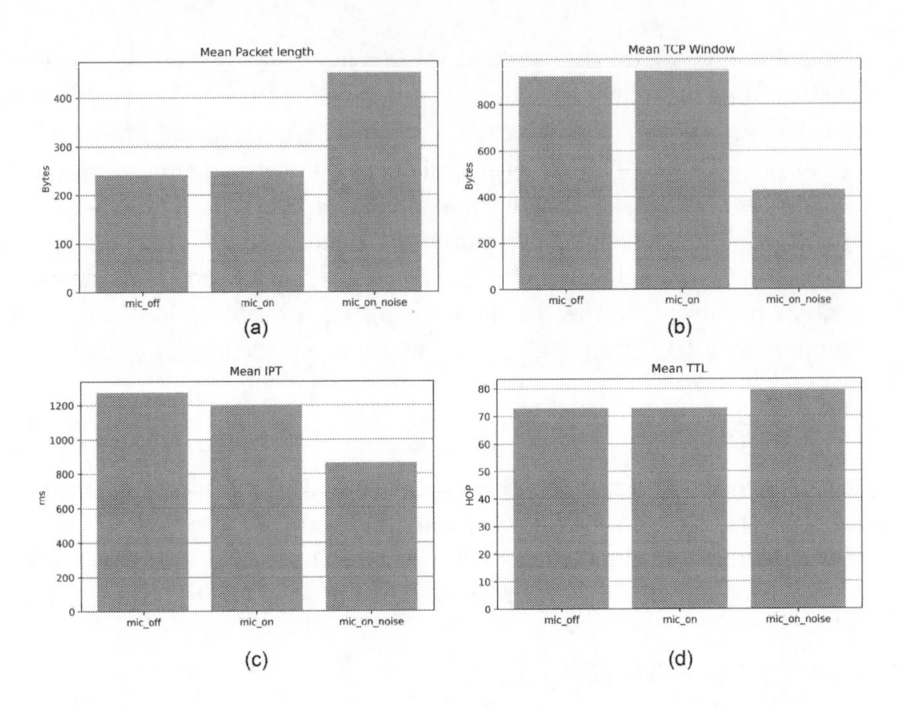

Figure 7.4 Average values for the packet length, TCP window, IPT, and TTL computed over the entire dataset. (a) Mean packet length, (b) mean TCP window, (c) mean IPT, and (d) mean TTL.

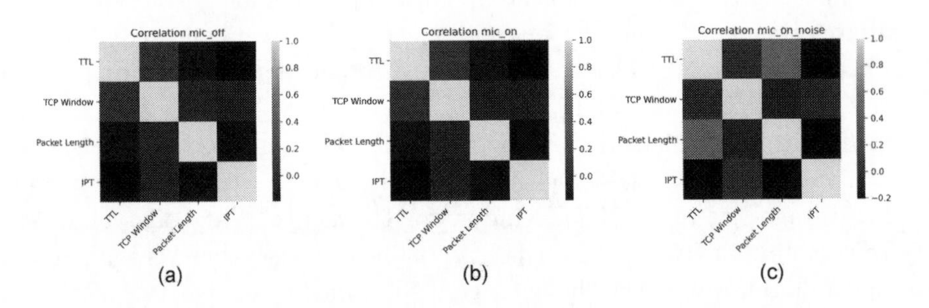

Figure 7.5 Correlation analysis on all the performed measurements for the TTL, IPT, TCP window, and packet length. (a) Correlation of `mic_off`. (b) Correlation of `mic_on`. (c) Correlation of `mic_on_noise`.

The first experiment aimed at investigating whether it is possible to identify if the microphone of the smart speaker or the device hosting the IVA is turned on or off. We point out that this can be also viewed as a sort of side channel, where the attacker can identify if humans are in the proximity of the device. In this perspective, Figure 7.6 shows the accuracy of the various classifiers adopted to infer from the traffic whether the microphone is ON or OFF, i.e., discriminate among `mic_on` or `mic_off` cases. To better understand the performances, we also investigated when the different "grouping" strategies presented in Section 7.2.2 are used to feed the machine learning algorithms.

As shown, best results are achieved by using the AdaBoost algorithm (denoted as AB in the figure). However, it is important to note that, for identifying the state of the microphone with an acceptable level of accuracy, the attacker has to collect about 500 s of traffic or 500 packets. Therefore, a real-time classification could not be possible in the sense that the attacker has to wait a non-negligible amount of time before he/she has the knowledge to launch the attack (e.g., force the physical perimeter where the smart speaker is deployed).

The second experiment aimed at discriminating between the two different behaviors of the surrounding environment, i.e., the `mic_on` and `mic_on_noise` states. We recall that such states can be used by the attacker to infer if the smart speaker operates in a silent environment or in the presence of noise, e.g., people are talking to each other or the television is turned on. In both cases, there is not a direct interaction, that is, in the case of GoogleHome, any user did not issue the "Ok Google" phrase. Then, the malicious user cannot exploit "macro" features of the traffic, such as the number of TCP connections, the IP range, or the traffic volume [4,22].

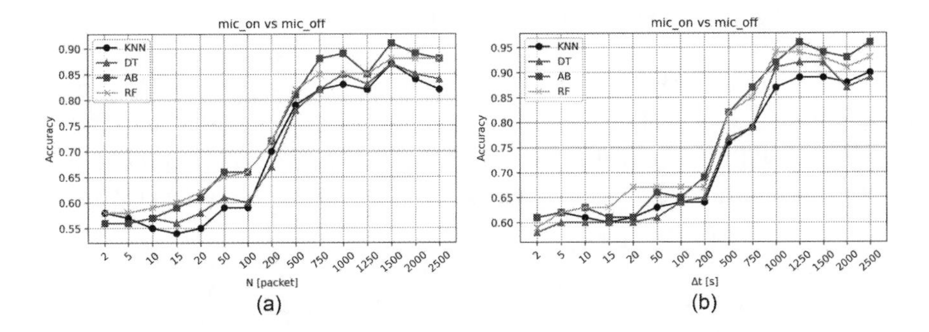

Figure 7.6 Accuracy of the classifiers for the `mic_off` and `mic_on` cases. (a) Grouping in a window of N packets. (b) Grouping in a window of Δt seconds.

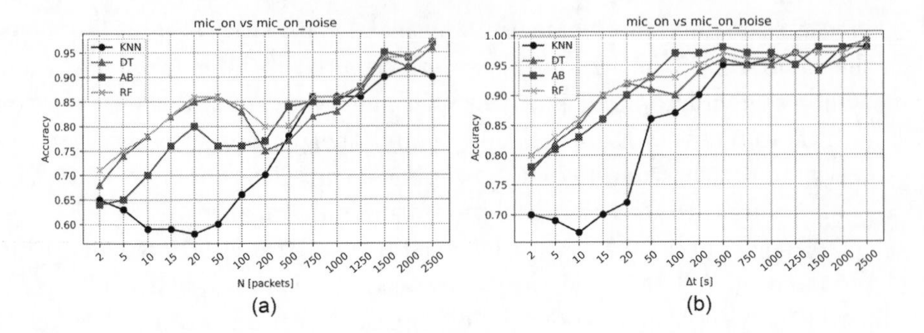

Figure 7.7 Accuracy of the classifiers for the `mic_on` and `mic_on_noise` cases. (a) Grouping in a window of N packets. (b) Grouping in window of Δt seconds.

Figure 7.7 depicts the obtained results. Compared to the previous experiment, to reach a good level of accuracy, it is sufficient to use a reduced amount of packets. As an example, for the case of the DT, good degrees of accuracy to decide whether the smart speaker is in the `mic_on` or `mic_on_noise` states are achieved by using time windows with $\Delta t = 15$ seconds or a burst of $N = 20$ packets. From the perspective of understanding the security and privacy of voice-based appliances, this result reveals a potential exploitable hazards. In fact, when the user does not directly interact with the smart speaker (e.g., the "Ok Google" phrase is not issued), the traffic generated towards the remote cloud should be the same for both the `mic_on` and `mic_on_noise` conditions. In other words, it is expected that the network traffic does not exhibit any signature. Even if we did not have access to the internals of the Google Home Mini used in our test bed, the different traffic behaviors could be due to the fact that the smart speaker is always in an "awake" mode and the selected stimuli are sent to the cloud to identify activation phrases like "Ok Google" or "Hey Siri." However, this could partially contradict the belief that such phrases are completely handled locally by the smart speaker or the IVA.

To assess the performances of the different classifiers in a comprehensive manner, Figure 7.8 shows the confusion matrices of the AB and DT classifiers when used to discriminate between the `mic_on` - `mic_on_noise` cases. It is possible to notice how the confusion matrices show the goodness of the chosen algorithms having the highest values distributed on the diagonal. Similar considerations can be done for the other techniques, but they have been omitted here for the sake of brevity.

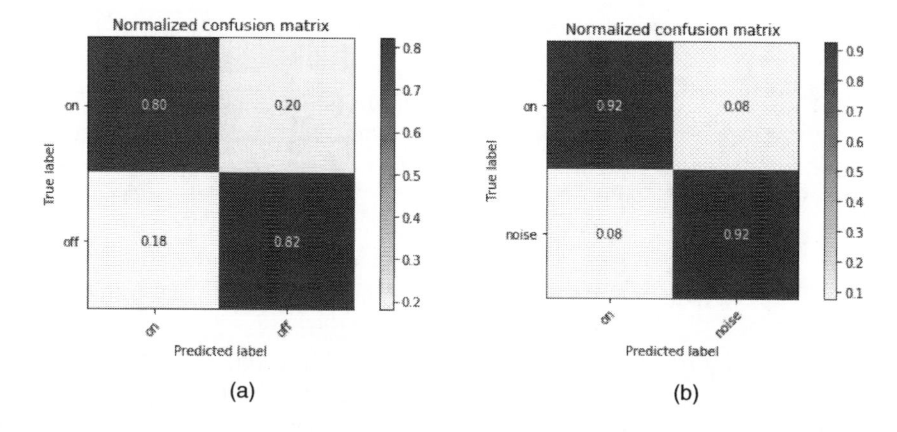

Figure 7.8 Confusion matrix for selected methods in different use cases. (a) AB: $\Delta t = 500$ - `mic_off` vs `mic_on`. (b) DT: $\Delta t = 15$ - `mic_on` vs `mic_on_noise`.

7.4.3 Analysis of the Training Time

As mentioned above, the ability for an attacker for performing a real-time classification is a key factor for successfully launching a physical attack. In this vein, the time needed to train from scratch a statistical tool is of paramount importance. Tables 7.1 and 7.2 report the training times

TABLE 7.1 Training Times for the Used Classifiers with Different Windows of Length Δt

	Training Time [s]							
	`mic_on - mic_off`				`mic_on - mic_on_noise`			
Δt [s]	kNN	DT	AB	RF	kNN	DT	AB	RF
2	1.516	0.471	11.039	44.009	1.549	0.411	13.283	41.164
5	0.710	0.290	8.354	29.012	0.582	0.276	9.902	26.203
15	0.073	0.153	4.456	15.214	0.045	0.127	5.256	11.946
50	0.010	0.052	1.675	5.293	0.012	0.037	1.960	4.557
200	0.003	0.014	0.514	1.448	0.004	0.014	0.611	1.089
500	0.002	0.006	0.269	0.550	0.002	0.004	0.312	0.463
1,000	0.002	0.004	0.188	0.287	0.002	0.003	0.214	0.295
1,500	0.002	0.003	0.159	0.212	0.002	0.002	0.192	0.241
2,000	0.001	0.002	0.146	0.190	0.002	0.005	0.168	0.212
2,500	0.002	0.002	0.137	0.178	0.001	0.002	0.163	0.193

TABLE 7.2 Training Times for the Used Classifiers with Different Windows of Length N

	Training Time [s]							
N	`mic_on - mic_off`				`mic_on - mic_on_noise`			
[Packet]	**kNN**	**DT**	**AB**	**RF**	**kNN**	**DT**	**AB**	**RF**
2	21.074	1.462	38.137	158.809	13.267	1.010	25.844	101.262
5	1.466	0.617	13.966	63.169	1.819	0.783	21.076	78.305
15	0.042	0.176	4.757	18.157	0.053	0.221	6.911	21.546
50	0.030	0.046	1.444	4.784	0.014	0.046	2.017	5.624
200	0.004	0.010	0.423	1.035	0.005	0.017	0.581	1.325
500	0.002	0.005	0.233	0.451	0.003	0.005	0.307	0.575
1,000	0.002	0.003	0.168	0.252	0.002	0.003	0.218	0.326
1,500	0.002	0.003	0.146	0.194	0.002	0.003	0.182	0.236
2,000	0.002	0.002	0.135	0.179	0.002	0.002	0.165	0.215
2,500	0.001	0.002	0.129	0.167	0.001	0.002	0.159	0.198

for each classifier when the different "grouping" strategies described in Section 7.3.1 are used.

As shown, by increasing the window (i.e., the value for Δt and N) the training time decreases. This behavior is justified by the fact that the classifiers are trained by using statistic information about the traffic provided by each single window. In other words, the larger the window, the more packets are contained, thus less information is provided to the machine learning algorithm. This can be also viewed by comparing the tables. In fact, despite the used window scheme, when the resulting amount of information is comparable, the algorithms behave in the same manner. We point out that, high values for Δt have been reported only for the sake of comparison, as they can be hardly used in real-world attacks. Yet, this allows to better understand that the traffic produced by smart speakers when in idle or with the microphone in an inactive state is limited. Specifically, the amount of time needed to collect a suitable amount of data to have statistical relevance could approach one hour.

In general, classifiers such as DT and kNN turn out to be faster in training already for small values of Δt and N. Therefore, in the perspective of launching an attack less complex algorithms, such as DT, are the preferred choice, In fact, the higher speed accounts for reduced times to launch the attack, thus reducing the chance of being detected. This is why, being able to inject additional information in the traffic to inflate the time needed to train the machine learning mechanism could be considered as a valuable countermeasure.

7.5 DEVELOPMENT OF COUNTERMEASURES

To develop suitable countermeasures and mitigation techniques, we performed a set of trials to identify the most important traffic features that can be exploited to feed machine-learning-capable threats. To this aim, we considered all the high-level traffic features introduced in Section 7.3.1. The outcome is summarized in Figure 7.9: for the sake of brevity, we only report results when using the two best classifiers (i.e., AB with $\Delta t = 500$ and DT with $N = 15$) as they represent the worst-case scenario for the victim (i.e., the attacker uses his/her best tools). To rank the importance of the features, we used the Gini Impurity and Gini Gain metrics [17]. Specifically, the Gini Impurity is a measurement of the likelihood of an incorrect classification of a new instance of a random variable, considering that the new instance is randomly classified according to the distribution of class labels of the dataset. Instead, the Gini Gain is calculated when building a DT, and it helps to determine the attribute that gives the highest amount of information about the class containing new data.

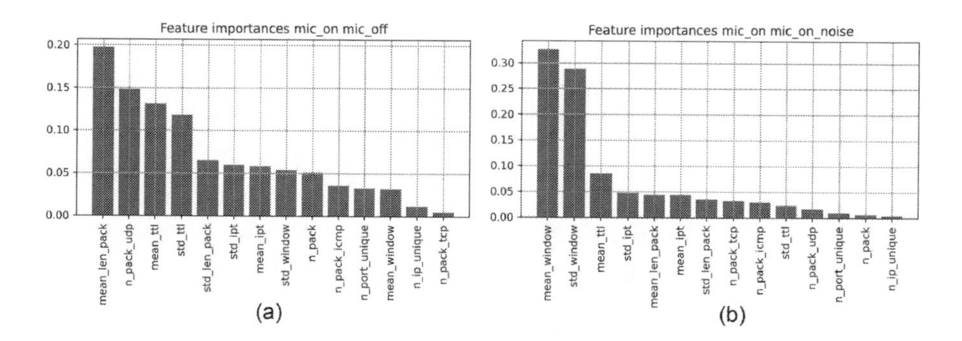

Figure 7.9 Importance of traffic features that can be used to leak information about smart speakers. (a) AB: $\Delta t = 500$ - `mic_off` vs `mic_on`. (b) DT: $\Delta t = 15$ - `mic_on` vs `mic_on_noise`.

Concerning the case aiming at discriminating between the `mic_on` and `mic_off` states, according to Figure 7.9a, we can notice that the features that affect the performances of the prediction more are the average packet length and the number of UDP datagram present within the considered window. Therefore, a possible countermeasure could be the insertion of suitable padding inside the packets so as to normalize the average length as well as the standard deviation. As regards the number of UDP packets, using a unique protocol for the transport could add another layer of privacy. Besides, Figure 7.9b reports results when a statistical tool has to recognize the `mic_on` and `mic_on_noise` states. In this case, the best features to be taken into account are the average value and the standard deviation of the TCP window, the average TTL, and standard deviation of the packet length. Consequently, a possible mitigation technique could be the alteration of the traffic with the insertion of appropriate "noise," for instance, by exploiting some form of traffic camouflage or morphing [15,24].

Another possible countermeasure could exploit a sort of security-by-design approach. Specifically, as shown in Table 7.1, the reduced amount of traffic produced by the protocol architecture used for the communications between the device and the remote datacenter could make it difficult to collect the needed amount of data in a limited time frame. Therefore, further reducing the traffic could not only represent an improvement in terms of bandwidth optimization but can also partially void the effectiveness of machine-learning-capable attacks. At the same time, avoiding fixed traffic patterns, for instance, by deploying port hopping or the aforementioned morphing mechanisms, can be an additional countermeasure built in the software layers responsible of implementing the network services for exchanging data with the IVA.

Lastly, we underline the importance of this kind of analysis also in the perspective of understanding the risks of machine-learning-capable privacy threats. In fact, as shown in Figures 7.4 and 7.9, the investigation of this class of threats can be used to anticipate the attacker. For instance, it turns out that the TCP window can be considered a sort of privacy-leaking side channel, which transmits a bit describing the state of the surrounding environment. Consequently, such a behavior should be taken into account to prevent attacks or to deploy proactive defense mechanisms to poison the database of the attacker (e.g., in an adversarial machine learning flavor) or to limit the effectiveness of reconnaissance campaigns.

7.6 CONCLUSIONS AND FUTURE WORKS

In this chapter, we investigated the feasibility of adopting machine-learning-based techniques to breach the privacy of users interacting with smart speakers or voice assistants. Different from other works discovering the presence of the user via intrinsically privacy-leaking activities (e.g., the activation of an IoT node and the related traffic flow), we concentrated on discriminating how the internal microphone is used. Results indicate the effectiveness of our approach, thus making the management of silence and noise *époque* as major privacy concerns. Therefore, suitable traffic morphing or protocol manipulation techniques should be put in place within the device or, at least, in home routers as to reduce the attack surface that can be exploited by malicious entities.

Future work will aim at refining our framework by considering smart speakers from other vendors. Besides, we are working towards the implementation of a sort of "warden" able to normalize traffic generated towards the IVA cloud.

Notes

[1]https://voicebot.ai/2019/04/15/smart-speaker-installed-base-to-surpass-200-million-in-2019-grow-to-500-million-in-2023-canalys/.

[2]https://store.google.com/product/google_home.

[3]https://www.apple.com/homepod/.

[4]https://www.amazon.com/echodot.

[5]https://developer.amazon.com/alexa.

[6]https://developer.apple.com/siri/.

[7]https://assistant.google.com/.

[8]https://store.google.com/it/product/google_home_mini.

[9]https://play.google.com/store/apps/details?id=com.google.android.apps.chromecast.app.

[10]https://www.wireshark.org/docs/man-pages/tshark.html.

[11]http://scikit-learn.org/.

Bibliography

[1] Abbas Acar, Hossein Fereidooni, Tigist Abera, Amit Kumar Sikder, Markus Miettinen, Hidayet Aksu, Mauro Conti, Ahmad-Reza Sadeghi, and A. Selcuk Uluagac. Peek-a-Boo: I See Your Smart Home Activities, Even Encrypted! 2018, http://arxiv.org/abs/1808.02741.

[2] Efthimios Alepis and Constantinos Patsakis. Monkey Says, Monkey Does: Security and Privacy on Voice Assistants. *IEEE Access*, 5:17841–17851, 2017.

[3] Yousef Amar, Hamed Haddadi, Richard Mortier, Anthony Brown, James Colley, and Andy Crabtree. An Analysis of Home IoT Network Traffic and Behaviour. 2018, http://arxiv.org/abs/1803.05368.

[4] Noah Apthorpe, Dillon Reisman, and Nick Feamster. A Smart Home is No Castle: Privacy Vulnerabilities of Encrypted IoT Traffic. 2017, http://arxiv.org/abs/1705.06805.

[5] Lei Bai, Lina Yao, Salil S. Kanhere, Xianzhi Wang, and Zheng Yang. Automatic Device Classification From Network Traffic Streams of Internet of Things. In *43rd Conference on Local Computer Networks*, pages 1–9. IEEE, 2018.

[6] Prasenjeet Biswal and Omprakash Gnawali. Does QUIC Make the Web Faster? In *2016 IEEE Global Communications Conference*, pages 1–6. IEEE, 2016.

[7] Leo Breiman. Random Forests. *Machine Learning*, 45(1):5–32, 2001.

[8] Andrea Cardaci, Luca Caviglione, Alberto Gotta, and Nicola Tonellotto. Performance Evaluation of SPDY Over High Latency Satellite Channels. In *International Conference on Personal Satellite Services*, pages 123–134. Springer, 2013.

[9] Luca Caviglione. A First Look at Traffic Patterns of Siri. *Transactions on Emerging Telecommunications Technologies*, 26(April):664–669, 2015.

[10] Luca Caviglione, Mauro Coccoli, and Alessio Merlo. A Taxonomy-based Model of Security and Privacy in Online Social Networks. *International Journal of Computer Sciences and Engineering*, 9(4):325–338, 2014.

[11] Luca Caviglione, Maciej Podolski, Wojciech Mazurczyk, and Massimo Ianigro. Covert Channels in Personal Cloud Storage Services: The case of Dropbox. *IEEE Transactions on Industrial Informatics*, 13(4):1921–1931, 2016.

[12] Mauro Conti, Nicola Dragoni, and Viktor Lesyk. A Survey of Man in the Middle Attacks. *IEEE Communications Surveys & Tutorials*, 18(3):2027–2051, 2016.

[13] Bogdan Copos, Karl Levitt, Matt Bishop, and Jeff Rowe. Is Anybody Home? Inferring Activity from Smart Home Network Traffic. In *IEEE Security and Privacy Workshops*, pages 245–251. IEEE, 2016.

[14] Wenrui Diao, Xiangyu Liu, Zhe Zhou, and Kehuan Zhang. Your Voice Assistant is Mine: How to Abuse Speakers to Steal Information and Control Your Phone. In *Proceedings of the 4th ACM Workshop on Security and Privacy in Smartphones & Mobile Devices*, pages 63–74. ACM, 2014.

[15] Kevin P. Dyer, Scott E. Coull, Thomas Ristenpart, and Thomas Shrimpton. Peek-a-Bboo, I Still See You: Why Efficient Traffic Analysis Countermeasures Fail. In *IEEE Symposium on Security and Privacy*, pages 332–346. IEEE, 2012.

[16] Marcia Ford and William Palmer. Alexa, Are You Listening to Me? An Analysis of Alexa Voice Service Network Traffic. *Personal and Ubiquitous Computing*, 23(1):67–79, 2019.

[17] Trevor Hastie, Robert Tibshirani, and Jerome Friedman. *The Elements of Statistical Learning: Data Mining, Inference, and Prediction.* Springer, New York, 2009.

[18] Shancang Li, Shancang Li, Kim-Kwang Raymond Choo, Qindong Sun, William J. Buchanan, and Jiuxin Cao. IoT Forensics: Amazon Echo as a Use Case. *IEEE Internet of Things Journal*, 14(8):1–1, 2015.

[19] Wojciech Mazurczyk and Luca Caviglione. Information Hiding as a Challenge for Malware Detection. *IEEE Security Privacy*, 13(2):89–93, 2015.

[20] Yair Meidan, Michael Bohadana, Asaf Shabtai, Juan David Guarnizo, Martín Ochoa, Nils Ole Tippenhauer, and Yuval Elovici. ProfilIoT: A Machine Learning Approach for IoT Device Identification Based on Network Traffic Analysis. In *Proceedings of the Symposium on Applied Computing*, pages 506–509. ACM, 2017.

[21] Richard Mitev, Markus Miettinen, and Ahmad-Reza Sadeghi. Alexa Lied to Me: Skill-based Man-in-the-Middle Attacks on Virtual

Assistants. In *Proceedings of the 2019 ACM Asia Conference on Computer and Communications Security*, pages 465–478. ACM, 2019.

[22] Antônio J. Pinheiro, Jeandro de M. Bezerra, Caio A. P. Burgardt, and Divanilson R. Campelo. Identifying IoT Devices and Events Based on Packet Length from Encrypted Traffic. *Computer Communications*, 144:8–17, 2019.

[23] S. Rezaei and X. Liu. Deep Learning for Encrypted Traffic Classification: An Overview. *IEEE Communications Magazine*, 57(5):76–81, May 2019.

[24] Sabine Schmidt, Wojciech Mazurczyk, Radoslaw Kulesza, Jörg Keller, and Luca Caviglione. Exploiting ip telephony with silence suppression for hidden data transfers. *Computers & Security*, 79:17–32, 2018.

[25] Mustafizur R. Shahid, Gregory Blanc, Zonghua Zhang, and Hervé Debar. IoT Devices Recognition Through Network Traffic Analysis. In *EEE International Conference on Big Data*, pages 5187–5192. IEEE, 2018.

[26] Siraj A. Shaikh, Howard Chivers, Philip Nobles, John A. Clark, and Hao Chen. Network Reconnaissance. *Network Security*, 2008(11):12–16, 2008.

[27] Charles V. Wright, Lucas Ballard, Fabian Monrose, and Gerald M. Masson. Language Identification of Encrypted VoIP Traffic: Alejandra y Roberto or Alice and Bob? In *USENIX Security Symposium*, volume 3, pages 43–54, 2007.

[28] Yuchen Yang, Longfei Wu, Guisheng Yin, Lijie Li, and Hongbin Zhao. A Survey on Security and Privacy Issues in Internet-of-Things. *IEEE Internet of Things Journal*, 4(5):1250–1258, 2017.

CHAPTER 8

Hardware Security in the Context of Internet of Things: Challenges and Opportunities

Pranesh Santikellur and Rajat Subhra Chakraborty
Indian Institute of Technology Kharagpur

Jimson Mathew
Indian Institute of Technology Patna

CONTENTS

THE INTERNET OF THINGS (IoT) promises to revolutionize the quality of human life in the near future through the integration of billions of smart, interconnected devices ("things") into the fabric of daily life. The primary enabling technologies for IoT are sensing, wireless communication, computing, and actuation, all of which have their associated challenges. However, the most formidable obstacle envisaged in the way of widespread deployment of IoTs is the lack of sufficient security measures in most IoT infrastructure. Since most IoT devices have severe energy and computational resource constraints, it becomes difficult or infeasible to implement full-scale cryptographic algorithms and protocols. Also, many cryptographic schemes while being mathematically secure, often fall prey to implementation-specific attacks, such as side-channel attacks (SCAs), and untrusted electronic components. In this chapter, we explore the unique challenges of secure IoT implementations, especially from the perspective of hardware security, and their solutions. We survey the reported attacks and vulnerabilities, describe the proposed solutions, and provide directions to the most important open problems. We put particular focus on *Physically Unclonable Function* (PUF) circuits, a promising hardware security primitive in the context of lightweight IoT security solutions.

8.1 INTRODUCTION

In computing and connected systems, in particular, significant new challenges are being faced because of the rapid evolution in *Cyber-Physical Systems* (CPS). One of the major thrusts towards CPS is the evolution of *Internet of Things* (IoT) [1]. IoT is a new computing paradigm that relies on pervasive and smart computing, transparently distributed across globally connected ultra-heterogeneous computing devices [1,2]. It provides seamless connectivity between devices, and processing of the generated data in a context-sensitive manner. The IoT will potentially revolutionize how the users exploit data, as well as the interaction between computing systems and the physical world.

The major challenge for networked systems is how to handle billions of devices generating volumes of real-time data and their security aspects. A taxonomy of IoT security is shown in Figure 8.1, covering multiple security considerations from the overall system perspective [1,3]. The following typical vulnerabilities can be envisaged:

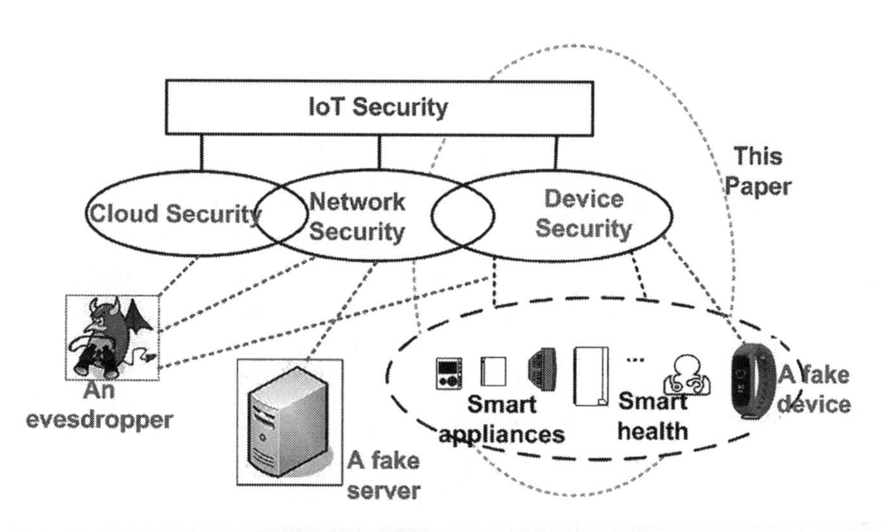

Figure 8.1 Taxonomy of IoT security.

1. An eavesdropper listening in on data or commands on the network/cloud can reveal confidential information about the operation of the infrastructure.

2. A fake server, in the absence of proper authentication mechanisms, can send incorrect key sequences to trigger random events at arbitrary IoT devices.

3. A fake IoT device, masquerading as an authentic device, can inject fake measurements to disrupt the overall measurement process, and trigger malfunctioning of the indented application [4].

The system designer has to be capable of dealing with the heterogeneity of interactions and information sharing between different participating devices and humans. In addition, the system has to be resistant against different major attack scenarios that may significantly vary in both motives and execution [1,3]. Adversaries could be very strategic in planning sequential attack actions. Finally, the defense mechanisms should react in real time and adaptive to the concrete actions of the attackers, simultaneously taking into account that there are typically limited resources (e.g., battery energy) available to execute each step, or decision made during the process. A motivational example is presented next to elucidate the complications involved in devising secure IoT systems.

8.1.1 Motivational Example

Most IoT devices work under severe energy availability constraints, but on the contrary, energy-expensive operations such as ubiquitous connectivity are indispensable. In addition, IoT devices are expected to have long life in potentially harsh environments, and many times, battery replacement is either not possible or extremely inconvenient. For example, consider a smart pacemaker implanted in a patient's heart which has the ability to monitor and, if necessary, to communicate abnormal heart muscle rhythm to the patient's physician over a radio link. The physician, in turn, can reprogram the pacemaker over the same radio link to generate pulses most suitable for the patient's detected heart condition. It is obvious that maintaining a radio link is essential, but changing the battery of the implanted pacemaker is extremely inconvenient. This type of severe energy availability constraint might force the designers of the pacemaker to compromise on the security aspect of the pacemaker's communication scheme. To save energy by avoiding often computation-intensive cryptographic operations, either the device will perform unencrypted (plaintext-mode) communication over the radio link, or use rudimentary encryption schemes (e.g., XOR-ing with a fixed key without any authentication of the communicating party), which make them vulnerable to many potentially life-threatening security issue. This attack has been known since at least 2008 [5], and one of the most widely reported solutions to this problem involves "friendly jamming" of radio signals corresponding to unauthorized commands [6]; however, there are several more challenges that need to be overcome to reach a satisfactory practical solution to this threat. One of the major hurdles has been the lack of security awareness among designers and manufacturers of smart medical electronics and many other IoT products – traditionally, security has always been an afterthought rather than being a design goal. Fortunately, the situation is slowly changing with security standards being developed for smart medical electronics/health monitoring and devices complying to these security guidelines being available in the market [5].

8.1.2 Contributions and Organization of This Chapter

The overarching goal of this chapter is to link research in hardware security with IoT security, concentrating on the challenges, and emerging solutions. IoT devices pose unique challenges related to security, trust, and privacy, and concentrating on the software-based protection schemes often leaves the hardware vulnerable, allowing for new attacks on the hardware and the hardware-software interface. Primarily, we focus on the following threats in the context of IoT security (Section 8.2) **Hardware Trojan Horse (HTH)** and **Counterfeit Integrated Circuits**. We describe different techniques to mitigate these threats, in particular, **testing techniques for HTH detection** and **image processing techniques for counterfeit integrated circuit detection**. We also describe the **role of PUF, a promising hardware security primitive, to develop secure lightweight communication protocols for IoT devices** (Sections 8.3–8.3.5). While several works have previously explored various aspects of IoT security [7–9], but the above-mentioned topics have not been considered in a consolidated form in a single chapter, to the best of our knowledge. We point to some open research issues and directions of future research in Section 8.4, and conclude in Section 8.5.

8.2 THREATS ON IoT IMPLEMENTATION

In this section, we discuss in detail two primary hardware-related threats on IoT implementations, each being the result of reduced trust levels in electronics design and manufacturing: (a) *HTHs* and (b) *Counterfeit Integrated Circuits.* We also discuss the state-of-the-art techniques to alleviate these threats.

8.2.1 HTH and Countermeasures

HTHs have attracted plenty of research interest recently [10–14]. HTHs are malicious modifications in integrated circuits (ICs)which can be inserted either during the circuit design or the IC fabrication phase. HTHs can be either free-running (e.g., a counter-based "time-bomb" HTH), or one which depends on a rare logic condition to activate. Once activated, the HTH can cause disastrous functional failure or cause leakage of sensitive information. Because of their rareness of activation, inserted HTH usually evades detection by traditional post-manufacturing testing.

Motivational Example: To appreciate the impact that an inserted but undetected HTH in an IoT device might have, consider the smart pacemaker considered in Section 8.1. The pacemaker, once implanted, behaves normally. However, an inserted HTH in the pacemaker might have a counter which, after counting till a pre-determined count, makes the pacemaker malfunction by modifying the pulses generated by it, an effect that can be fatal to the person in whom the pacemaker is implanted. A secondary effect is the fast draining of battery power of the implanted pacemaker due to constant switching of the HTH counter circuit. The inserted HTH might also be activated by a pre-determined sequence of commands sent over its live radio link, in which case an adversary with knowledge of this sequence can potentially gravely harm or kill the person with the implanted pacemaker.

Till date, extensive research works have been done on the design, detection, and prevention of HTHs. The aim of this section is to make the reader accustomed to the state of the art through a comprehensive survey. We start with an elaborate study of the works addressing the classification of HTHs depending on various aspects. Understanding the possible threats due to HTHs as well as their proper classification is important as it eventually helps to devise proper countermeasures against them. A survey of state-of-the-art countermeasures will be presented subsequently. Although, in this chapter, we mainly focus on the detection of HTHs inserted through the untrusted fabrication facilities, brief reviews of other possible attack models are also presented for the sake of completeness.

8.2.1.1 Modern Integrated Circuit Design and Manufacturing Practices

It is argued that virtually every step of the IC life cycle is vulnerable to HTH attacks. At the specification phase, HTH may make its way through altered functional specification or design and security constraints. The design phase incorporates several steps, namely: behavioral description, register-transfer level (RTL) synthesis, logic synthesis, technology mapping, placement and routing, and the mask preparation. HTHs can be inserted at any of these steps by principle. However, it is reasonable to assume that the in-house specification, design, and integration teams are trusted and attacks may happen mostly through the third-party components, tools, and facilities. Figure 8.2 shows the potentially vulnerable points in a supply chain with relatively reasonable assumptions. One of the possible means of the injection of HTH in the design phase is through the third-party intellectual property (3PIP) cores and cell

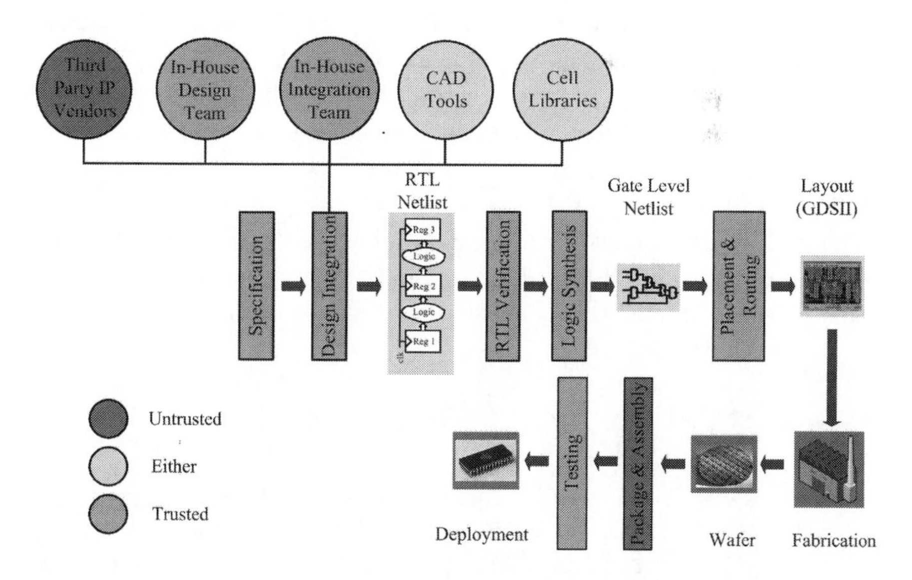

Figure 8.2 A typical IC design and supply chain with its potentially vulnerable locations. Although, in principle, all steps can be vulnerable to HTH attacks with some reasonable assumptions the in-house team and stages are considered as trusted [15]. (Adapted with modifications.)

libraries which are made available in different level of abstractions. Also, malicious computer-aided design (CAD) tools may be devised, with the power to insert HTHs automatically [16]. Unlike the design phase, where the team is assumed to be in-house and trusted, the fabrication facility is considered as untrusted in most of the cases. The adversary is assumed to modify or even replace the mask to insert the Trojan. Finally, in the assembly phase, the *Printed Circuit Board* (PCB) may be maliciously modified. The testing phase, unlike other phases, cannot be used to insert HTH. But considering multi-level attacks [17], where one or more attackers may have a malicious collusion, the testing phase may also be exploited, for example by skipping the testing of the infected chips.

8.2.2 HTH Classification Based on Triggering Mechanism

Hardware Trojans can be classified in many ways, and we describe one of the most widely accepted classifications based on HTH structure and impact. Generally, HTH comprises of two sections: a *trigger* for its initiation and a *payload* to cause the undesired function. But few types of HTHs exist without trigger [18]. The triggers can be launched from the internals of circuits such as the output of on-chip sensor [19], as well as external such as particular input. However, it is reasonable to assume that the attacker will retain some control over the HTH, and in the process, she will make it externally triggered. To be undetectable during the test phase and most of the deployment phase, HTHs are designed to get activated at extremely rare conditions [20,21]. Based on structure, classification of HTHs was mentioned in [15] (refer to Figure 8.3). This classification considers digital, analog, and hybrid (combination of digital and analog) HTHs, e.g., analog trigger mechanism with digital payload [22]. Digital HTHs constitute the most common threats.

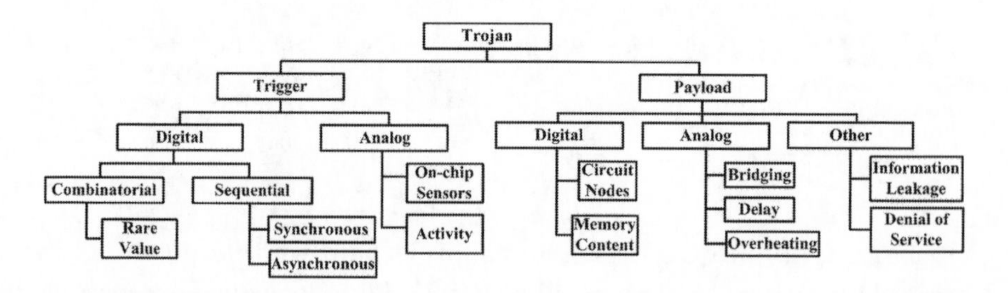

Figure 8.3 Trojan designs: a structural classification [15]. (Adapted with modifications.)

Figure 8.4 shows several possible models of Trojan triggers. The digital trigger mechanism can be either combinational or sequential. To ensure that triggers for the HTH are adequately rare, HTHs are combined with low controllable nets in circuits, so that it launches in synchrony with low controllability values. Figure 8.4a shows the model of such a

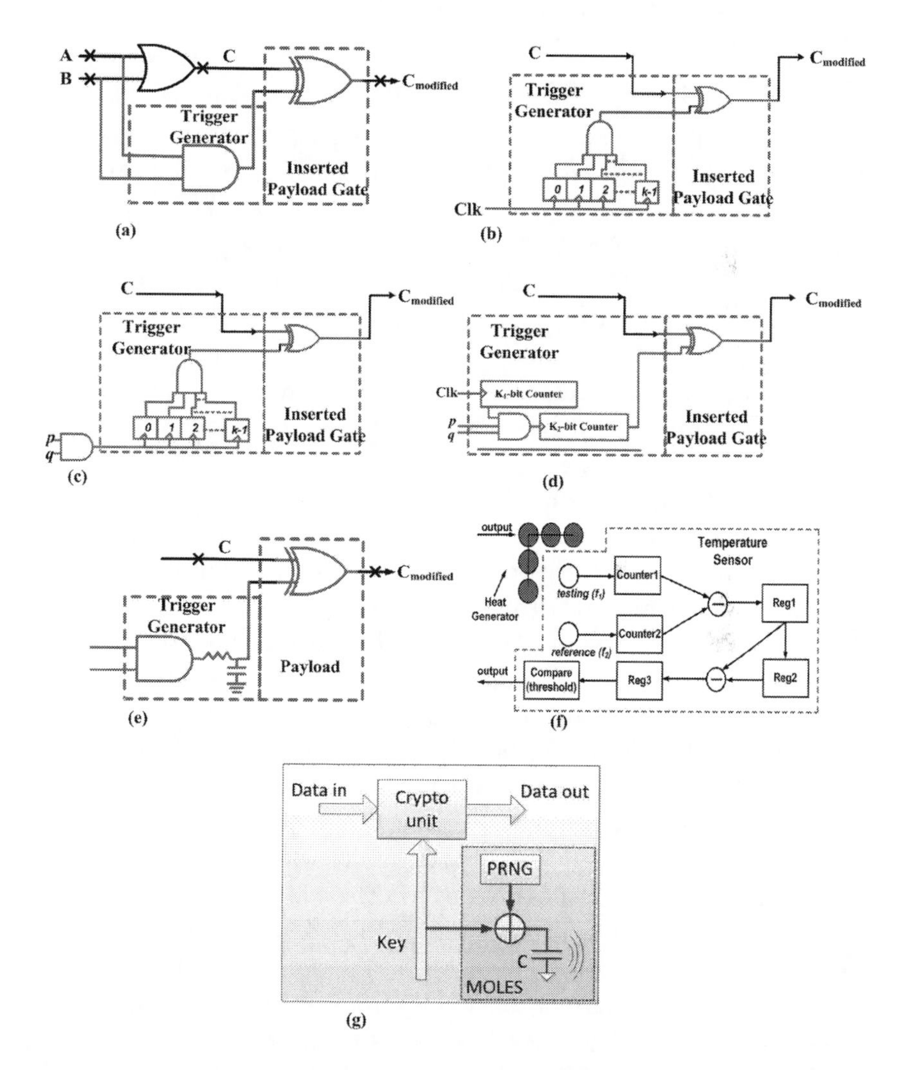

Figure 8.4 Structural models of HTHs [15,18]. (a) Combinational Trojan, (b) sequential Trojan (synchronous), (c) sequential Trojan (asynchronous), (d) hybrid Trojan, (e) analog Trojan triggered based on logic values, (f) analog Trojan triggered based on circuit activity, and (g) side-channel Trojan.

Trojan. On the other hand, Sequential HTHs may be classified into two general types: synchronous and asynchronous. Figure 8.4b shows synchronous HTHs which involves large counters, and Figure 8.4c shows asynchronous HTHs where the trigger activates only on the occurrence of one or more rare events. Figure 8.4d, e, and g show examples of hybrid Trojan and a triggerless Trojan, respectively.

8.2.3 Impact of Undetected HTH Insertion

Most of the HTH effects can be abstracted in the form of some hardware faults. Faults caused by HTHs may result in catastrophic functional failure or temporary or permanent denial of service (DoS) [23]. Uncontrolled parametric degradation or drainage of battery is also feasible by means of analog HTHs. Analog Trojans, also referred to as *reliability Trojans*, attack on different process steps trading off the reliability of the chips [24,25]. The activation of analog HTHs usually happens with the aging of the chips, and analog HTHs in turn cause wear-out failures in complementary metal–oxide–semiconductor (CMOS) ICs. The common wear-out failure mechanisms which the HTHs leverage include hot carrier-induced fluctuation, electromigration, time-dependent dielectric breakdown, and negative bias temperature instability. In [25], the authors discuss the list of critical process parameters that can be used for reliability attack and also the manifestation of reliability attacks. The chapter also discusses stress testing coupled with increased excitation for identification of reliability tampering. In work described in [26] proposes HTH requiring the change in the dopant masks.

HTH payloads can also be engineered specially for the target application classes. The most prominent examples are the cryptographic algorithms, where the target is to leak the secret key. Leakage of cryptographic keys is possible by injecting faults with a HTH [17]. However, leaking of secret is still possible by HTHs without injecting any fault. Example of such HTH can be found in [18], where a HTH was utilized to leak the secret key through its power dissipation signature ("power side-channel"). The structure of this HTH circuit is shown in Figure 8.4g. In another example, a secret key was leaked through wireless channels [27]. HTHs inserted in general purpose processors can be exploited by malicious software [28–32]. For example, most modern processors use "no-execute" (NX) bit technology to mark certain segments of memory as non-executable; the processor will then decline to execute any code living in these regions of memory. This NX bit thus serves as protection bit for execution, utilized to keep certain sorts of malicious programs away to execute from its space. Attacks like "Rakshasa" [33] disables this NX bit, enabling the free access to protected areas. Other similar backdoors such ad disabling the secure boot with rare inputs are prevalent [29].

Devices in every IoT network domain would usually employ a suitable Identity Management framework [34]. Assume that one of the devices has a built-in hardware-based authentication mechanism. e.g., based on a PUF circuit. If the device authentication mechanism gets compromised for a single device, then the security of the entire connected system is at risk. Therefore, the identity of the devices in an IoT domain must not be compromised. In other words, any impersonation, tampering, or HTH attack on one of the IoT devices that is not properly protected by may compromise the system security.

8.2.3.1 Trojans on IoT Infrastructure

Field Programmable Gate Arrays (FPGA) are gradually getting popular in the field of embedded systems. Like the application-specific integrated circuits (ASICs), FPGAs are also found vulnerable against HTHs. However, the attack model significantly varies between these two platforms. Unlike ASICs, FPGA HTHs are assumed to be inserted at the user side through the manipulation of the configuration bitstream, which is the means of mapping a design on this platform. Although leading FPGA vendors like *Xilinx* and *Altera* provide bitstream encryption as a security measure, recently they have been found to be vulnerable [35,36]. The bitstream encryption key can be compromised utilizing side-channel analysis. However, the bitstream formats are often kept proprietary providing an inherent obfuscation. Nevertheless, in recent works, such obfuscated bitstreams were successfully reverse-engineered [37]. Once the bitstream format gets revealed it is not difficult to insert a Trojan by manipulating it as shown in [38]. In this work, the Trojan is inserted into unused logic blocks. Till date, no successful countermeasures have been proposed for such attacks.

Another possible attack model exploits the *Dynamic Partial Reconfiguration* (DPR) primitive in most of the FPGAs to insert an HTH [39]. DPR is an extremely useful feature supported by modern FPGAs that allow modifications to be made to a deployed FPGA, without taking it offline. However, indiscriminate application of this procedure might allow HTHs to be easily inserted to security-sensitive FPGA-based circuitry. An attacker may reset the FPGA after the attack just by resetting the

DPR block removing the Trojan and eventually bypassing all the diagnosis events. In [39], this technique was successfully utilized to launch a fault attack on a Advanced Encryption Standard (AES) crypto-core over an Ethernet connection. However, careful access control on the DPR facility may minimize the risk of such attacks [40], which will be discussed again later in the chapter.

The work in [41] proposes a case study on implanting an HTH in a commercial product that weakens the overall system security. The target device is *Data Traveler 5000*, a FIPS-140-2 level 2 certified Universal Serial Bus (USB) flash drive from Kingston. The target device uses Advanced RISC Machine (ARM) central processing unit (CPU) to perform the device authentication and the Xilinx FPGA for encryption and decryption of the user data. The device uses ARM CPU as master and Xilinx FPGA as a slave device. The complete attack involved bitstream file format reverse engineering, IP core analysis, and hardware configuration manipulation. The bitstream manipulation alters the exploited AES-256 algorithm to convert it into a linear mapping, thereby making cryptanalysis possible. The trick is to patch or reprogram the bitstream without disturbing the integrity check. The readers are requested to refer the work [41] for physical attack to reveal the bitstream and modification of AES-256 core.

A relatively recent work [42] has explored HTH attack on a neural network hardware. With the increasing popularity of sophisticated machine learning (ML) frameworks like Deep Learning (DL), DL-enabled IoT devices hold immense promise in many application domains. IoTs being resource-constrained, and DL being an extremely resource-demanding framework, a lot of studies have been carried out to make DL suitable for IoT devices. The main success of deep learning in recent years is credited to their image and audio recognition capabilities. Deep Convolutional Neural Network (CNN) is the main workhorse behind audio and visual recognition tasks. Several works [43–45] discuss implementation of a fast CNN inference in IoTs. The neural network approximates the function using weights and biases. The authors of [42] propose the idea to insert the HTH into the neural network model by maliciously changing the model weights, so that the neural network will malfunction when the HTH is triggered. One of the methods to maliciously change the weights is to use adversarial examples. The author introduces a variant of the Jacobian-based Saliency Map Attack (JSMA) [46] which requires minimum hardware changes for an effective HTH. Several other works [47–50] discuss HTH insertion on neural networks under different conditions. For the interest of the reader, few more recent Trojan-based attacks on IoT are mentioned with references below.

The work in [37] presents and FPGA HTH attack targeting the S-box implemented in the look-up table without the knowledge of internal routing. The other USB-related attacks are reported here [51–53]. The work in [54] describes a new HTH-enabled DoS attack on power budgeting scheme in network-on-chip. The power request packet is made to travel through a HTH-infected router, and hence to allow the modification of request packet. The authors [55] carried out an HTH attack on an FPGA-based random number generator. HTH attacks on cryptographic IP cores are well studied by the authors of [27,37,56,57]. The attacks on wireless cryptographic ICs using silicon and mixed-signal System On a Chips (SoCs) are detailed in [58,59], respectively. Static random-access memory (SRAM) attacks and embedded memory attacks using HTH are studied in [60,61], respectively.

8.2.4 Countermeasures against HTHs

There are several challenges related to the formulation and implementation of a proper detection scheme. HTHs are carefully engineered to evade all such standard pre-silicon and post-silicon tests. Also, as it can be seen from the discussions of the previous subsection, the feasible space of HTHs is prohibitively large. Even if one considers only the combinational Trojans in a circuit with say n gates, the total possible number of Trojans is $\mathcal{O}(2^n)$. This makes deterministic and exhaustive testing infeasible. Again, due to their negligibly small circuit footprints to remain stealthy, HTHs do not add significant effects to the physical parameters (e.g., circuit propagation delay, supply current, leakage power, dynamic power, etc.) of the circuits. Further, such effects, even if they exist, get masked due to the presence of random manufacturing process variation induced noise. Prevention and runtime detection of HTHs are also being currently researched. However, considering the enormous versatility of the adversary, it is extremely difficult to derive a truly secure on-chip countermeasure, as the added hardware resources implementing the scheme may themselves get tampered.

8.2.4.1 HTH Attack Models and Countermeasure Classification

Research efforts for the protection of HTHs involve multiple disciplines. Before going into the details of the HTH countermeasures, it is necessary to fix the proper attack models for HTHs. Virtually every step of the

IC manufacturing process may become a target of HTH attack. In literature most of the detection mechanisms assume either the foundry or the procured 3PIPs to be malicious, which give rise to two prominent attack models, namely *malicious foundry model* and *malicious 3PIP model*. These two attack models are significantly different. The malicious foundry model often assumes the existence of a *golden model* of the IC under test. Such golden models may exist in various levels of abstractions, for example RTL, gate-level netlist, or even in the form of an IC instance with its trust verified. However, certain golden models, especially the trusted IC instances can only be acquired through destructive reverse engineering or exhaustive testing which is almost practically infeasible. As a result, detection techniques relying on such golden ICs often utilize detailed simulation models which are assumed to mimic the behavior of the IC well [62]. Also, recently golden model free detection techniques are getting prominent in the malicious foundry model of HTH detection [63]. On the contrary, the 3PIP-based attack model does not assume the availability of any golden model. Only the availability of the IP in some soft form is assumed (RTL, gate-level netlist etc.) with some standard specifications. However, slight variations of this attack model exist.

Based on attack models, the classification of various HTH countermeasures is presented in Figure 8.5. First, we shall describe the mechanisms in the malicious foundry model. Gradually, survey of the countermeasures for the 3PIP attack model will be presented.

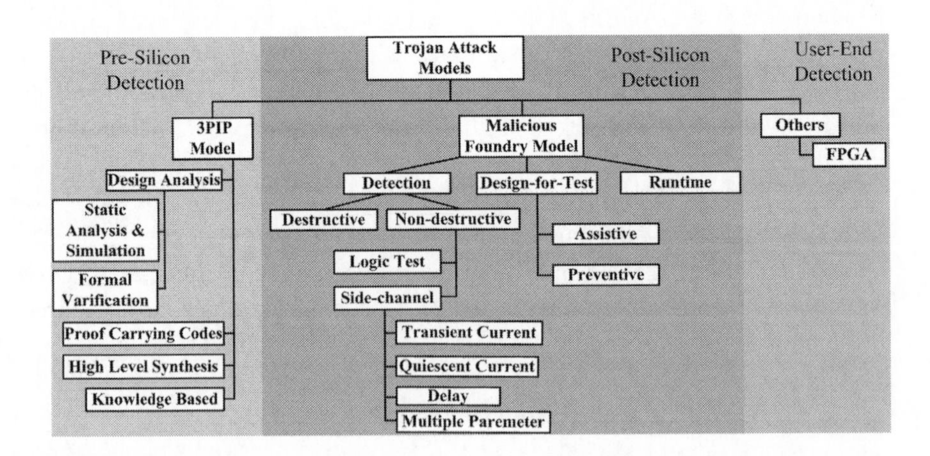

Figure 8.5 Classification of HTH countermeasures with respect to their attack models.

8.2.5 Countermeasures in Malicious Foundry Model

The possible countermeasures in the malicious foundry model can be broadly classified into three classes: (a) detection approaches; (b) Design-for-Test (DFT) circuit insertion, and (c) runtime approaches. Detection approaches in the malicious foundry model typically attempt to detect HTHs in packaged ICs. Detection techniques are fundamentally of two types: *destructive* and *non-destructive*. Non-destructive HTHs detection approaches are useful for the identification of HTHs during post-manufacturing test, and fall into two main categories: (a) logic testing based and (b) side-channel analysis based. Logic testing based concentrates on the generation of test vectors to activate rare activation conditions for likely HTHs and to propagate the anomalous effect to the primary output of the circuits. On the other hand, side-channel analysis based approaches focus on analysis of physical parameters like power consumption and circuit propagation delay to identify the possible existence of HTH. The DFT techniques attempt to prevent HTH insertion or help in the detection schemes. Runtime validation approaches are based on online checking methods where changes in the consistency of circuit functionality is monitored to detect and possibly shield against HTHs.

8.2.5.1 Destructive Detection Approaches

Destructive techniques for HTH detection [15,64] selects a small sample of manufactured ICs, and then each sample IC is de-metallized using chemical mechanical polishing (CMP), followed by image reconstruction and analysis with scanning electron microscope (SEM) [11]. Although such a method possibly can detect any inserted HTH, the time and computational effort required for a reasonable complex IC may be significant, from several weeks to months. A major drawback of these techniques is that the IC gets destroyed at the end of this invasive process, and yet it does not guarantee HTH detection, since it is possible that the adversary will only modify the design of a few IC instances in a batch of manufactured ICs.

8.2.5.2 Non-Destructive Detection Approaches

Two prominent classes of HTH detection approaches are the logic testing and side-channel analysis. In the following we survey both of these paradigms.

A. Logic Testing

Logic testing based HTH detection attempts to activate injected HTHs by automatic test pattern generation (ATPG), applying the generated test vectors and observing the impact. Besides, a significant number of the side-channel based methodologies expect the availability of such rare test vectors to amplify their precision in the detection. Hence, logic testing tools are also used by side-channel based methods. For example, in [65], a guided test generation strategy for small but critical parts of a design was devised, which was found to increase the toggling rate of the embedded HTH circuit, hence increasing its detection probability through power side-channel. With the same target of amplifying HTH effects in power signals, the sustained vector technique in [66] attempts to minimize the original circuit activity, so that any additional circuit activity due to an inserted HTH can be highlighted. In a slightly different flavor, the work in [67] addressed a similar issue. Here, the circuit was first partitioned into smaller sub-circuits, called "regions," restricted across clock boundaries. Once the regions were identified, the technique attempted to generate a minimal test set to maximize the circuit activity in each region. In another work [68], a scalable side-channel approach was developed, in conjunction with a HTH-targeted test generator.

In spite of the fact that ATPG for HTH detection is an extremely useful and prime line of research, there are a few major difficulties involved with it. As discussed before, because of the enormously large HTH design space, it is infeasible to consider the entire HTH design space and then to create a reasonably compact set of effective test vectors. Alternatively, a statistical ATPG approach which uses feasible HTH sampling for generation of a compact set of test vectors was proposed in [20], referred to as MERO ("Multiple Excitation of Rare Occurrence"). The principle line of thought behind MERO was to produce a set of compact test vectors that can maximize the Trojan detection scope with least time and cost. Results were reported for both Trojan trigger coverage and Trojan detection coverage for a large number of test cases. Logic testing approach is ineffective in detecting large HTHs. However, as mentioned previously, in such a scenario, they can be combined with side-channel based techniques.

B. Side-Channel Analysis

Side-channel analyses for Trojan detection is the most widely explored area in this community [69–71]. These methods are generic in nature and can be easily utilized for analog HTHs too. The main assumption behind these methods is that the malicious inclusion of extra circuitry during design or fabrication will definitely affect the physical characteristics of the circuit, e.g., power, delay, electromagnetic (EM) emission etc. Measurable deviations in both the idle (static) and transient (switching) current profiles of the circuit can be caused by the inserted HTH. Similar arguments are true for the EM radiations. A delay impact may result due to either incorporation of extra logic level(s) in a circuit path or increase in capacitive load on a circuit path. The effect caused by obscure HTH in physical parameters can be watched and validated with golden circuit model to detect the presence of HTH. The effectiveness of side-channel based HTH detection techniques depends on two major factors: the environmental or measurement noise and the process variation noise. The first one is a time-varying quantity whereas the second one varies with the location inside a die or among dies. Both of these noises can mask the HTH effect making it hard-to-detect. The detection sensitivity of side-channel techniques is often quantified by two quantities, namely Signal-to-Noise Ratio (SNR) and Trojan-to-Circuit Ratio (TCR). The "signal" here is the impact of the HTH on a side-channel parameter, and the "noise" is the side-channel parameter corresponding to rest of the circuit activities, added to environmental and process noise. The TCR is a measure of how different the Trojan affected parameters are compared to the original – the precision to detect is determined by the TCR. It is difficult to accurately model the noise components involved in side-channel based HTH detection scheme. In general, the environmental and measurement noise can be removed utilizing similar setup and environmental conditions during characterization and testing. However, process variation noise is even more difficult to model accurately. Till date, many works have addressed the issue of process noise calibration, one notable among which is is [72], where a virtual probe framework was proposed to get an exact profiling of spatial variations inside the chip with low overhead cost. It estimates that parameter estimation was performed using *maximum a posteriori* (MAP) and linear programming.

Next we present notable side-channel approaches as well as their challenges and effectiveness.

1. *Static Current Analysis*: When idle, each static CMOS logic gate draws a small amount of "leakage current," which is a non-ideal effect more pronounced for sub-micron technologies. Any extra

addition of gates due to HTH might get captured by leakage current analysis as it is bound to add an extra leakage current even if it is in a dormant state. This observation gave rise to several leakage current based HTH detection techniques [70,73]. But the extremely small additional leakage current caused by the few extra HTH gates are not easy to observe due to the low TCR. At the same time, the exponential dependence of the leakage current on the transistor threshold voltage (V_T) makes the SNR even worse, as V_T is a parameter which gets largely affected by the process variation noise. Many of the proposed methods try to mitigate such unwanted effects by means of multiple supply pads on the chip [70]. Other approaches include gate-level characterization [74] and scalable versions of it based on intelligent circuit segmentation [73], eventually leading to a golden model free self-consistency analysis [75,76]. Although claimed to detect ultra small Trojans with single gates, such approaches may have limited accuracy considering the increasing effect of process variation on V_T with device scaling.

2. *Transient Current Analysis*: Transient current (I_{DDT}) analysis based HTH detection techniques assume that the switching activity caused in Trojan gates will lead to detectable extra power consumption inside the circuit. However, to make such power consumption detectable, one has to minimize both the environmental and process noise effects. The second challenge involved is to either fully or partially activate the HTH circuitry to improve the TCR. Region-based partitioning [67] and directed test vector generation [65] have been proposed to serve this purpose. On the other hand, to improve detection sensitivity further, process calibration is done by most of the techniques using statistical averaging techniques [13,68,71,77–80].

3. *Path Delay Analysis*: The main challenge in path delay analysis based HTH detection schemes is again to mitigate the effect of noise, especially process variation noise [71,81–84]. The selection of proper test vectors to excite suitable paths is also crucial. The accurate measurement of path delays often requires the addition of extra test structures inside the circuit which often increases the hardware overhead.

4. *Electromagnetic Radiation Analysis*: Recently, some works started addressing the Trojan detection problem with the EM radiation based side-channel perspective. Setting an EM probe over the chip the circuit activity dependent EM traces can be collected which are assumed to be sensitive towards HTHs. However, the research in this direction is still at a primitive stage, and only a few works on FPGA platforms can be found [85].

5. *Multiple Parameter Analysis*: In order to increase the reliability of the HTH detection process, multiple testing mechanisms are often combined together. The work in [86] proposed such a scheme, where the I_{DDT} side-channel parameter was utilized along with the maximum operating frequency ($F_{\max}$) to enhance the detection reliability and sensitivity. In [87], the problem of multiple parameter analysis was formulated as a multimodal sub-modular problem, to systematically analyze the effect of different parameters. Further enhancements of such methods are possible with the choice of proper test vectors.

6. *Golden Model Free Analysis*: Although most of the detection methods in the malicious foundry model assumes the existence of a golden model of the design, in practice it is hard to identify such a model. In most of the cases, destructive reverse engineering and exhaustive testing will be required to obtain such a model. Even then, the trustworthiness of such a golden model remains under question. Recently, several golden model free detection techniques have emerged, removing this shortcoming of the detection methods. One of such methods, called *Temporal Self-Referencing* (TeSR) [63], compared the transient current signatures form an IC with itself in different time windows, hence eliminating the need for a golden model. The assumption was that when a Trojan-free IC passes through the same set of state transitions in multiple time windows, the I_{DDT} signature will remain the same. Otherwise, the IC was assumed to be infected. This approach was however limited for sequential Trojans only, and was further enhanced by exploiting the structural self similarities of different design sub-blocks (spacial self referencing) [68]. Another self-referencing approach utilized the path delays of similar paths to detect a Trojan [88].

8.2.5.3 DFT Insertion Approaches

Recently design modifications for the sake of HTH detection or prevention are getting popular. Such DFT measures, also known as Design-for-Security (DFS) or Design-for-Trust measures can be broadly classified into two classes. Methodologies from the first class mainly try to prevent the insertion of HTHs, whereas the techniques in the second class mainly try to assist certain Trojan detection schemes – mainly side-channel analysis based schemes. We now shed light on some of the notable preventive and assistive DFT measures.

1. *Preventive DFT Measures*: Complete control of end-to-end supply chain of IC is the most ideal way to prevent HTH insertion. But having such control with recent trends of distributed design and manufacturing in semiconductors is difficult. Hence, DFT methodologies have to be employed for protection against HTHs. The methods in this class can be broadly classified into three subclasses: (a) netlist obfuscation; (b) layout filling, and (c) split manufacturing. Also, some approaches considering the modification in the supply chain itself are getting popular.

   (a) *Netlist Obfuscation*: Circuit obfuscation modifies the structure and functionality of a circuit and makes it difficult to reverse engineer. As a consequence, it becomes difficult for an attacker to insert an HTH unless he/she understands the functionality of the circuit properly. The main idea is to have two different modes of operation for a circuit – *obfuscated* and *normal*, with a one-way transition from the obfuscated to the normal mode possible only through the application of a particular secret key sequence as input. On applying the incorrect key sequence, the circuit continues to operate with incorrect functionality in the obfuscated mode, making it hard for an adversary (who either has access to the circuit netlist or has reverse-engineered the layout to extract the circuit netlist) to identify the rare nets in the normal mode of operation, which are the likely trigger nodes for HTH insertion. The state transition behavior of the circuit are modified, usually achieved by structural modification of the circuit netlist. Circuit obfuscation can be used both as a countermeasure to reverse-engineering, as well as to prevent HTH insertion [89–92]. Notable among them are the

HARPOON [89], where special combinational logic primitives called "modification cells" were inserted inside the circuit to obfuscate it. In a recent work [92], a fault analysis based obfuscation technique was proposed which effectively achieves maximum obfuscation with reasonable circuit overheads. In spite of the fact that obscurity strategies are very successful, they include additional hardware overheads. Additionally, it requires the secure storage of the necessary secret key or key sequence. Furthermore, it has been shown that reverse-engineered can be successfully applied for such obfuscated circuits [93], using Boolean satisfiability solvers (SAT solvers).

(b) *Layout Filling*: The main idea is to fill up vacant spaces in the layout of an IC so that additional gates cannot be inserted. In the *Built-in Self-Authentication* (BISA) technique [94], empty layout space is filled up with circuits to prevent layout tampering without modifying functionality. But BISA is ineffective against malicious tampering of some transistors in one or more standard cells.

(c) *Split Manufacturing*: To prevent Trojan attack, it has been proposed that the manufacturing process should be split among different foundries [95], with devices and interconnects being processed at different fabrication facilities with different levels of process sophistication and trust levels. The technique is costly and cumbersome, and has also been attacked [96].

Protective DFT measures may be achieved in some other forms too. For example, in [16], it was proposed that keeping the account of all the hardware resources in disposal for entire clock cycles may prevent the insertion of a HTH, which is utilizing the already existing logic. However, ensuring that a design always uses all its hardware resources does not imply that the resource utilization is not malicious. Moreover, such a high resource utilization is against the standard low-power design practices in principle. Another interesting approach is to use reconfigurable logic along with the standard logic [97,98] to keep the attacker unaware about some parts of the logic. These reconfigurable blocks are to be programmed post-fabrication by a trusted party. Such a technique also achieves obfuscation in some sense.

2. *Assistive DFT Measures*: Similar to the conventional fault testing, HTH detection measures may be facilitated with the help of certain embedded test structures. These structures are meant to assist the detection methods by addressing the challenges related to the HTH detection. In the following, we shall discuss some notable assistive DFT measures found in the literature.

    (a) *On-Chip Security Monitors*: The major challenge related to side-channel analysis based Trojan detection is the proper mitigation of the environmental, measurement and process noise effects. However, even with the judicious vector generation and best available measurement setup, this is often found to be a critical task. To assist the state-of-the-art delay side-channel based methods, in [99,100] a solution was proposed where the circuit paths were configured as ring oscillators to precisely detect even small delay variations. In [101], on-chip transient current sensors were applied improve HTH detection sensitivity compared to off-chip current monitoring. Two more area-efficient HTH detection mechanisms proposed are TRUSTNET and DATAWATCH [102]. However, all such security measures themselves are vulnerable against the adversary in the foundry. For example, in [103], a successful attack against the ring oscillator based scheme was proposed.

    (b) *Removal of Rare-Triggered Nets*: Low signal transition probability nets are likely to be utilized by an adversary to trigger HTHs. In [80], through dummy flip-flop insertion, the signal probabilities of the low singal probabilty nets inside a circuit were increased, increasing the possibility of HTH activation. This helps in their easy detection through logic values at the primary output or through transient current signals. In [104], the scheme was enhanced by intelligent flip-flop insertion to reduce the hardware overhead. However, sophisticated attacks have also been proposed against this scheme [105].

    (c) *Improving the TCR*: The TCR factor can be improved significantly by appropriate test vector generation as shown previously. However, similar effects can be achieved by means of DFT. In [106], a post-layout scan chain reordering for localizing the switching activities inside the circuit was presented. Such a method is found to be highly effective even when the HTH is placed in a distributed manner inside the circuit.

This is because the reduction of circuit switching is significantly higher than that of the HTH switching, which in turn improves the TCR as well as the SNR. Another alternative technique is the "voltage inversion" scheme described in [107]; however, it suffers from scalability issues for large. circuits.

8.2.5.4 Runtime Monitoring Approaches

Online monitoring approaches can be considered to be the last resort against HTHs undetected pre-deployment, some of which we describe below.

a. *Configurable Security Monitors*: In this approach, reconfigurable logic is added to ASICs, with possible customization for real-time functionality monitoring [19,108]. But an inherent assumption is that the monitors are perfectly trustable, which decreases their usefulness.

b. *Variant-Based Parallel Execution*: This approach utilizes multi-core architectures to ensure in-field trusted execution [109]. This countermeasure incorporates scheduling and execution of functionally equivalent variants of codes from different competitors on different processing elements (PE). The results are compared and new PEs are engaged while a mismatch happens. The PE assignment and computation process is then continued until the match is found, and in this process all Trojan infected PEs are expected to be identified. However, such a protection mechanism involves high performance and power overheads and is highly architecture specific.

c. *Software-Assisted Approach*: In [29], a verifiable "hardware guard" module external to the processor was used for execution monitoring, to identify attempts of DoS and privilege escalation attacks. A "memory guard" module to monitor the memory bus was proposed in [110].

8.2.5.5 Other Emerging Detection Approaches

Emerging solutions such as split manufacturing [96,111] are recently getting popular for HTH detection and prevention. Also, optical measurement based non-destructive reverse-engineering [112] has recently been introduced for these purposes. Such methodologies often try to make the detection approaches more scalable. At the architecture level, various data guarding mechanisms, such as bus scrambling, fully homomorphic encryption of the data, introduction of random time delays and insertion of dummy instructions are getting prominent [113]. The primary target of such methods is to prevent the HTH from getting its activation signal or to prevent its access to sensitive data. A secured bus architecture was proposed in this context at [114], targeting HTHs which enable bus locking attack. However, most of such approaches require further research for getting standardized.

8.2.6 Countermeasures in 3PIP Model

Trojan detection approaches in the 3PIP model vary significantly from that of the malicious foundry model. Unavailability of the golden model makes the detection critical in this model, and only suspicious behaviors can be traced. However, the existence of the design in a soft form along with some specifications also make the detection easier in some sense. The detection schemes in this paradigm can be broadly classified into two classes depending on the availability of specifications. If detailed specification as well as a secured agreement between the IP vendor and the consumer exists, trusted IP acquisition can be achieved through a secured design paradigm. One of such methods is known as *Proof-Carrying Codes* (PCC) [115,116]. A list of security-related properties is detailed, and the vendor creates the formal proofs of these properties. The designer is supposed to build the circuit in a way so that the security proofs can be verified easily from it at the consumer side, to detect if the IP has been modified. However, at the current state of the art, it is really difficult to define the security properties in the presence of intelligent adversaries.

In the absence of detailed specifications and secured agreements among the two parties, analysis of the procured netlist or RTL code is usually performed. A set of suspicious signals is expected to be returned after analysis containing many false positives. It is considered that the lesser the number of false positives better is the method. Such methods again can be classified into three subclasses, namely, signal and circuit based methods, redundancy-based methods, and knowledge-based methods. Methods from the first subclass usually exploit formal methods such as model checking, circuit simulation, and static analysis techniques. The circuit simulation based techniques usually do not take the internal circuit structures into account and just rely on the statistics of individual signals. One of such methods, proposed in [117], utilizes N-detect ATPG followed by a sequential equivalence checking on a gate-level netlist to detect certain suspicious signals. Sequential circuits are unrolled upto certain time stages. However, the existence of a trusted behavioral level specification is assumed in this scheme. In another scheme from this sub-genre [118], standard code coverage metrics (i.e., line coverage, toggle coverage, finite-state machine (FSM) coverage, etc.) were utilized to detect uncovered regions of the code. A significantly different approach was proposed in [30], where the concept of *Unused Circuit Identification* (UCI) was proposed. With graph-based circuit analysis, UCI effectively detects HTHs which are otherwise undetectable by code coverage or simulation-based methods. Relations among each pair of wires are exploited in this scheme with the assumption that if the values of any pair do not differ with the change of inputs, they are suspicious. However, the scheme was shown to be evadable with MUX-based logic in [119]. In another scheme [120], it was shown that systematic hardware description language (HDL) coding can bypass UCI. Such drawbacks of the UCI scheme were successfully overcome with the K-Map based VeriTrust method [121]. VeriTrust utilized inactive minterms and their corresponding logic expressions to detect suspicious circuits. Another method with similar features but significantly different approach was FANCI [122], where complex circuit properties were utilized through a simulation-based method. Both FANCI and VeriTrust methods were evaded with the introduction of DeTrust [123]. Breaking the Trojan activation paths between different time windows, DeTrust successfully bypasses both the schemes. In some recent works, structural features of Trojan structures were exploited to identify them [124]. All of the methods in this subclass have different false positive rates and different scalability levels.

A significantly different approach of 3PIP Trojan detection was proposed in [125], which procures the same IP from different vendors [126], and then securely assembles design through high-level synthesis [127]. Scalability might be an issue in such schemes as they require multiple copies of the same IP.

Recently several ML models have been proposed to detect the hardware Trojan using RTL and gate level in IoT devices. The works [128–130] analyze gate-level netlists to differentiate the HTH-infected circuit netlists from normal ones. The classifiers used was SVM, Random-forest and multi-layer perceptron. The method [131] uses features extracted from RTL source code, and uses 21 circuits from the Trust-Hub [132] database identify the Trojans present. Another method [133] uses controllability and transition probability characteristics of design as features for the deep learning along with *k-means clustering*, and the effectiveness of the technique was evaluated on ISCAS'85 combinational benchmark circuits. True negative rate (TNR) and True positive rate (TPR) are used as evaluation criteria to assess the methods. Combining both RTL and gate-level based approaches, the authors proposes [134] *RG-Secure* algorithm for IoT system that can classify multi-type HTHs; the evaluation was performed using 15 benchmark circuits from Trust-hub. We see that there is a lot of scope to conduct the research with ML specifically for silicon and FPGA-based IoT devices.

8.2.7 Counterfeit ICs

Counterfeit ICs constitute a major hindrance in realizing trusted electronic systems [136], decreasing reliability and increasing the possibility of HTH infection. Data provided by Information Handling Services (IHS) [137] shows that reports of counterfeit parts in 2012 have increased four times since 2009 [136]; the seriousness of the problem can be envisaged to have increased many fold since then.

Counterfeit ICs have been classified into seven major classes [136], based on the illegal or unauthorized aspect that makes them "counterfeit": **Recycled**, **Remarked**, **Overproduced**, **Out of spec/-damaged**, **Cloned**, **Forged Documentation**, and **Tampered**, with the last two having the possibility of inserted HTHs or malicious software/firmware.

8.2.8 Counterfeit IC Detection

Physical inspection based detection is possible based on characterization of several external properties of ICs, as shown in Figure 8.6 [135]. Of these, texture mismatch identification requires help of a microscope,

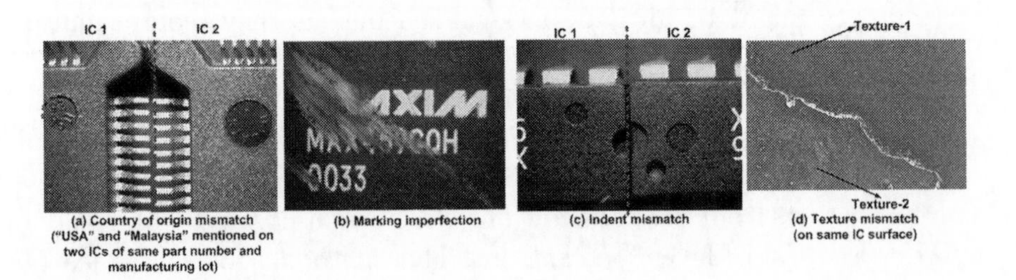

(a) Country of origin mismatch ("USA" and "Malaysia" mentioned on two ICs of same part number and manufacturing lot) (b) Marking imperfection (c) Indent mismatch (d) Texture mismatch (on same IC surface)

Figure 8.6 Examples of different types of counterfeit ICs [135].

while the others are detectable by naked eye. Over the past few years, a taxonomy of detection methods has been created as shown in Figure 8.7 [136,138]. Many of these techniques are inspired by technologies used in the domain of IC and PCB reverse-engineering [139,140]. An IC should ideally undergo each of these tests before being released in the market and being deployed in a system. But most of these tests are required to be done manually, sometimes by highly skilled engineers and technicians, and require a *Subject Matter Expert* (SME) [136,138,141]. This makes the process costly, time-consuming, and clearly unscalable for large deployment volumes.

As shown in Figure 8.7, counterfeit IC detection methods are classified into three broad categories, namely: (a) physical inspection; (b) electrical characterization, and (c) aging-based fingerprinting. Image processing based counterfeit IC detection can be characterized to be a physical inspection [142]. Physical inspection methods investigate various external properties of the components like texture, indents, imperfections, etc. Image processing helps in studying these physical properties without the help of any SME, and thus is amenable to automation of detection.

In spite of its promise, relatively few works exist in the current research literature on this topic. One of the earliest works was [143], which employed *SEM*, X-Ray microscopy, and *Energy Dispersive Spectroscopy* (EDS). The work described in [142] applies Artificial Neural Networks (ANN) to automatically detect scratches present on its surface, using images from the database at [144] for training. It is quite computationally expensive. Another recent work [145] has analyzed texture of IC package surface images to distinguish between counterfeit and non-counterfeit ICs, using three different texture analysis techniques. The advantage of this scheme is that expensive image acquisition systems (e.g. SEM or X-ray imaging) are not required, and images acquired using optical microscopes suffice. The technique was improved in [146] to combine texture comparison with indent shape and position comparison, thereby achieving greater detection accuracy. However, the computational effort necessary and scalability of these techniques pose similar challenges as other techniques proposed previously.

8.2.9 Trends of HTH and Counterfeit Electronics Research

From the above it is clear that the threats of HTH and counterfeit electronics are very potent, and the vulnerability of IoT devices from these

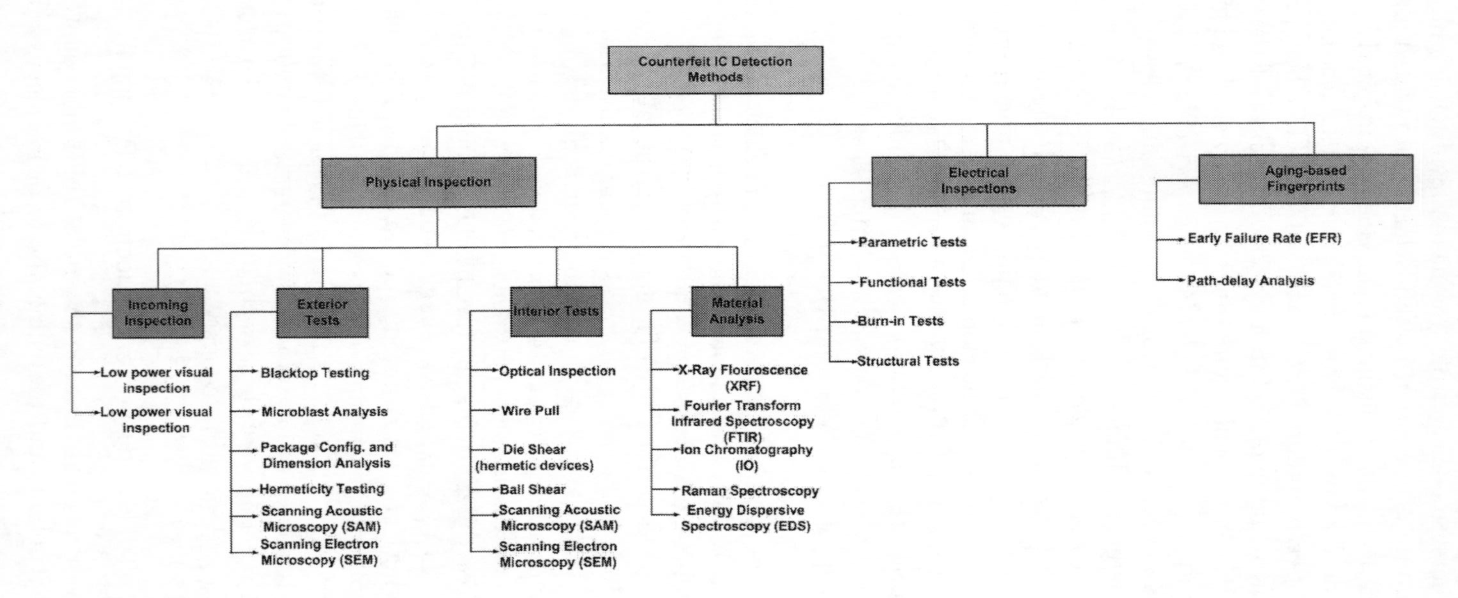

Figure 8.7 Taxonomy of counterfeit IC detection [136].

threats are far from being mitigated. Split manufacturing and its secure variants are currently the most widely investigated techniques, holding great promise. Sophisticated attacks against split manufacturing are being proposed regularly, along with countermeasures against them [147], so this is a dynamic topic of research where the last word is yet to be heard. An added challenge is the difficulty of adopting such a specialized manufacturing flow at commercial scales for mass-market products such as IoTs. Counterfeit electronic components constitute an even older threat (as old as mass-scale IC manufacturing itself), but unfortunately research for their detection is at a nascent stage. Image processing (optical and X-ray based) based techniques described above are the only techniques known currently that have shown some promise in detecting counterfeit ICs, to the best of our knowledge. It might happen that IC counterfeiters develop techniques to evade the current image processing techniques, and then newer sophisticated techniques would be required for counterfeit IC detection.

8.3 PUFs FOR IoT SECURITY

As mentioned previously, the unique challenges posed by resource constraints of IoT devices necessitates non-traditional, lightweight schemes for secure communication and authentication. PUF circuits have recently attracted plenty of attention as a promising hardware security primitive with diverse applications, including IoT devices [148]. In this section, we explore in detail the structure and workings of common PUF variants, their vulnerabilities, and solutions to these threats. In the next section, we describe proposed applications of PUF circuits to IoT security.

8.3.1 Physically Unclonable Function (PUF)

Biometrics is the common method that uses human attributes for authentication and identification. *PUF* [149] circuits provide comparable functionality by assigning fingerprints to ICs. During IC manufacturing, the random physical process variations impose unique characteristics to each IC. Traditionally, such process variation adversely affects circuit performance and functionality and is considered counter-productive. On the other hand, PUF circuits exploit these process variations to generate a unique identifier for that device which can be used in cryptographic protocols, primarily authentication. An n-bit input, m-bit output PUF instance can be thought to be a Boolean function $f : \{0,1\}^n \rightarrow \{0,1\}^m$. The inputs to the PUF are called "challenges," and the corresponding outputs generated from the PUFs are termed "responses." While silicon-based electrical PUFs have been most widely studied, there are other PUFs utilizing other materials and physical characteristics, e.g., Optical PUF [150], Microelectromechanical systems (MEMS) PUF [151,152], Memristor PUF [153,154], and Spintronics PUF [155]. Although silicon-based electrical PUFs have been widely studied, there are other PUF designs using different manufacturing processes and materials (sometimes hybrid with silicon technology), e.g., Optical PUF [150], MEMS PUF [151,152], Memristor PUF [153,154], and Spintronics PUF [155].

8.3.1.1 PUF Classification

Based on the size of their challenge-response space, PUFs are usually classified into the following classes [156]:

- *Strong PUF*: The CRP space of these PUFs are large enough to evade exhaustive characterization of the PUF circuit. This kind of PUFs is secure against challenge-replication based attacks if the same challenge is not applied twice during protocol execution. The *Arbiter PUF*(APUF) [149] is a widely studied Strong PUF.

- *Weak PUF*: This type of PUFs generate a comparatively smaller number of CRPs, and are susceptible to replication attack. The attacker with all the challenges will be able to emulate this PUF. Hence the responses of such PUFs are not exposed out of the hardware device. The examples of Weak PUF are SRAM PUF and Ring Oscillator PUF (ROPUF). For this reason, responses of this classes of PUFs are not allowed to leave out the device. In the extreme case, a weak PUF might not have any challenge, and generate a constant instance-specific response, e.g. the *MECCA* PUF [157], which can be used for key generation and authentication.

Note that the above notion of "strength" of PUFs is not based on the security level of PUF. Majority of primitive strong PUFs are susceptible to modeling attacks though they are secured to emulation attacks. In the rest of the chapter, we would concentrate on strong PUFs and their applications for IoT security.

Based on the way responses of PUF are generated, the PUF can be classified into the following categories:

- *Delay PUF*: Delay PUFs are principally based on the inherent delay characteristics of the physical elements that vary from chip-to-chip. The difference in delay caused by an identically laid out pair of electrical paths generates the response. Delay PUF can be strong PUF or Weak PUF based on its architecture. The widely studied examples of delay PUF are APUF and ROPUF.
- *Memory PUF*: This is based on the unpredictable power-up state values of bistable memory elements which occur due to random variations in transistor parameter values. The well-known example of Memory PUF is SRAM PUF.
- *Delay + Memory PUF*: Here the response is based on two possible stable states like memory PUFs, whereas the notion of delay PUF is utilized to build bistable delay components. The Bistable Ring PUF (BRPUF) [158] is an example of such a PUF, where a bistable ring comprises an even number of inverting stages.

In spite of the fact that majority of the memory PUFs don't need challenges to generate responses and considered as weak PUFs, the combination of some memory PUFs are used as strong PUF [159]. For example, consider SRAM cells which are organized in $l \times k$ grid to form $l \times k$ SRAM PUF. To make it use as a strong PUF, the challenge bits are used as addresses of SRAM.

We next describe the major reported attacks on PUFs.

8.3.2 Attacks on PUFs

Exploration of attacks on PUFs is as old as the concept of PUF itself [149]. PUF attacks are mainly classified into two main categories: *statistical attacks* and *modeling attacks*. Statistical attacks find the correlation between CRPs, and the same is used for prediction of the related set of challenges. On the other hand, model building attack focuses on constructing an accurate mathematical model of the PUF by using a relatively small set of known CRPs. After the successful model is built, the prediction is carried out for unknown challenges. The following discussion concentrates briefly on two attack categories.

In [160], authors had reported that challenge bit positions does not satisfy *Strict Avalanche Criterion* (SAC), i.e., change in any challenge bit should flip the response bit with probability 0.5. The attacker can use poor SAC to perform differential attack [160,161], or to successfully predict the response for related challenges, if the responses for some challenges are known [162]. The larger parts of strong PUF designs are susceptible to differential attack.

Modeling Attacks: The modeling attacks are the most potent threats to strong PUFs. Model building is typically achieved through ML algorithms [163] e.g., Logistic Regression (LR), Support Vector Machine (SVM) and ANN, evolutionary algorithms [164,165] (e.g., Evolutionary Strategy (ES)), and more recently, Probably Approximately Correct (PAC) Learning [166–168]. ML algorithms usually rely on the availability of a database of *challenge-response pairs* (CRPs), obtained by either directly evaluating a PUF instance, or eavesdropping an unencrypted communication session between a PUF-enabled entity and verifier during authentication. Some other source of information which can be used for modeling include the side-channel information leaked through PUF implementations. Various modeling attacks presented in the literature has been shown in Figure 8.8.

Hybrid Modeling Attacks: Side-channel based modeling attacks use the information which is leaked from the execution of implemented PUF circuits. The work [169] examines two side-channels, i.e., power traces of latch element in APUF and timing signatures. The side-channel information along with the CRP data boosts ML modeling performance. The paper presents the modeling of 512-bit 16-XOR PUF with an accuracy of 96.5% with just 4.1 million CRPs. The result is remarkable using the side-channel features compared to the work [170] where 64-bit 8-XOR required 350 million of CRPs without using any side-channel information.

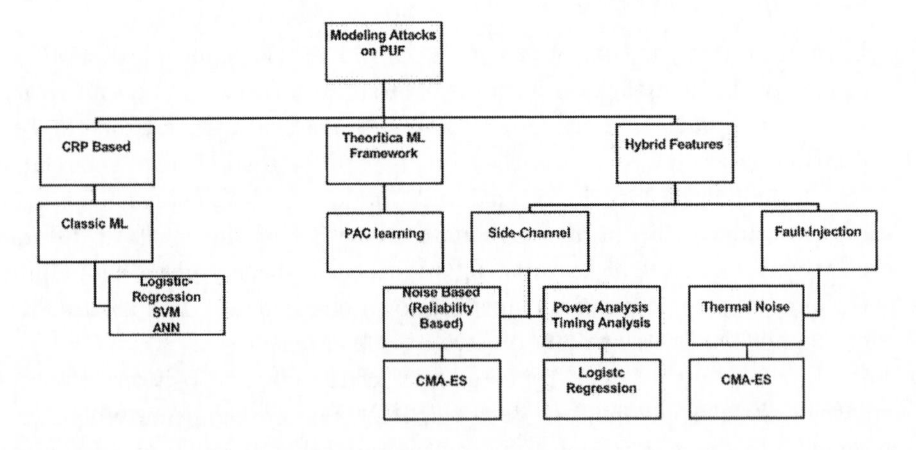

Figure 8.8 Modeling attacks taxonomy.

Similarly, the paper [171] shows that the power traces information of APUF using ML can be used to attack controlled PUF.

Reliability-Based Attacks: In [172], the author presents a new ML attack using the imperfect reliability of PUF instances. The attack has been tried on XOR PUFs and Lightweight Secure PUF (LSPUFs). Reliability information can be captured using the same challenge with the repeated observation of responses. The adversary can exploit this information to understand the variation of responses with respect to environmental factors. The author proposes an ES-based attack. Fault injection attacks induce deliberate fault in the process which leads to a change in the behavior. The change in the behavior is used for understanding the internals of the system. The paper [171] uses thermal noise as fault injector to the controlled Arbiter PUF. The thermal noise adds as a Gaussian noise and can flip the response bit.

We now describe various modeling attempts on three widely studied delay PUFs: APUF, FF-APUF, and XOR APUF.

8.3.2.1 Modeling Attacks on Delay PUF Variants

In [149], the authors constructed a linear additive delay model of the APUF circuit, that can be learned using SVM [149], LR [173], and ANN [174], with LR achieving the best modeling accuracy (close to 100%). To enhance the modeling robustness, Feed-Forward APUF (FF-APUF) [149] and XOR APUF [175] were proposed to add the non-linearity to the delay model. ES was initially reported in [173] to attack FF-PUF, with the practical complexity of the attack depending on various structural properties of the PUF being analyzed. Along with this, [176] reported various side-channel based model building attacks.

In an x-XOR APUF, the responses of x individual APUFs are XOR-ed to compute the final response. It can be shown [173] that the XOR-ing operation increases the complexity of the learning problem. On the theoretical aspect of the problem of XOR PUF modeling, [177] establishes the PAC learning framework stating that XOR APUF can be learned with given levels of accuracy and confidence. The paper also establishes that considering the theoretical limit on the number of arbitrary chains, the learning of XOR can happen in polynomial time. On the practical modeling attack, earlier work [173] had shown LR to be effective for XOR APUF modeling without using any side-channel information, for $x < 6$; however, for $x \geq 6$, the computational effort required is substantial. A later work [170] analyzed the scalability of ML-based modeling of XOR APUF using LR, using extensive computational infrastructure. The recent known tensor regression based modeling building attack [178] uses least number of CRPs to achieve the comparable accuracy for higher order XOR PUFs ($x < 9$). Deep learning based attack on various arbiter-based compositions was shown with less computational resource in the work [179]. For the benefit of the research community, the authors have released the dataset and ML software code online.

8.3.2.2 Cryptanalytic Attacks on Delay PUFs

The *LSPUF* [180] generates multibit response, where each output bit is defined by an XOR APUF. Unlike XOR APUF that applies the same challenge to all APUFs, different challenges derived from a common challenge are applied to individual APUFs in LSPUF, using an input logic network. Majority of modeling attacks developed for XOR APUF are also applicable to LSPUF; but in general, the LSPUF was thought to be more secure than the XOR APUF, because of its multi-bit output and the presence of an input layer. However, in [181], cryptanalytic attacks on LSPUFs were proposed, whereby with certain probability an adversary can predict the value of an output bit for a given challenge. This cryptanalytic attack was used in turn to aid in the ML-based modeling attack accuracy for LSPUF outputs, taking advantage of the interdependence of the LSPUF output bits. An attack of similar flavor on an *Enhanced ROPUF* [182] was also described in [183].

In [184], the authors introduced an abstract model for strong PUFs based on their response generation technique as $sign(f(\mathbf{a})-f(\mathbf{b}))$, where vector $\mathbf{a}$ and $\mathbf{b}$ are random parameter vectors specific to the applied challenge $\mathbf{c}$. Function f maps its input physical parameter vector to some measurable behaviors of circuit such as delay, voltage, and current. They classified the function f as (a) fully continuous random function (FCRF) and (b) mixed continuous-discrete random function (MCDRF). The composition of primitive PUFs fall into the class MCDRF, whereas primitive PUFs are in FCRF. Modeling robustness of PUF depends on the non-linearity of the function f. Vijayakumar et al. [185] have extended this abstract model to demonstrate the designing requirements for strong PUF with high modeling robustness. Through extensive experimental results, they have shown that modeling robustness of primitive strong PUF increases with increasing non-linearity of function blocks f_i. Design of these functional blocks with higher non-linearity is a challenging task. It is obvious that further modeling robustness can be enhanced with an additional non-linear composition of primitive strong PUFs.

8.3.3 Attacks on PUF Protocols

In addition to the attacks on bare PUF design with unrestricted access to PUF, there are many attacks developed for PUF- based protocols. In [186,187], the authors proposed modeling attack on slender PUF and fuzzy extractor based authentication protocols. The fuzzy extractor based authentication protocol uses helper data, and the adversary exploits these helper data for modeling of PUF instead of its response. An attack on PUF based key-generation protocol was reported in [187] by manipulating helper data. There are a few attacks on ROPUF-based key generation protocols by using (or manipulating) helper data [183,188]. Thus, to ensure the security of PUF-based protocols, we need to use strong PUFs with modeling robustness and higher reliability, and helper data should be protected with proper access control mechanisms.

8.3.4 Philosophy Behind Modeling Resistant Strong PUF Designs

Modeling robustness of a strong PUF design relies on the complexity in its challenge-response behavior. It is more difficult to learn a non-linear behavior compared to a linear one using ML algorithms. We have experienced this fact through the modeling of non-linear design variants of APUF, namely FF-APUFs and XOR APUFs, where non-linearity is introduced by some digital means like feed-forward loop and XOR of multiple outputs. The major problem with primitive delay-based digital PUFs is that delay of a path-segment varies linearly in threshold voltage variation, and the cascaded architecture of delay-based PUF makes the resulting path delay linear and additive in nature. In addition to these digital strong PUF designs, there are a few attempts on modeling resistant analog PUF designs where exploited non-linearity sources are (a) sub-threshold current in metal–oxide–semiconductor field-effect transistor (MOSFET) [184], (b) non-linear current mirror [189], and (c) circuit block with non-linear voltage transfer characteristics [190]. These primitive PUF designs have better modeling robustness than primitive digital strong PUFs such as APUF.

8.3.5 Application of PUF for IoT Security

Although PUFs have been considered an important hardware security primitive for a long time, their applicability to IoT devices or other resource-constrained devices such as *Radio Frequency Identification* (RFID) chip has been explored in relatively few research papers. One of the first published works on this topic was [191], where the authors described a mutual authentication protocol based on PUFs, suitable for RFIDs. The protocol is *ultra-lightweight*, in a sense that it is suitable for RFID tags where the only arithmetic/logical operations supported are bitwise AND, OR, XOR, etc., and left and right shifts, but due to hardware resource constraints, random number generation or hash functions are not available. The proposed protocol is based on a challenge-response information exchange between an RFID reader and the RFID tag, assuming a backend database stores all the RFID golden CRPs. One of the implementation issues of the proposed protocol is that the PUF circuitry in the RFIDs are required to generate a relatively high number of output bits (e.g., 96 bits); hence, the hardware footprint of the PUF circuitry might be unacceptable in an RFID context. Another such work is [148]. For the interest of readers, we provide appropriate references on the recent research with a brief description below.

In [192], the authors survey the research work on PUF-based key generation and analyse the protocols focused on strong PUF. The work defines the protocol requirements for the token-server entity authentication and analyses 19 different proposals under a common framework for comparison. In [193] the authors focused on the difficulty of storing a PUF CRP database, due to the high cardinality of the CRP set of each device, and the high count of the IoT devices. Hence, as an alternative, the authors proposed storing an encrypted accurate mathematical model that substitutes the PUF at the verifier (authentication server). The authors implemented the scheme using an *Enhanced Anderson PUF* [194], and 128-bit AES on Xilinx FPGA. The modeling ease of the Enhanced Anderson PUF might be a concern in this scheme, if the adversary gets sufficient access to characterize the PUF in the deployed IoT nodes.

More sophisticated PUF-based security protocols suitable for IoT applications were proposed in [195–198]. In [195], the authors proposed a PUF-based authentication protocol which combined the unclonability of PUFs with the security of cryptographic hash functions in the context of *Message Authentication Code* (MAC), to argue about the security of the proposed scheme. However, the authors did not provide any formal proof of security, or any experimental validation, of the proposed protocol. Another concern might be the fact that the protocol follows a fixed sequence of CRPs for multiple authentication sessions for the same device. In [196], the author proposed a protocol that combines elliptic curve cryptography with PUFs for device enrollment, authentication, decryption, and digital signature generation. Elliptic curve cryptography is ideal for resource-constrained IoT devices because it provides security levels equivalent to other cryptographic schemes based on different hard problems, with reduced storage and computational requirements. An FPGA-based implementation of the protocol was also provided.

An *identity-based cryptosystem* providing secure authentication and message exchange, while being lightweight and thus suitable for IoT devices, was proposed in [197]. A PUF is used to generate the identity of each device, which is also its public key used to encrypt messages (by *Elliptic Curve Cryptography*). The work also provides formal proofs of security of the proposed protocol. Comparison with several PUF-based authentication protocols (many of which have fallen prey to attacks) was performed, and an experimental evaluation proved the low hardware software footprints of the proposed protocol. A mutual authentication protocol on a PUF of a different flavor was reported in [199]. This paper proposes the application of a *Hardware Embedded Delay PUF* (HELP) implemented on FPGA.

The suitability of PUF as a hardware security primitive suitable for IoT applications was explored in [200] in relative detail. In this paper, the author described a practical attack on a *Philips Hue* smart lighting system, a typical commercial off-the-shelf IoT system. Taking advantage of weaknesses in the authentication protocol used by this smart lighting system, which used the physical address as an identifier to authenticate the device trying to control the smart lighting system, a practical attack was demonstrated using widely available tools such as [201]. The author also proposes a PUF-based authentication scheme based on the physical unclonability of PUF instances which can prevent such an attack. A recent work in [202] proposes the certificateless PUF-based authentication protocol which blends *Identity based Encryption* (IBE) with PUF. It does not require any storage of an explicit CRP database. The main idea is to map the challenges and responses in such a way that they can be stored publicly ensuring their integrity. The verifier is assumed to the root of trust and can verify the prover using the public key. However, the computation overhead at the Private Key Generator (PKG) inside the IBE described in this work needs to be analyzed and optimized.

A PUF-based authentication protocol for access control to prevent malicious DPR attempts on FPGAs connected over a network was presented in [40]. In this work, the authors considered the modeling attack of XOR PUF to be a difficult computational problem, and proved the security of the scheme by formally reducing the problem of modeling a XOR PUF to the problem of breaking the protocol. This protocol can be envisaged to be extended for IoT applications also, especially for small-scale IoT networks with relatively few devices.

8.4 OPEN RESEARCH ISSUES AND FUTURE RESEARCH DIRECTIONS

Many IoT implementations follow the sensing-communication-actuation model of working, where the sensor and communication units (and sometimes the actuators) are expected to be working autonomously placed in locations where they are not continuously surveyed. An adversary can take advantage of this fact to launch powerful, non-disruptive attacks termed as *SCAs* to gain information about the internal workings of the system. All integrated circuits and computing devices leak information about the operations they are performing, and if the power dissipation of such a chip can be faithfully recorded (often by tracing the electromagnetic signature in its vicinity by sensitive antenna probes), much information can be gathered through mathematical analyses of such information. A famous example is the *Differential Power Analysis* [203] of cryptographic hardware implementation, which enables relatively easy computation of the secret key from the power side-channel. Similarly, by carefully measuring the time required to perform arithmetic operations ("timing side-channel"), several software implementations of cryptographic algorithms can be attacked ("timing attack") [204]. Another powerful timing attack is the so-called *Cache Timing Attack* [205], which is an SCA based on the minute differences in the execution time of cryptographic algorithms resulting from cache hit or cache miss. Countermeasures to prevent or to increase the computational complexity of SCAs can be applied at the algorithmic [206] or circuit levels [207]. However, typically such countermeasures result in substantial (often ∼200%) hardware and/or performance overhead. It is an open research problem to develop lightweight design techniques to prevent SCAs, suitable for resource-constrained IoT devices. Also, SCA for hardware security primitives like PUFs have been relatively less explored, but it has been shown that SCA can help to reduce the complexity of modeling attack on PUFs [208–210].

Reliability of IoT devices, which might often be deployed in harsh environmental conditions, and the related security vulnerabilities also constitute a great concern. *Fault Attacks* [211] constitute a class of attacks whereby the incorrect execution of cryptographic algorithms, through the introduction or occurrence of circuit-level faults, results in faulty ciphertext to be produced, which can then be analyzed to reveal the secret key. Attacks following similar philosophy have also been described for PUFs [212]. Other attacks relying on the auxiliary information ("helper data") typically employed to correct the response of an unreliable PUF [172] have also been reported. The development of hardware security primitives and protocols based on them, which are robust to the threat of attacks on unreliable IoT devices and security primitives, is another open challenge. PUF circuits are not perfectly reliable; hence, to ensure they are usable as hardware security primitives, low-cost reliability enhancement techniques are required, e.g., the one described in [213].

The heterogeneity of IoT devices implies that it is infeasible to procure all IoT devices from a single or a small number of "trusted" vendors, a technique that is more practical for ICs or system components. Hence, the issue of untrusted vendors and all the vulnerabilities resulting from them as described earlier in the paper are exacerbated in IoTs. Ensuring trust in a large collection of heterogeneous devices is another open research problem, especially because of the diversity and scaling issues associated with IoT deployment.

8.5 CONCLUSIONS

The different threats related to hardware implementation of security features in IoT devices are formidable enough to prevent the IoT realizing its full potential. Hardware Trojan Horses and counterfeit ICs can lead to a variety of potent attacks, causing disruption of functionality and serious compromise of security and privacy. The challenges of defending against these threats are exacerbated by the severe resource constraints of IoT devices, which prevent the adoption of traditional cryptographic algorithms and protocols for secure communication and secure system implementation. We discussed several techniques, ranging from circuit testing and novel hardware security primitives (PUFs), to image processing algorithms to mitigate these threats. In particular, PUFs show great promise in the development of secure lightweight communication protocols suitable for IoT devices, in spite of several threats against them such as ML-based modeling attacks. Hardware security aspects for IoT devices can be envisaged to be an exciting and rewarding field of research in the near future.

Bibliography

[1] Jun Zhou, Zhenfu Cao, Xiaolei Dong, and Athanasios V. Vasilakos. Security and Privacy for Cloud-Based IoT: Challenges. *IEEE Communications Magazine*, 55(1):26–33, January 2017.

[2] Hong-Linh Truong and Schahram Dustdar. Principles for Engineering IoT Cloud Systems. *IEEE Cloud Computing*, 2(2):68–76, March 2015.

[3] Alessio Botta, Walter de Donato, Valerio Persico, and Antonio Pescapé. On the Integration of Cloud Computing and Internet of Things. In *Proceedings of the International Conference on Future Internet of Things and Cloud (FiCloud 2014)*, pages 23–30, August 2014.

[4] Bonnie Zhu, Anthony Joseph, and Shankar Sastry. A Taxonomy of Cyber Attacks on SCADA Systems. In *Proceedings of International Conference on Internet of Things and 4th International Conference on Cyber, Physical and Social Computing (iThings/CPSCom 2011)*, pages 380–388, Oct 2011.

[5] Steven Cherry. Hacking Pacemakers, 2013. https://spectrum.ieee.org/podcast/biomedical/devices/hacking-pacemakers, Accessed: June 2020.

[6] Shyamnath Gollakota, Haitham Hassanieh, Benjamin Ransford, Dina Katabi, and Kevin Fu. They Can Hear Your Heartbeats: Non-invasive Security for Implantable Medical Devices. In *Proceedings of the ACM SIGCOMM 2011 Conference*, SIGCOMM '11, pages 2–13, 2011.

[7] Mohammad Abdullah Al Faruque and Korosh Vatanparvar. Energy Management-as-a-Service Over Fog Computing Platform. *IEEE Internet of Things Journal*, 3(2):161–169, April 2016.

[8] Tanzeem Choudhury et al. The Mobile Sensing Platform: An Embedded Activity Recognition System. *IEEE Pervasive Computing*, 7(2):32–41, April 2008.

[9] Geng Wu, Shilpa Talwar, Kerstin Johnsson, Nageen Himayat, and Kevin D. Johnson. M2M: From mobile to embedded internet. *IEEE Communications Magazine*, 49(4):36–43, April 2011.

[10] Jean Kumagai. Chip detectives [reverse engineering]. *Spectrum, IEEE*, 37(11):43–48, 2000.

[11] DARPA. *TRUST in Integrated Circuits (TIC)*. https://apps.dtic.mil/dtic/tr/fulltext/u2/a503809.pdf, 2007. Accessed: June 2020.

[12] Sally Adee. The Hunt For The Kill Switch. *IEEE Spectrum*, 45(5):34 –39, May 2008.

[13] Dakshi Agrawal, Selcuk Baktir, Deniz Karakoyunlu, Pankaj Rohatgi, and Berk Sunar. Trojan detection using IC Fingerprinting. In *Proceedings of the IEEE Symposium on Security and Privacy*, pages 296–310, Washington, DC, USA, 2007.

[14] Mohammad Tehranipoor and Farinaz Koushanfar. A Survey of Hardware Trojan Taxonomy and Detection. *IEEE Design Test of Computers*, 27(1):10–25, Jan.–Feb. 2010.

[15] Rajat Subhra Chakraborty, Seetharam Narasimhan, and Swarup Bhunia. Hardware Trojan: Threats and emerging solutions. In *Proceedings of the IEEE International High Level Design Validation and Test Workshop (HLDVT'09)*, pages 166–171. IEEE, 2009.

[16] Miodrag Potkonjak. Synthesis of trustable ICs using untrusted CAD tools. In *Proceedings of the 47th Design Automation Conference*, pages 633–634. ACM, 2010.

[17] Sk. Subidh Ali, Rajat Subhra Chakraborty, Debdeep Mukhopadhyay, and Swarup Bhunia. Multi-level attacks: An emerging security concern for cryptographic hardware. In *Design, Automation & Test in Europe Conference & Exhibition (DATE), 2011*, pages 1–4. IEEE, 2011.

[18] Lang Lin, Wayne Burleson, and Christof Paar. MOLES: Malicious off-chip leakage enabled by side-channels. In *Proceedings of the*

2009 International Conference on Computer-Aided Design, pages 117–122. ACM, 2009.

[19] Swarup Bhunia, Miron Abramovici, Dakshi Agrawal, Paul Bradley, Michael Hsiao, Jim Plusquellic, and Mohammad Tehranipoor. Protection Against Hardware Trojan Attacks: Towards a Comprehensive Solution. *IEEE Design & Test*, 30(3):6–17, 2013.

[20] Rajat Subhra Chakraborty, Francis Wolff, Somnath Paul, Christos Papachristou, and Swarup Bhunia. MERO: A statistical approach for hardware Trojan detection. In *Cryptographic Hardware and Embedded Systems-CHES 2009*, pages 396–410. Springer, 2009.

[21] Susmit Jha and Sumit Kumar Jha. Randomization Based Probabilistic Approach to Detect Trojan Circuits. In *2008 11th IEEE High Assurance Systems Engineering Symposium*, pages 117–124, 2008.

[22] Zhimin Chen, Xu Guo, Raghunandan Nagesh, Anand Reddy, Michael Gora, and Abhranil Maiti. Hardware Trojan designs on BASYS FPGA board. *Embedded System Challenge Contest in Cyber Security Awareness Week-CSAW*, 2008.

[23] Swarup Bhunia, Michael S Hsiao, Mainak Banga, and Sriram Narasimhan. Hardware Trojan Attacks: Threat Analysis and Countermeasures. *Proceedings of the IEEE*, 102(8):1229–1247, 2014.

[24] Aswin Sreedhar, Sandip Kundu, and Israel Koren. On Reliability Trojan Injection and Detection. *Journal of Low Power Electronics*, 8(5):674–683, 2012.

[25] Yuriy Shiyanovskii et al. Process reliability based Trojans through NBTI and HCI effects. In *2010 NASA/ESA Conference on Adaptive Hardware and Systems (AHS)*, pages 215–222. IEEE, 2010.

[26] Georg Becker, Francesco Regazzoni, Christof Paar, and Wayne Burleson. Stealthy dopant-level hardware Trojans. In *Cryptographic Hardware and Embedded Systems-CHES 2013*, pages 197–214. Springer, 2013.

[27] Yier Jin and Yiorgos Makris. Hardware Trojans in Wireless Cryptographic ICs. *IEEE Design & Test of Computers*, 27(1):26–35, 2010.

[28] Samuel T King, Joseph Tucek, Anthony Cozzie, Chris Grier, Weihang Jiang, and Yuanyuan Zhou. Designing and Implementing Malicious Hardware. *USENIX LEET*, 8:1–8, 2008.

[29] Gedare Bloom, Bhagirath Narahari, and Rahul Simha. OS support for detecting Trojan circuit attacks. In *HOST'09. IEEE International Workshop on Hardware-Oriented Security and Trust*, pages 100–103. IEEE, 2009.

[30] Matthew Hicks, Murph Finnicum, Samuel T King, Milo MK Martin, and Jonathan M Smith. Overcoming an untrusted computing base: Detecting and removing malicious hardware automatically. In *2010 IEEE Symposium on Security and Privacy (SP)*, pages 159–172. IEEE, 2010.

[31] Jean-François Gallais, Johann Großschädl, Neil Hanley, Markus Kasper, Marcel Medwed, Francesco Regazzoni, Jörn-Marc Schmidt, Stefan Tillich, and Marcin Wójcik. Hardware Trojans for inducing or amplifying side-channel leakage of cryptographic software. In *Trusted Systems*, pages 253–270. Springer, 2011.

[32] Xinmu Wang, Tatini Mal-Sarkar, Aswin Krishna, Seetharam Narasimhan, and Swarup Bhunia. Software exploitable hardware Trojans in embedded processor. In *2012 IEEE International Symposium on Defect and Fault Tolerance in VLSI and Nanotechnology Systems (DFT)*, pages 55–58. IEEE, 2012.

[33] Jonathan Brossard and Florentin Demetrescu. Hardware backdooring is practical, 2012. Presented in Black Hat, USA 2012.

[34] Anders Fongen. Identity Management and Integrity Protection in the Internet of Things. In *Proceedings of the International Conference on Emerging Security Technologies (EST 2012)*, pages 111–114, September 2012.

[35] Amir Moradi, Markus Kasper, and Christof Paar. Black-box side-channel attacks highlight the importance of countermeasures. In *Topics in Cryptology–CT-RSA 2012*, pages 1–18. Springer, 2012.

[36] Amir Moradi, David Oswald, Christof Paar, and Pawel Swierczynski. Side-channel attacks on the bitstream encryption mechanism of Altera Stratix II: Facilitating black-box analysis using software reverse-engineering. In *Proceedings of the ACM/SIGDA*

International Symposium on Field Programmable Gate Arrays, pages 91–100. ACM, 2013.

[37] Pawel Swierczynski, Marc Fyrbiak, Philipp Koppe, and Christof Paar. FPGA Trojans through detecting and weakening of cryptographic primitives. *IEEE Transactions on Computer-Aided Design of Integrated Circuits and Systems*, 34(8):1236–1249, 2015.

[38] Rajat Subhra Chakraborty, Indranil Saha, Ayan Palchaudhuri, and Gowtham Kumar Naik. Hardware Trojan insertion by direct modification of FPGA configuration bitstream. *Design & Test, IEEE*, 30(2):45–54, 2013.

[39] Anju P. Johnson, Sayandeep Saha, Rajat Subhra Chakraborty, Debdeep Mukhopadhyay, and Sezer Gören. Fault attack on AES via hardware Trojan insertion by dynamic partial reconfiguration of FPGA over ethernet. In *Proceedings of the 9th Workshop on Embedded Systems Security*, page 1. ACM, 2014.

[40] Anju P. Johnson, Rajat Subhra Chakraborty, and Debdeep Mukhopadhyay. A PUF-enabled Secure Architecture for FPGA-based IoT Applications. *IEEE Transactions on Multi-Scale Computing Systems*, 1(2):110–122, 2015.

[41] Pawel Swierczynski, Marc Fyrbiak, Philipp Koppe, Amir Moradi, and Christof Paar. Interdiction in practice? Hardware Trojan against a high-security USB flash drive. *Journal of Cryptographic Engineering*, 7(3):199–211, 2017.

[42] Joseph Clements and Yingjie Lao. Hardware Trojan Attacks on Neural Networks. *arXiv preprint arXiv:1806.05768*, 2018.

[43] Mohammad Motamedi, Daniel Fong, and Soheil Ghiasi. Fast and Energy-Efficient CNN Inference on IOT Devices. *arXiv preprint arXiv:1611.07151*, 2016.

[44] Partha Maji, Daniel Bates, Alex Chadwick, and Robert Mullins. ADaPT: optimizing CNN inference on IoT and mobile devices using approximately separable 1-D kernels. In *Proceedings of the 1st International Conference on Internet of Things and Machine Learning*, page 43. ACM, 2017.

[45] Motamedi, Mohammad and Portillo, Felix and Saffarpour, Mahya and Fong, Daniel and Ghiasi, Soheil. Resource-scalable cnn synthesis for iot applications. *arXiv preprint arXiv:1901.00738*, 2018.

[46] Nicolas Papernot, Patrick McDaniel, Somesh Jha, Matt Fredrikson, Z Berkay Celik, and Ananthram Swami. The Limitations of Deep Learning in Adversarial Settings. In *2016 IEEE European Symposium on Security and Privacy (EuroS&P)*, pages 372–387. IEEE, 2016.

[47] Tao Liu, Wujie Wen, and Yier Jin. SIN 2: Stealth infection on neural network? A low-cost agile neural Trojan attack methodology. In *2018 IEEE International Symposium on Hardware Oriented Security and Trust (HOST)*, pages 227–230. IEEE, 2018.

[48] Jing Ye, Yu Hu, and Xiaowei Li. Hardware Trojan in FPGA CNN Accelerator. In *2018 IEEE 27th Asian Test Symposium (ATS)*, pages 68–73. IEEE, 2018.

[49] Yang Zhao, Xing Hu, Shuangchen Li, Jing Ye, Lei Deng, Yu Ji, Jianyu Xu, Dong Wu, and Yuan Xie. Memory Trojan Attack on Neural Network Accelerators. In *2019 Design, Automation & Test in Europe Conference & Exhibition (DATE)*, pages 1415–1420. IEEE, 2019.

[50] Wenshuo Li, Jincheng Yu, Xuefei Ning, Pengjun Wang, Qi Wei, Yu Wang, and Huazhong Yang. Hu-fu: Hardware and software collaborative attack framework against neural networks. In *2018 IEEE Computer Society Annual Symposium on VLSI (ISVLSI)*, pages 482–487. IEEE, 2018.

[51] John Clark, Sylvain Leblanc, and Scott Knight. Risks associated with USB Hardware Trojan devices used by insiders. In *2011 IEEE International Systems Conference*, pages 201–208. IEEE, 2011.

[52] John Clark, Sylvain Leblanc, and Scott Knight. Compromise through USB-based Hardware Trojan Horse device. *Future Generation Computer Systems*, 27(5):555–563, 2011.

[53] Pravin Phule. A low cost hardware Trojan horse device based on unintended USB channels and a solution. *International Journal of Advanced Computer Research*, 2(7):114–118, 2012.

[54] Yiming Zhao, Xiaohang Wang, Yingtao Jiang, Yang Mei, Amit Kumar Singh, and Terrence Mak. On a New Hardware Trojan Attack on Power Budgeting of Many Core Systems. In *2018 31st IEEE International System-on-Chip Conference (SOCC)*, pages 1–6. IEEE, 2018.

[55] Vidya Govindan, Rajat Subhra Chakraborty, Pranesh Santikellur, and Aditya Kumar Chaudhary. A Hardware Trojan Attack on FPGA-Based Cryptographic Key Generation: Impact and Detection. *Journal of Hardware and Systems Security*, 2(3):225–239, 2018.

[56] Yu Liu, Yier Jin, and Yiorgos Makris. Hardware Trojans in wireless cryptographic ICs: silicon demonstration & detection method evaluation. In *2013 IEEE/ACM International Conference on Computer-Aided Design (ICCAD)*, pages 399–404. IEEE, 2013.

[57] Maryam Jalalitabar, Marco Valero, and Anu G Bourgeois. Demonstrating the Threat of Hardware Trojans in wireless sensor networks. In *2015 24th International Conference on Computer Communication and Networks (ICCCN)*, pages 1–8. IEEE, 2015.

[58] Yu Liu, Yier Jin, and Yiorgos Makris. Hardware Trojans in wireless cryptographic ICs: Silicon demonstration amp; detection method evaluation. In *Proceedings of IEEE/ACM ICCAD*, pages 399–404, 2013.

[59] Hassan Salmani. Hardware Trojans in Analog and Mixed-Signal Integrated Circuits. In *Trusted Digital Circuits*, pages 121–131. Springer, 2018.

[60] Tamzidul Hoque, Xinmu Wang, Abhishek Basak, Robert Karam, and Swarup Bhunia. Hardware Trojan attacks in embedded memory. In *2018 IEEE 36th VLSI Test Symposium (VTS)*, pages 1–6, April 2018.

[61] Roghayeh Saeidi and Hossein Gharaee Garakani. SRAM hardware Trojan. In *2016 8th International Symposium on Telecommunications (IST)*, pages 719–722, September 2016.

[62] Reza Rad, Jim Plusquellic, and Mohammad Tehranipoor. A sensitivity analysis of power signal methods for detecting hardware Trojans under real process and environmental conditions. *IEEE*

Transactions on Very Large Scale Integration (VLSI) Systems, 18(12):1735–1744, 2010.

[63] Seetharam Narasimhan, Xinmu Wang, Dongdong Du, Rajat Subhra Chakraborty, and Swarup Bhunia. TeSR: A robust temporal self-referencing approach for hardware Trojan detection. In *2011 IEEE International Symposium on Hardware-Oriented Security and Trust (HOST)*, pages 71–74. IEEE, 2011.

[64] Jeffrey A Kash, James C Tsang, and Daniel R Knebel. Method and apparatus for reverse engineering integrated circuits by monitoring optical emission, 2002. US Patent 6,496,022.

[65] Mainak Banga, Maheshwar Chandrasekar, Lei Fang, and Michael S Hsiao. Guided test generation for isolation and detection of embedded Trojans in ICs. In *Proceedings of the 18th ACM Great Lakes symposium on VLSI*, pages 363–366. ACM, 2008.

[66] Mainak Banga and Michael Hsiao. A novel sustained vector technique for the detection of hardware Trojans. In *2009 22nd International Conference on VLSI Design*, pages 327–332. IEEE, 2009.

[67] Mainak Banga and Michael Hsiao. A region based approach for the identification of hardware Trojans. In *Proceedings of International Symposium on HOST*, pages 40–47, 2008.

[68] Dongdong Du, Seetharam Narasimhan, Rajat Subhra Chakraborty, and Swarup Bhunia. Self-referencing: A Scalable Side-channel Approach for Hardware Trojan Detection. In *Proceedings of the International Workshop on Cryptographic Hardware and Embedded Systems (CHES'11)*, pages 173–187, Berlin, Heidelberg, 2010.

[69] Reza Rad, Jim Plusquellic, and Mohammad Tehranipoor. Sensitivity analysis to hardware Trojans using power supply transient signals. In *Proceedings of International Symposium on HOST*, pages 3–7, 2008.

[70] Jim Aarestad, Dhruva Acharyya, Reza Rad, and Jim Plusquellic. Detecting Trojans through leakage current analysis using multiple supply pad IDDQs. *IEEE Transactions on Information Forensics and Security*, 5(4):893–904, 2010.

[71] Yier Jin and Yiorgos Makris. Hardware Trojan detection using path delay fingerprint. In *HOST 2008. IEEE International Workshop on Hardware-Oriented Security and Trust*, pages 51–57. IEEE, 2008.

[72] Wangyang Zhang, Xin Li, Frank Liu, Emrah Acar, Rob A Rutenbar, and Ronald D Blanton. Virtual probe: A statistical framework for low-cost silicon characterization of nanoscale integrated circuits. *Computer-Aided Design of Integrated Circuits and Systems, IEEE Transactions on*, 30(12):1814–1827, 2011.

[73] Sheng Wei and Miodrag Potkonjak. Scalable hardware Trojan diagnosis. *IEEE Transactions on Very Large Scale Integration (VLSI) Systems*, 20(6):1049–1057, 2012.

[74] Miodrag Potkonjak, Ani Nahapetian, Micheal Nelson, and Tammara Massey. Hardware Trojan horse detection using gate-level characterization. In *Proceedings of the 46th ACM/IEEE DAC*, pages 688–693, 2009.

[75] Yousra Alkabani and Farinaz Koushanfar. Consistency-based characterization for IC Trojan detection. In *Proceedings of the 2009 International Conference on Computer-Aided Design*, pages 123–127. ACM, 2009.

[76] Sheng Wei and Miodrag Potkonjak. Self-Consistency and Consistency-Based Detection and Diagnosis of Malicious Circuitry. *IEEE Transactions on Very Large Scale Integration (VLSI) Systems*, 22(9):1845–1853, 2014.

[77] Dhruva Acharyya and Jim Plusquellic. Calibrating power supply signal measurements for process and probe card variations. In *DBT 2004. Proceedings. 2004 IEEE International Workshop on Current and Defect Based Testing*, pages 23–30. IEEE, 2004.

[78] Devendra Rai and John Lach. Performance of delay-based Trojan detection techniques under parameter variations. In *HOST'09. IEEE International Workshop on Hardware-Oriented Security and Trust*, pages 58–65. IEEE, 2009.

[79] Charles Lamech, Rza Rad, Mohammad Tehranipoor, and Jim Plusquellic. An Experimental Analysis of Power and Delay Signal-to-Noise Requirements for Detecting Trojans and Methods for

Achieving the Required Detection Sensitivities. *IEEE Transactions on Information Forensics and Security*, 6(3):1170–1179, 2011.

[80] Hassan Salmani, Mohammad Tehranipoor, and Jim Plusquellic. A novel technique for improving hardware Trojan detection and reducing Trojan activation time. *IEEE Transactions on Very Large Scale Integration (VLSI) Systems*, 20(1):112–125, 2012.

[81] Sheng Wei and Miodrag Potkonjak. Malicious circuitry detection using fast timing characterization via test points. In *2013 IEEE International Symposium on Hardware-Oriented Security and Trust (HOST)*, pages 113–118. IEEE, 2013.

[82] Jie Li and John Lach. At-speed delay characterization for IC authentication and Trojan horse detection. In *2008 IEEE International Workshop on Hardware-Oriented Security and Trust*, pages 8–14. IEEE, 2008.

[83] Byeongju Cha and Sandeep K. Gupta. Trojan detection via delay measurements: A new approach to select paths and vectors to maximize effectiveness and minimize cost. In *Design, Automation Test in Europe Conference Exhibition (DATE), 2013*, pages 1265–1270, March 2013.

[84] Seyed Mohammad Hossein Shekarian and Morteza Saheb Zamani. Improving hardware Trojan detection by retiming. *Microprocessors and Microsystems*, 39(3):145–156, 2015.

[85] Oliver Soll, Thomas Korak, Michael Muehlberghuber, and Marcus Hutter. EM-based detection of hardware Trojans on FPGAs. In *2014 IEEE International Symposium on Hardware-Oriented Security and Trust (HOST)*, pages 84–87. IEEE, 2014.

[86] Seetharam Narasimhan, Dongdong Du, Rajat Subhra Chakraborty, Somnath Paul, Francis Wolff, Christos Papachristou, Kaushik Roy, and Swarup Bhunia. Multiple-parameter side-channel analysis: A non-invasive hardware Trojan detection approach. In *Proceedings of International Symposium on HOST*, pages 13–18, 2010.

[87] Farinaz Koushanfar and Azalia Mirhoseini. A unified framework for multimodal submodular integrated circuits Trojan detection.

IEEE Transactions on Information Forensics and Security, 6(1):162–174, 2011.

[88] Norimasa Yoshimizu. Hardware Trojan detection by symmetry breaking in path delays. In *2014 IEEE International Symposium on Hardware-Oriented Security and Trust (HOST)*, pages 107–111. IEEE, 2014.

[89] Rajat Subhra Chakraborty and Swarup Bhunia. HARPOON: An Obfuscation-Based SoC Design Methodology for Hardware Protection. *IEEE Transactions on CAD of Integrated Circuits and Systems*, 28(10):1493–1502, 2009.

[90] Rajat Subhra Chakraborty and Swarup Bhunia. Security Against Hardware Trojan through a Novel Application of Design Obfuscation. In *Proceedings of the IEEE/ACM International Conference on Computer-Aided Design (ICCAD)'09*, pages 113–116, New York, 2009.

[91] Sophie Dupuis, Papa-Sidi Ba, Giorgio Di Natale, Marie-Lise Flottes, and Bruno Rouzeyre. A novel hardware logic encryption technique for thwarting illegal overproduction and Hardware Trojans. In *On-Line Testing Symposium (IOLTS), 2014 IEEE 20th International*, pages 49–54. IEEE, 2014.

[92] Jeyavijayan Rajendran, Huan Zhang, Chi Zhang, Garrett S Rose, Youngok Pino, Ozgur Sinanoglu, and Ramesh Karri. Fault Analysis-Based Logic Encryption. *IEEE Transactions on Computers*, 64(2):410–424, 2015.

[93] Pramod Subramanyan, Sayak Ray, and Sharad Malik. Evaluating the security of logic encryption algorithms. In *2015 IEEE International Symposium on Hardware Oriented Security and Trust (HOST)*, pages 137–143. IEEE, 2015.

[94] Kan Xiao and M. Tehranipoor. BISA: Built-in self-authentication for preventing hardware Trojan insertion. In *Proceedings of International Symposium on HOST*, pages 45–50, 2013.

[95] Frank Imeson, Ariq Emtenan, Siddharth Garg, and Mahesh V Tripunitara. Securing Computer Hardware Using 3D Integrated Circuit (IC) Technology and Split Manufacturing for Obfuscation. In *USENIX Security*, volume 13, 2013.

[96] Jeyavijayan Rajendran, Ozgur Sinanoglu, and Ramesh Karri. Is split manufacturing secure? In *Design, Automation & Test in Europe Conference & Exhibition (DATE), 2013*, pages 1259–1264. IEEE, 2013.

[97] Bao Liu and Brandon Wang. Embedded reconfigurable logic for ASIC design obfuscation against supply chain attacks. In *Proceedings of the conference on Design, Automation & Test in Europe*, page 243. European Design and Automation Association, 2014.

[98] Bao Liu and Brandon Wang. Reconfiguration-Based VLSI Design for Security. *IEEE Journal on Emerging and Selected Topics in Circuits and Systems*, 5(1):98–108, 2015.

[99] Jeyavijayan Rajendran, Vinayaka Jyothi, Ozgur Sinanoglu, and Ramesh Karri. Design and analysis of ring oscillator based Design-for-Trust technique. In *VLSI Test Symposium (VTS), 2011 IEEE 29th*, pages 105–110. IEEE, 2011.

[100] Xuehui Zhang and Mohammad Tehranipoor. RON: An on-chip ring oscillator network for hardware Trojan detection. In *Design, Automation & Test in Europe Conference & Exhibition (DATE), 2011*, pages 1–6. IEEE, 2011.

[101] Seetharam Narasimhan, Wen Yueh, Xinmu Wang, Saibal Mukhopadhyay, and Swarup Bhunia. Improving IC security against Trojan attacks through integration of security monitors. *IEEE Design & Test of Computers*, 29(5):37–46, 2012.

[102] Adam Waksman and Simha Sethumadhavan. Tamper evident microprocessors. In *2010 IEEE Symposium on Security and Privacy (SP)*, pages 173–188. IEEE, 2010.

[103] Yier Jin and Yiorgos Makris. Is single Trojan detection scheme enough? In *Proceedings of the IEEE International Conference on Computer Design (ICCD)*, pages 305–308, 2011.

[104] Bin Zhou, Wei Zhang, Srikanthan Thambipillai, Jason Teo Kian Jin, Vivek Chaturvedi, and Tao Luo. Cost-efficient Acceleration of Hardware Trojan Detection Through Fan-Out Cone Analysis and Weighted Random Pattern Technique. *IEEE Transactions on Computer-Aided Design of Integrated Circuits and Systems*, 35(5):792–805, May 2016.

[105] Seyed Mohammad Hossein Shekarian, Morteza Saheb Zamani, and Shirin Alami. Neutralizing a design-for-hardware-trust technique. In *2013 17th CSI International Symposium on Computer Architecture and Digital Systems (CADS)*, pages 73–78. IEEE, 2013.

[106] Hassan Salmani, Mohammad Tehranipoor, and Jim Plusquellic. A layout-aware approach for improving localized switching to detect hardware Trojans in integrated circuits. In *2010 IEEE International Workshop on Information Forensics and Security (WIFS)*, pages 1–6, Dec 2010.

[107] Mainak Banga and Michael Hsiao. VITAMIN: Voltage Inversion Technique to Ascertain Malicious Insertions in ICs. In *Proceedings of HOST*, pages 104 –107, Washington, DC, USA, 2009.

[108] Miron Abramovici and Paul Bradley. Integrated circuit security: new threats and solutions. In *Proceedings of the 5th Annual Workshop on Cyber Security and Information Intelligence Research: Cyber Security and Information Intelligence Challenges and Strategies*, page 55. ACM, 2009.

[109] D. McIntyre, F. Wolff, C. Papachristou, Swarup Bhunia, and D. Weyer. Dynamic evaluation of hardware trust. In *2009 IEEE International Workshop on Hardware-Oriented Security and Trust*, pages 108–111. IEEE, 2009.

[110] Gedare Bloom, Bhagirath Narahari, Rahul Simha, and Joseph Zambreno. Providing secure execution environments with a last line of defense against Trojan circuit attacks. *Computers & Security*, 28(7):660–669, 2009.

[111] Richard Wayne Jarvis and Michael G McIntyre. Split manufacturing method for advanced semiconductor circuits, 2007. US Patent 7,195,931.

[112] Franco Stellari et al. Functional block extraction for hardware security detection using time-integrated and time-resolved emission measurements. In *2014 IEEE 32nd on VLSI Test Symposium (VTS)*, pages 1–6. IEEE, 2014.

[113] Adam Waksman and Simha Sethumadhavan. Silencing hardware backdoors. In *2011 IEEE Symposium on Security and Privacy (SP)*, pages 49–63. IEEE, 2011.

[114] Lok-Won Kim, John D Villasenor, and Cetin K Koç. A Trojan-resistant system-on-chip bus architecture. In *Military Communications Conference, 2009. MILCOM 2009. IEEE*, pages 1–6. IEEE, 2009.

[115] Yier Jin and Yiorgos Makris. Proof carrying-based information flow tracking for data secrecy protection and hardware trust. In *2012 IEEE 30th on VLSI Test Symposium (VTS)*, pages 252–257. IEEE, 2012.

[116] Eric Love, Yier Jin, and Yiorgos Makris. Proof-carrying hardware intellectual property: A pathway to trusted module acquisition. *IEEE Transactions on Information Forensics and Security*, 7(1):25–40, 2012.

[117] Banga, Mainak and Hsiao, Michael. Trusted RTL: Trojan detection methodology in pre-silicon designs. In *2010 IEEE International Symposium on Hardware-Oriented Security and Trust (HOST)*, pages 56–59. IEEE, 2010.

[118] Xuehui Zhang and Mohammad Tehranipoor. Case study: Detecting hardware Trojans in third-party digital IP cores. In *2011 IEEE International Symposium on Hardware-Oriented Security and Trust (HOST)*, pages 67–70. IEEE, 2011.

[119] Cynthia Sturton, Matthew Hicks, David Wagner, and Samuel T King. Defeating UCI: Building stealthy and malicious hardware. In *2011 IEEE Symposium on Security and Privacy (SP)*, pages 64–77. IEEE, 2011.

[120] Jie Zhang and Qiang Xu. On hardware Trojan design and implementation at register-transfer level. In *2013 IEEE International Symposium on Hardware-Oriented Security and Trust (HOST)*, pages 107–112. IEEE, 2013.

[121] Jie Zhang, Feng Yuan, Lingxiao Wei, Zelong Sun, and Qiang Xu. VeriTrust: Verification for Hardware Trust. In *Proceedings of the 50th Annual Design Automation Conference*, page 61. ACM, 2013.

[122] Adam Waksman, Matthew Suozzo, and Simha Sethumadhavan. FANCI: Identification of stealthy malicious logic using boolean functional analysis. In *Proceedings of the 2013 ACM*

SIGSAC conference on Computer & communications security, pages. 697–708. 2013.

[123] Jie Zhang, Feng Yuan, and Qiang Xu. DeTrust: Defeating hardware trust verification with stealthy implicitly-triggered hardware Trojans. In *Proceedings of the 2014 ACM SIGSAC Conference on Computer and Communications Security*, pages 153–166. ACM, 2014.

[124] Masaru Oya, Youhua Shi, Masao Yanagisawa, and Nozomu Togawa. A score-based classification method for identifying hardware-Trojans at gate-level netlists. In *Proceedings of the 2015 Design, Automation & Test in Europe Conference & Exhibition*, pages 465–470. EDA Consortium, 2015.

[125] Jeyavijayan Rajendran, Huan Zhang, Ozgur Sinanoglu, and Ramesh Karri. High-level synthesis for security and trust. In *2013 IEEE 19th International On-Line Testing Symposium (IOLTS)*, pages 232–233. IEEE, 2013.

[126] Chen Liu, Jeyavijayan Rajendran, Chengmo Yang, and Ramesh Karri. Shielding heterogeneous MPSoCs from untrustworthy 3PIPs through security-driven task scheduling. *IEEE Transactions on Emerging Topics in Computing*, 2(4):461–472, 2014.

[127] Xiaotong Cui, Kun Ma, Liang Shi, and Kaijie Wu. High-level synthesis for run-time hardware Trojan detection and recovery. In *Proceedings of the 51st Annual Design Automation Conference*, pages 1–6. ACM, 2014.

[128] Kento Hasegawa, Masaru Oya, Masao Yanagisawa, and Nozomu Togawa. Hardware Trojans classification for gate-level netlists based on machine learning. In *2016 IEEE 22nd International Symposium on On-Line Testing and Robust System Design (IOLTS)*, pages 203–206, July 2016.

[129] Kento Hasegawa, Masao Yanagisawa, and Nozomu Togawa. Trojan-feature extraction at gate-level netlists and its application to hardware-Trojan detection using random forest classifier. In *2017 IEEE International Symposium on Circuits and Systems (ISCAS)*, pages 1–4. IEEE, 2017.

[130] Kento Hasegawa, Masao Yanagisawa, and Nozomu Togawa. Hardware Trojans classification for gate-level netlists using multi-layer neural networks. In *2017 IEEE 23rd International Symposium on On-Line Testing and Robust System Design (IOLTS)*, pages 227–232. IEEE, 2017.

[131] Tao Han, Yuze Wang, and Peng Liu. Hardware Trojans Detection at Register Transfer Level Based on Machine Learning. In *2019 IEEE International Symposium on Circuits and Systems (ISCAS)*, pages 1–5, May 2019.

[132] Mohammad Tehranipoor, Ramesh Karri, Farinaz Koushanfar, and Miodrag Potkonjak. TrustHub. Available: https://www.trust-hub.org, 2016.

[133] K Reshma, M Priyatharishini, and M Nirmala Devi. Hardware Trojan Detection Using Deep Learning Technique. In *Soft Computing and Signal Processing*, pages 671–680. Springer, 2019.

[134] Chen Dong, Guorong He, Ximeng Liu, Yang Yang, and Wenzhong Guo. A Multi-Layer Hardware Trojan Protection Framework for IoT Chips. *IEEE Access*, 2019.

[135] Robb Hammond. Counterfeit Electronic Component Detection. https://www.aeri.com/counterfeit-electronic-component-detection/, Accessed: April 2020.

[136] Ujjwal Guin et al. Counterfeit Integrated Circuits: A Rising Threat in the Global Semiconductor Supply Chain. *Proceedings of the IEEE*, 102(8):1207–1228, August 2014.

[137] IHS Markit. Reports of Counterfeit Parts Quadruple Since 2009, Challenging US Defense Industry and National Security, 2012. https://technology.informa.com/389481/reports-of-counterfeit-parts-quadruple-since-2009-challenging-u_4 Accessed: June 2020.

[138] Ujjwal Guin, Daniel DiMase, and Mohammad Tehranipoor. A Comprehensive Framework for Counterfeit Defect Coverage Analysis and Detection Assessment. *Journal of Electronic Testing*, 30(1):25–40, 2014.

[139] Randy Torrance and Dick James. The State-of-the-Art in IC Reverse Engineering. In *Proceedings of the International Workshop on Cryptographic Hardware and Embedded Systems (CHES)*, pages 363–381, 2009.

[140] Navid Asadizanjani, Mark Tehranipoor, and Domenic Forte. PCB Reverse Engineering Using Nondestructive X-ray Tomography and Advanced Image Processing. *IEEE Transactions On Components, Packaging and Manufacturing Technology*, 1–8, 2017.

[141] Ujjwal Guin, Daniel DiMase, and Mohammad Tehranipoor. Counterfeit Integrated Circuits: Detection, Avoidance, and the Challenges Ahead. *Journal of Electronic Testing and Test Applications*, 30(1):9–23, Feburary 2014.

[142] Navid Asadizanjani, Mark Tehranipoor, and Domenic Forte. Counterfeit Electronics Detection Using Image Processing and Machine Learning. *Journal of Physics: Conference Series*, 787(1):1–6, 2017.

[143] Sina Shahbazmohamadi, Domenic Forte, and Mark Tehranipoor. Advanced Physical Inspection Methods for Counterfeit Detection. In *Proceedings of International Symposium for Testing and Failure Analysis (ISFTA)*, pages 55–64, 2014.

[144] Nathan Dunn, Navid Asadizanjani, Sachin Gattigowda, Mark Tehranipoor, and Domenic Forte. A Database for Counterfeit Electronics and Automatic Defect Detection Based on Image Processing and Machine Learning, 2016. In *Proceedings of the 42nd International Symposium for Testing and Failure Analysis, Texas, USA*, pages 1–8. 2016.

[145] Pallabi Ghosh and Rajat Subhra Chakraborty. Counterfeit IC Detection By Image Texture Analysis. In *2017 Euromicro Conference on Digital System Design (DSD)*, pages 283–286, 2017.

[146] Pallabi Ghosh and Rajat Subhra Chakraborty. Recycled and Remarked Counterfeit Integrated Circuit Detection by Image Processing based Package Texture and Indent Analysis. *IEEE Transactions on Industrial Informatics*, pages 1–1, 2018.

[147] Yujie Wang, Pu Chen, Jiang Hu, and Jeyavijayan (JV) Rajendran. The Cat and Mouse in Split Manufacturing. In *Proceedings of the 53rd Annual Design Automation Conference*, DAC '16, pages 165:1–165:6, 2016.

[148] Yongming Jin, Wei Xin, Huiping Sun, and Zhong Chen. PUF-Based RFID Authentication Protocol against Secret Key Leakage. In Quan Z. Sheng, Guoren Wang, Christian S. Jensen, and Guandong Xu, editors, *APWeb*, volume 7235 of *Lecture Notes in Computer Science*, pages 318–329. Springer, 2012.

[149] Daihyun Lim. *Extracting Secret Keys from Integrated Circuits.* Master's thesis, MIT, USA, 2004.

[150] Ravikanth S. Pappu. *Physical one-way functions.* PhD thesis, Massachusetts Institute of Technology, March 2001.

[151] Oliver Willers, Christopher Huth, Jorge Guajardo, and Helmut Seidel. MEMS-based Gyroscopes as Physical Unclonable Functions. *IACR Cryptology ePrint Archive*, 2016:261, 2016.

[152] Aydin Aysu, Nahid Farhady Ghalaty, Zane R. Franklin, Moein Pahlavan Yali, and Patrick Schaumont. Digital fingerprints for low-cost platforms using MEMS sensors. In *Proceedings of the Workshop on Embedded Systems Security (WESS)*, pages 2:1–2:6, 2013.

[153] Wenjie Che, Jim Plusquellic, and Swarup Bhunia. A Non-volatile Memory Based Physically Unclonable Function Without Helper Data. In *Proceedings of the 2014 IEEE/ACM International Conference on Computer-Aided Design*, ICCAD '14, pages 148–153, Piscataway, NJ, USA, 2014. IEEE Press.

[154] Jimson Mathew, Rajat Subhra Chakraborty, Durga Prasad Sahoo, Yuanfan Yang, and Dhiraj K. Pradhan. A Novel Memristor-Based Hardware Security Primitive. *ACM Transactions on Embedded Computing Systems*, 14(3):60, 2015.

[155] Swaroop Ghosh. Spintronics and Security: Prospects, Vulnerabilities, Attack Models, and Preventions. *Proceedings of the IEEE*, 104(10):1864–1893, 2016.

[156] Roel Maes. *Physically Unclonable Functions - Constructions, Properties and Applications.* Springer, 2013.

[157] Aswin Raghav Krishna, Seetharam Narasimhan, Xinmu Wang, and Swarup Bhunia. MECCA: A Robust Low-Overhead PUF Using Embedded Memory Array. In Bart Preneel and Tsuyoshi Takagi, editors, *Cryptographic Hardware and Embedded Systems – CHES 2011: 13th International Workshop, Nara, Japan, September 28–October 1, 2011. Proceedings*, pages 407–420, 2011.

[158] Qingqing Chen, György Csaba, Paolo Lugli, Ulf Schlichtmann, and Ulrich Rührmair. The Bistable Ring PUF: A new architecture for strong Physical Unclonable Functions. In *Proceedings of IEEE International Symposium on Hardware-Oriented Security and Trust (HOST)*, pages 134 –141, June 2011.

[159] Daniel E. Holcomb and Kevin Fu. Bitline PUF: Building Native Challenge-Response PUF Capability into Any SRAM. In *Proceedings of Cryptographic Hardware and Embedded Systems (CHES) 2014*, pages 510–526, 2014.

[160] Mehrdad Majzoobi, Farinaz Koushanfar, and Miodrag Potkonjak. Testing Techniques for Hardware Security. In *Proceedings of IEEE International Test Conference (ITC)*, pages 1–10, October 2008.

[161] Dai Yamamoto, Masahiko Takenaka, Kazuo Sakiyama, and Naoya Torii. Security Evaluation of Bistable Ring PUFs on FPGAs using Differential and Linear Analysis. In *Proceedings of the Federated Conference on Computer Science and Information Systems (FedCSIS)*, pages 911–918, 2014.

[162] Phuong Ha Nguyen, Durga Prasad Sahoo, Rajat Subhra Chakraborty, and Debdeep Mukhopadhyay. Security Analysis of Arbiter PUF and Its Lightweight Compositions Under Predictability Test. *ACM Transactions on Design Automation of Electronic Systems*, 22(2):20:1–20:28, December 2016.

[163] Christopher M. Bishop. *Pattern Recognition and Machine Learning (Information Science and Statistics)*. Springer, 1 edition, 2007.

[164] Thomas Bäck. *Evolutionary Algorithms in Theory and Practice: Evolution Strategies, Evolutionary Programming, Genetic Algorithms*. Oxford University Press, 1996.

[165] Hans-Paul Schwefel. *Evolution and Optimum Seeking*. Sixth-generation computer technology series. Wiley, 1995.

[166] Fatemeh Ganji, Shahin Tajik, and Jean-Pierre Seifert. PAC learning of Arbiter PUFs. *Journal of Cryptographic Engineering*, 6(3):249–258, 2016.

[167] Dana Angluin. Learning regular sets from queries and counterexamples. *Information and Computation*, 75(2):87–106, 1987.

[168] Fatemeh Ganji. *On the Learnability of Physically Unclonable Functions*. Springer, 2018.

[169] Ulrich Rührmair, Xiaolin Xu, Jan Sölter, Ahmed Mahmoud, Mehrdad Majzoobi, Farinaz Koushanfar, and Wayne P. Burleson. Efficient Power and Timing Side Channels for Physical Unclonable Functions. In *Proceedings of 16th International Workshop on Cryptographic Hardware and Embedded Systems (CHES)*, pages 476–492, 2014.

[170] Johannes Tobisch and Georg Becker. On the Scaling of Machine Learning Attacks on PUFs with Application to Noise Bifurcation. In *Proceedings of 11th International Workshop on Radio Frequency Identification: Security and Privacy Issues (RFIDsec)*, pages 17–31, 2015.

[171] Georg Becker et al. Active and Passive Side-Channel Attacks on Delay Based PUF Designs. *IACR Cryptology ePrint Archive*, 2014:287, 2014.

[172] Georg Becker. The Gap Between Promise and Reality: On the Insecurity of XOR Arbiter PUFs. In *Cryptographic Hardware and Embedded Systems (CHES)*, pages 535–555, Springer, Berlin, Heidelberg, 2015.

[173] Ulrich Rührmair, Frank Sehnke, Jan Sölter, Gideon Dror, Srinivas Devadas, and Jürgen Schmidhuber. Modeling Attacks on Physical Unclonable Functions. In *Proceedings of 17th ACM Conference on Computer and Communications Security (CCS)*, pages 237–249, New York, ACM, 2010.

[174] Gabriel Hospodar, Roel Maes, and Ingrid Verbauwhede. Machine learning attacks on 65nm Arbiter PUFs: Accurate Modeling Poses Strict Bounds on Usability. In *Proceedings of the 4th IEEE International Workshop on Information Forensics and Security (WIFS)*, pages 37–42, 2012.

[175] G. Edward Suh and Srinivas Devadas. Physical Unclonable Functions for Device Authentication and Secret Key Generation. In *Proceedings of Design Automation Conference (DAC)*, pages 9–14, New York, ACM, 2007.

[176] Raghavan Kumar and Wayne Burleson. Side-Channel Assisted Modeling Attacks on Feed-Forward Arbiter PUFs Using Silicon Data. In *Proceedings of 11th International Workshop on Radio Frequency Identification: Security and Privacy Issues (RFIDsec)*, pages 53–67, 2015.

[177] Fatemeh Ganji, Shahin Tajik, and Jean-Pierre Seifert. Why Attackers Win: On the Learnability of XOR Arbiter PUFs. In *Proceedings of International Conference on Trust and Trustworthy Computing (TRUST)*, pages 22–39, 2015.

[178] Pranesh Santikellur, Lakshya, Shashi Ranjan Prakash and Rajat Subhra Chakraborty. A Computationally Efficient Tensor Regression Network Based Modeling Attack on XOR APUF. In *2019 Asian Hardware Oriented Security and Trust Symposium (AsianHOST)*, pages 1–6, 2019.

[179] Pranesh Santikellur, Aritra Bhattacharyay, and Rajat Subhra Chakraborty. Deep Learning based Model Building Attacks on Arbiter PUF Compositions. Cryptology ePrint Archive, Report 2019/566, 2019. Available: https://eprint.iacr.org/2019/566.

[180] Mehrdad Majzoobi, Farinaz Koushanfar, and Miodrag Potkonjak. Lightweight Secure PUFs. In *Proceedings of the 2008 IEEE/ACM International Conference on Computer-Aided Design (ICCAD)*, pages 670–673, Piscataway, NJ, USA, IEEE Press, 2008.

[181] Durga Prasad Sahoo, Phuong Ha Nguyen, Debdeep Mukhopadhyay, and Rajat Subhra Chakraborty. A Case of Lightweight PUF Constructions: Cryptanalysis and Machine Learning Attacks. *IEEE Transactions on Computer-Aided Design of Integrated Circuits and Systems*, 34(8):1334–1343, August 2015.

[182] Abhranil Maiti, Inyoung Kim, and Patrick Schaumont. A Robust Physical Unclonable Function With Enhanced Challenge-Response Set. *IEEE Transactions on Information Forensics and Security*, 7(1):333–345, Feburary 2012.

[183] Phuong Ha Nguyen, Durga Prasad Sahoo, Rajat Subhra Chakraborty, and Debdeep Mukhopadhyay. Efficient Attacks on Robust Ring Oscillator PUF with Enhanced Challenge-Response Set. In *Proceedings of Design, Automation Test in Europe Conference Exhibition (DATE)*, 2015.

[184] Mukund Kalyanaraman and Michael Orshansky. Novel strong PUF based on Nonlinearity of MOSFET Subthreshold Operation. In *2013 IEEE International Symposium on Hardware-Oriented Security and Trust, HOST 2013, Austin, TX, USA, June 2–3, 2013*, pages 13–18, 2013.

[185] Arunkumar Vijayakumar, Vinay C. Patil, Charles B. Prado, and Sandip Kundu. Machine Learning Resistant Strong PUF: Possible or a pipe dream? In *Proceedings of International Symposium on Hardware Oriented Security and Trust (HOST)*, pages 19–24, 2016.

[186] Georg Becker. On the Pitfalls of Using Arbiter-PUFs as Building Blocks. *IEEE Transactions on Computer-Aided Design of Integrated Circuits and Systems*, 34(8):1295–1307, 2015.

[187] Jeroen Delvaux and Ingrid Verbauwhede. Attacking PUF-Based Pattern Matching Key Generators via Helper Data Manipulation. In *Proceedings of Topics in Cryptology - CT-RSA - The Cryptographer's Track at the RSA Conference*, pages 106–131, San Francisco, CA, USA, February 2014.

[188] Jeroen Delvaux and Ingrid Verbauwhede. Key-recovery Attacks on Various RO PUF Constructions via Helper Data Manipulation. In *Design, Automation and Test in Europe (DATE)*, 2014.

[189] Raghavan Kumar and Wayne Burleson. On Design of a Highly Secure PUF based on Non-linear Current Mirrors. In *2014 IEEE International Symposium on Hardware-Oriented Security and Trust (HOST)*, pages 38–43, 2014.

[190] Arunkumar Vijayakumar and Sandip Kundu. A Novel Modeling Attack Resistant PUF Design based on Non-linear Voltage Transfer Characteristics. In *Proceedings of Design, Automation & Test in Europe Conference & Exhibition (DATE)*, pages 653–658, 2015.

[191] Ramzi Bassil, Wissam El-Beaino, Ayman Kayssi, and Ali Chehab. A PUF-based ultra-lightweight mutual-authentication RFID protocol. In *2011 International Conference for Internet Technology and Secured Transactions*, pages 495–499, 2011.

[192] Jeroen Delvaux, Roel Peeters, Dawu Gu, and Ingrid Verbauwhede. A Survey on Lightweight Entity Authentication with Strong PUFs. *ACM Computing Surveys*, 48(2):26:1–26:42, October 2015.

[193] Mario Barbareschi, Pierpaolo Bagnasco, and Antonino Mazzeo. Authenticating IoT Devices with Physically Unclonable Functions Models. In *2015 10th International Conference on P2P, Parallel, Grid, Cloud and Internet Computing (3PGCIC)*, pages 563–567, 2015.

[194] Miaoqing Huang and Shiming Li. A delay-based PUF design using multiplexer chains. In *2013 International Conference on Reconfigurable Computing and FPGAs (ReConFig)*, pages 1–6, 2013.

[195] Muhammad N. Aman, Kee Chaing Chua, and Biplab Sikdar. Physical Unclonable Functions for IoT Security. In *Proceedings of the 2Nd ACM International Workshop on IoT Privacy, Trust, and Security*, IoTPTS '16, pages 10–13, 2016.

[196] John Ross Wallrabenstein. Practical and Secure IoT Device Authentication Using Physical Unclonable Functions. In *2016 IEEE 4th International Conference on Future Internet of Things and Cloud (FiCloud)*, pages 99–106, 2016.

[197] Urbi Chatterjee, Rajat Subhra Chakraborty, and Debdeep Mukhopadhyay. A PUF-Based Secure Communication Protocol for IoT. *ACM Transactions on Embedded Computing Systems*, 16(3):67:1–67:25, April 2017.

[198] Wenjie Che, Mitchell Martin, Goutham Pocklassery, Venkata K. Kajuluri, Fareena Saqib, and Jim Plusquellic. A Privacy-Preserving, Mutual PUF-Based Authentication Protocol. *Cryptography*, 1(1), 2016.

[199] Jim Aarestad, Philip Ortiz, Dhruva Acharyya, and Jim Plusquellic. HELP: A Hardware-Embedded Delay PUF. *IEEE Design & Test*, 30(2):17–25, April 2013.

[200] Debdeep Mukhopadhyay. PUFs as Promising Tools for Security in Internet of Things. *IEEE Design & Test*, 33(3):103–115, 2016.

[201] Offensive Security. Kali Linux: Penetration Testing. Available: http://www.kali.org, 2016.

[202] Urbi Chatterjee, Vidya Govindan, Rajat Sadhukhan, Debdeep Mukhopadhyay, Rajat Subhra Chakraborty, Debashis Mahata, and Mukesh Prabhu. Building PUF based Authentication and Key Exchange Protocol for IoT without Explicit CRPs in Verifier Database. *IEEE Transactions on Dependable and Secure Computing*, pages 1–1, 2018.

[203] Paul Kocher, Joshua Jaffe, and Benjamin Jun. Differential Power Analysis. In Michael Wiener, editor, *Advances in Cryptology – Proceedings of Annual International Cryptology Conference (CRYPTO'99)*, pages 388–397, Springer, Berlin, Heidelberg, 1999.

[204] Paul Kocher. Timing Attacks on Implementations of Diffie-Hellman, RSA, DSS, and Other Systems. In Neal Koblitz, editor, *Proceedings of Advances in Cryptology — Annual International Cryptology Conference (CRYPTO'96)*, pages 104–113, Springer, Berlin, Heidelberg, 1996.

[205] Daniel J. Bernstein. Cache-timing attacks on AES. Available: https://cr.yp.to/antiforgery/cachetiming-20050414.pdf, 2005.

[206] Standaert, François-Xavier and Petit, Christophe and Veyrat-Charvillon, Nicolas. *Masking with Randomized Look Up Tables*, pages 283–299, Springer, Berlin Heidelberg, 2012.

[207] Kris Tiri and Ingrid Verbauwhede. A Logic Level Design Methodology for a Secure DPA Resistant ASIC or FPGA Implementation. In *Proceedings of the Conference on Design, Automation and Test in Europe - Volume 1 (DATE'04)*, page 10246, 2004.

[208] Ahmed Mahmoud, Ulrich Ruhrmair, Mehrdad Majzoobi, and Farinaz Koushanfar. Combined Modeling and Side Channel Attacks on Strong PUFs. Available: https://eprint.iacr.org/2013/632.pdf, 2013.

[209] Xiaolin Xu and W. Burleson. Hybrid Side-Channel/Machine-Learning Attacks on PUFs: A New Threat? In *Proceedings of Design, Automation and Test in Europe Conference and Exhibition (DATE)*, pages 1–6, March 2014.

[210] Ulrich Rührmair and Jan Sölter. PUF modeling attacks: An introduction and overview. In *Proceedings of Design, Automation and Test in Europe Conference and Exhibition (DATE)*, pages 1–6, 2014.

[211] Michael Tunstall, Debdeep Mukhopadhyay, and Sk. Subidh Ali. Differential Fault Analysis of the Advanced Encryption Standard using a Single Fault. In *Information Security Theory and Practice. Security and Privacy of Mobile Devices in Wireless Communication*, pages 224–233. Springer, 2011.

[212] Jeroen Delvaux and Ingrid Verbauwhede. Fault Injection Modeling Attacks on 65 nm Arbiter and RO Sum PUFs via Environmental Changes. *IEEE Transactions on Circuits and Systems I: Regular Papers*, 61-I(6):1701–1713, 2014.

[213] Yingjie Lao, Bo Yuan, Chris H. Kim, and Keshab K. Parhi. Reliable PUF-based Local Authentication with Self-Correction. *IEEE Transactions on Computer-Aided Design of Integrated Circuits and Systems*, 36(2):201–213, Feburary 2016.

CHAPTER 9

Security Challenges in Hardware Used for Smart Environments

Sree Ranjani Rajendran

Indian Institute of Technology Madras

CONTENTS

9.1 INTRODUCTION TO IoT AND HARDWARE SECURITY ISSUES

Internet of Things (IoT) is a most emerging technology in which Internet connects everything from anywhere at all times using Internet Protocol (IP). This basic idea leads a root for the emerging smart homes, smart cities, and smart power grids, which together form a smart environment as shown in Figure 9.1. However, the smart environment or smart space is equipped with sensors, actuators, and various computing components [1]. Thus, the personalized services are identified and delivered by these devices in the IoT, when they are interacting and exchanging information with the environment. In 2020, it is expected that more than 50 billion "things" are connected to IoT [2]. Hence, such things/devices used in IoT to perform an application of smart environment should be secure, private, and trustworthy; 128-bit IPv6 protocol gives us the opportunity to assign ubiquitous IDs to trillions of devices

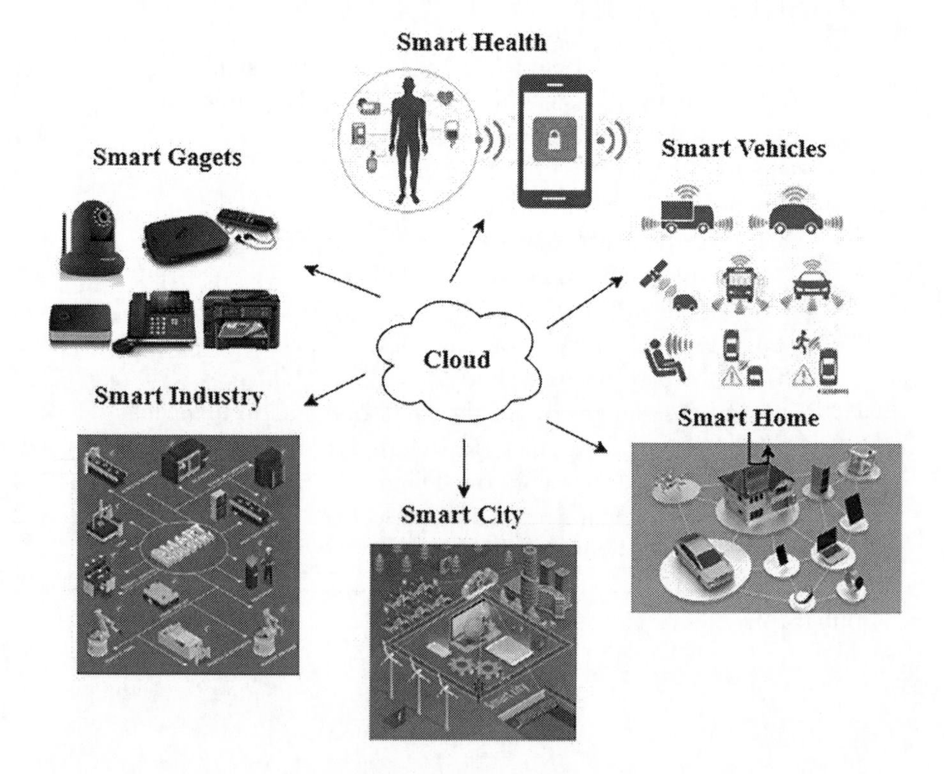

Figure 9.1 Examples of IoT applications.

in the network [3]. However, it is very difficult to manage the privacy protection, confidentiality, issues of information base due to the huge number of devices in a protocol, etc. But the devices developed by the nano-scale technology also provide security and piracy challenges to the great extent. This is because the hardware threats are becoming more vulnerable as software threats. This chapter describes various hardware security issues in the IoT devices and their countermeasures to overcome these threats.

Hardware security is a road ahead challenge for the integrated circuit (IC) manufacturers over a decade. Due to globalization, the fabrication process of IC manufacturing is outsourced, and this leads a back door for various threats [4–6]. Intellectual property (IP) piracy, malicious circuit insertion (aka hardware Trojans or HTs), IC overbuilding, and reverse engineering (RE) are the major threats in IC design flow, and semiconductor industry is losing $4 billion annually [5]. These hardware threats spoil the efficiency of any system as hardware plays a vital role in implementation of cryptographic applications [6], IoT, and embedded security applications [7].

9.1.1 Hardware Issues with Real-Time Examples and Their Countermeasures

A wide range of attacks from the cyber security domain can be mounted on real-time embedded system devices. Many of them are thwarted by IT-based countermeasures including antimalware programs, firewalls, intrusion detection systems, and anomaly detection tools. Recently, in the H2020 European CIPSEC (Enhancing Critical Infrastructure Protection with Innovative SECurity Framework) project, the above tools are considered as a unified whole, thus structuring a cyberattack protection framework that is applicable to many different Critical infrastructure environments that involve real-time embedded systems [8]. In real-time systems, hardware security challenges are addressed in [9]. The cache side-channel attacks are such case but also the newly discovered Rowhammer fault injection attack that can be mounted even remotely to gain full access to a device DRAM (dynamic random access memory). Under the light of the above dangers that are focused on the device hardware structure is vulnerable to hardware attacks [9]. Section 9.3 clearly describes the hardware threats in IoT devices with its countermeasures.

9.2 SECURITY CHALLENGES IN IoT DEVICES

Confidentiality, integrity, authentication, non-repudiation and availability are the main objectives of any IoT devices. These security characteristics are achieved by applying different cryptographic mechanism, whereas implementing these cryptographic algorithms to provide a security of IoT devices is more challenging. Many of IoT devices are of low-resource devices (LRDs) or constrained-resource devices (CRDs) category as they are extremely small in size. So it is more difficult to add different layers of security due to the physical limitations of IoT devices. Limited central processing unit (CPU) and memory are the constrained resource which limits their ability to process information and complex security algorithms. In IoT, there exists a serious trade between the constrained resource and security, and this may reflect in the confidentiality also [10]. However, confidentiality and integrity are provided by implementing public key infrastructure (PKI). In many IoT devices, the encryption process with public key demands computational and memory resources beyond their capabilities. Updating the IoT devices in small and remote devices is a great challenge, and some require a human interface [11]. Recent research focuses mainly on implementing encryption in the low-resource IoT devices. Various works to provide a secure IoT devices by the implementation of encryption in low resource are discussed in [12–15]. Whereas to achieve integrity, hardware security managers (HSM) is used with attestation [11]. These keys require human intervention as either a password or any other identification parameter. In unattended and inaccessible IoT devices, it is more difficult to implement, and it requires another method to protect authentication keys. IoT devices should authenticate and identify the correct users; otherwise, unauthorized users may attack the devices. Other than these security challenges, there also exists hardware-related threats in IoT devices. The hardware threats are deployed in IoT devices to access them physically. These vulnerabilities are more challenging to detect, and the following section describes the vulnerabilities in the hardware in detail.

9.3 OVERVIEW OF HARDWARE ATTACKS AND THREAT MODELS

Hardware used for IoT devices is vulnerable to various attacks at all stages of IC design flow. Figure 9.2 describes the hardware threats in IC design flow, which makes the end-users to re-evaluate their trust in

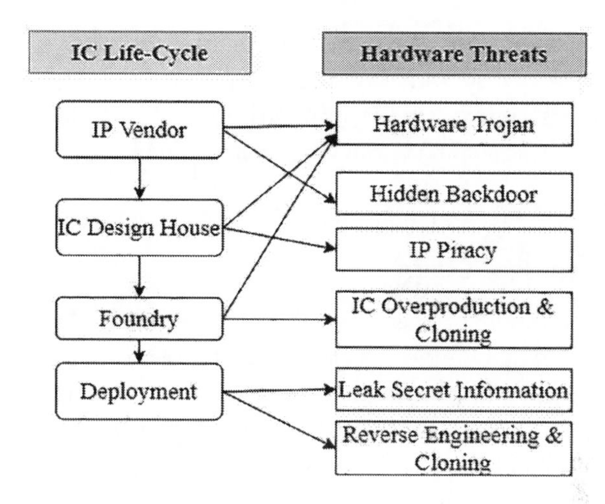

Figure 9.2 Hardware threats in IC life-cycle.

hardware. The attacker can access the design at any stage of design flow and add vulnerabilities to it. The mitigation of these security threats, with minimal overhead on design, and fabrication and testing of ICs, is challenging to the IC chip industries. IP/ IC piracy, insertion of malicious circuit (aka HTs), hidden back door, IC overproduction and cloning, and RE are some of the addressable hardware threats in the IC design industries. These threats will cause the end-users not to trust the IC chips, and there arise countermeasures to build a secured chips. Some of the threats are as follows:

9.3.1 Hardware Trojans

The outsourcing of fabrication process leads a way to insert a malicious circuitry in the design without the knowledge of the designer. These malicious circuits remain stealthy, and they may cause malfunction, leak secret information, cause denial of service, or degrade the performance of the IC as the moment they are triggered. These Trojans may be inserted at the specification, design, and fabrication stages of IC design flow, and their action may be bypassed at the testing, assembly, and package stages. These Trojans are triggered either externally or internally, some of the Trojans may always be ON, and they are most challenging to detect. The physical characteristics of the Trojan may vary depending on its size, distribution, type of Trojan (either parametric or functional), and its structure (either layout-same or layout-change). The location of Trojan insertion also plays a major role as they will remain stealthy until triggered. Hence, the Trojans are inserted at the rarely triggering nodes of any design so that the conventional test may fail to detect their presence [16–19].

9.3.2 Hidden Back Door

Hardware components may contain hidden back doors, which can be enabled with catastrophic effects or for ill-gotten profit [20]. The third-party IP provider or the malicious insider may insert a back doors on the core design component. A malfunction can be caused even by the change in few lines of hardware description language (HDL) of the core design. Use of third-party IPs also leaves its way to back door threats in today's hardware design. These back doors are dormant during testing, and hence, it is so hard to detect while validation. Verification fails because designs are too large to formally verify, and there are exponentially many different ways to express a hardware back door. A suspected actual case is the failure of Syrian radar [21], where the state-of-the-art Syrian military radar could not detect an enemy bombing due to the radar's temporary malfunction.

9.3.3 IP/IC Piracy

Outsourcing various stages of IC design flow leads a challenge to the major hardware threat known as IP/ IC piracy [22]. Electronics and defense industries are highly affected by this hardware threat. Global piracy of hardware and software IP is now approaching $1B per day, with a major share in computers, peripherals, and embedded systems [23]. Indeed, once a fab starts producing chips from client's masks, unauthorized copies can be made cheaply. As pointed out by the US Defense Science Board [22], masks can also be stolen by industrial and military spies.

9.3.4 IC Overproduction & Cloning

Overproduction of IC without the knowledge of the designer at the fabrication unit is a major source of concern in the hardware supply chain, which causes a reliability and security issues in all sectors including computers, telecommunications, and automotive electronics, and even in defense systems [24,25]. Detection and/or prevention of counterfeit

TABLE 9.1 Hardware Threat Scenarios in IC Design Flow

Threats	3PIP Vendor	SoC Integrator	Foundry	User
Hardware Trojans	✓	✓	✓	✗
Hidden Back Door	✓	✓	✓	✗
IP/ IC Piracy	✓	✓	✓	✗
IC overproduction and Cloning	✓	✓	✓	✓
Reverse Engineering	✓	✓	✓	✓

electronic components has become a major challenge in the electronic component supply chain [26].

9.3.5 Reverse Engineering

RE is a process used to identify the device technology, to extract the gate level netlist and inferring the implemented functionality [27–29]. To reverse engineer an IC, techniques and tools are developed [30,31]. The IC design can be stolen to identify the device technology and illegally fabricated by RE. Table 9.1 shows hardware threat scenarios in IC design flow. Among these hardware threats, this chapter focuses on the most challenging threat known as HTs and its countermeasures in the IoT devices used for smart environment.

9.4 SECURITY CHALLENGES OF HARDWARE TROJANS AT IoT DEVICES

Any malicious circuit or modification in the original design of IC during the design or manufacturing process is known as hardware Trojan or hardware malware. These types of hardware malware cause system failure or denial of service, or leak secret information when they are triggered. The electronic components of IoT devices are made up of ICs from different vendors, and they are manufactured at the untrusted foundries; hence, they are not trustable. Unlike software Trojans, these HTs cannot be eliminated by a firmware update, so once the IC is manufactured with HT, it cannot be repaired again. Thus, HTs are more challenging to remove and detrimental. Also the threat posed by HT exists throughout the lifetime of devices as it is inserted at the hardware level. HT can be embedded in any type of IC, such as application-specific integrated circuits (ASIC), field-programmable gate arrays (FPGAs), system-on-chip (SoC), and digital signal processors (DSPs) [32]. IC security measures will be impeded by the HT, and this results in faulty output of the infected chip. These vulnerable HT can even destroy the hardware or may leak any secret information or even bypass the existing security systems of the hardware. Hence, these HT will be a challenging threat to the already vulnerable IoT devices. Modern world is moving towards smart cities, smart environment, smart healthcare, smart transportation and smart farmlands, smart industries and smart home, etc. If devices connected in these types of smart hardware are vulnerable to most challenging threat HT, then entire system will be vulnerable, and this results in evaluating the trust of hardware in all the domain of IoT. Hence, there exists a countermeasure for the HT vulnerability, which is discussed in Section 9.5.

9.5 COUNTERMEASURES AGAINST HARDWARE TROJAN

HT is a most challenging hardware threat, and its countermeasures are discussed in the following section.

9.5.1 Trojan Detection and Diagnosis

Handling the HT attacks typically includes two steps: HT detection, which determines whether HT exists in the circuit or not, and HT diagnosis, which determines where the HT located in the circuit, its intrusion type, associated pins, etc. Although literature gives many HT detection approaches [33–40], HT diagnosis was rarely discussed in the previously proposed methods. The classification of HT detection approaches is discussed in [41]. Among the all HT detection approaches, the researchers still find the difficulty in detecting the presence of HTs, which are amid in between the millions of gates in the circuit. HT detection also has to consider cost of testing and running time as a major concern. In addition, Trojan diagnosis will cost more than the detection, since pointing the Trojan location among the millions of gates is not an easy process. Diagnosis is far more challenging because it requires an golden reference circuit for comparison [39]. Since the conventional methods of HT detection (logic testing or functional testing) fail to detect a stealthy Trojans, the side-channel parameters such as supply current or path delay measurements are widely used to detect HTs in the circuit design [33]. The parameter variation or process variation (PV) is the most important factor to be considered while undergoing side-channel analysis. Because the side-channel parameters are directly affected by PV, their affects may be wrongly distinguished as a presence of Trojan. The major source of parameter variation (PV) is two physical properties, (a) threshold voltage and (b) effective channel length, which induce an unintended design modification in the circuit design. These parameter variations will limit the effectiveness of side-channel analysis. Gate-level characterization (GLC) is the process in which each individual gate in the IC are characterized in terms of its physical and manifestation properties. These characterized gates are used as a means for comparing golden chip and the one under authentication [35,38]. This effect of Trojan on the extracted GLC are masked by the nano-metric feature-sized parameter variation [42]. Consistency analysis is carried out to develop a HT detection and diagnosis at a low cost at GLC. The power measurement at various gate levels was the major source of cost, and to overcome this, a self-consistency-based approach is developed in [43]. As a result, the entire circuit is divided into sub-segments or modules for power measurements. By reusing the existing measurements in various sub-segments, the cost of HT detection is reduced by the required number of power measurements. Further, the computational complexity of GLC is reduced by considering the minimum number of sub-segments generated by the algorithm [43]. In GLC, by applying thermal conditioning, the scaling factors of all the gates are calculated by linear programming. The statistical methods in linear programming and post-processing steps had improved the accuracy of GLC to some extent [44]. The existing self-referencing schemes [41,45] fail to assure the golden-chip-free detection methods. The idea that the current signature of the chip at different time windows is compared to detect the Trojans by eliminating the power noise was proposed. But this method fails to detect the small-sized sequential Trojans, as their effects of current signature are negligible [45]. Then emerges a self-referencing approach of HT detection and diagnosis method using power signature, which gives an efficient golden-chip-free Trojan detection at low cost and is the most effective and efficient method [46,47]. The post-manufacturing tests will detect small Trojans, since those Trojans will have a limited number of states to switch through. The inconsistency analysis among the same modules of the circuit for the set of same input vectors, at different time windows, describes the novelty of the proposed scheme. If the inconsistency metric value is above a specific threshold value, then the presence of Trojan is confirmed without demanding a golden chip. These power measurements are reused in various sub-segments, so that the cost of

HT detection is reduced by minimizing the required number of power measurements.

9.5.2 Online-Monitoring HT Detection

Detection of HT by conventional methods is more difficult due to their inherent surreptitious nature. This makes the end-users to associate lower levels of trust with the design. Detection and prevention of HTs at the circuit netlist level have thus emerged as a significance research area along with the investigation of newer threats. Hence, a design-for-trust (DFTr) technique is required to regain the trust of an end-user at the design phase. In this work, a trustable hardware is designed with low hardware overhead, which inserts online monitoring logical malfunction detection circuitry at specific sites of the netlist. The suitable sites of online monitoring circuit insertion are chosen based on signal probability reliability analysis (SPRA) algorithm, which uses an astute analysis of fault propagation through logic gates. System failure can be avoided by activating proactive system-level countermeasures upon the detection of HTs. The major advantages of the online monitoring approach [48,49] are that there is no specific fixed structure for the checkers and hence impediment to RE-based attacks. The EDA tools are used to implement the proposed schemes with automated design methodology. The main goal of this research is to provide a comprehensive solution for all the stages of the design flow. This is achieved by the promising online checking technique, to detect a hard-to-detect HTs with acceptable and controlled hardware overhead. The core idea of the proposed online monitoring technique is to monitor the internal logic values at the probable HT insertion sites of IC. The main focus is on the expected circuit functionality and to monitor the correlation between the sets of logic values of the selected sites to that of the neighbor internal nodes. If their internal logics are uncorrelated, then the checker will detect the presence of Trojan. The following are the primary objectives of the proposed technique to enhance Trojan detectability:

- To the given netlist, the internal nodes are identified to which the HTs are inserted probably.
- The low overhead checkers are inserted at a selected subset of internal nodes.
- The logic malfunctions are identified and reported by the online checkers.

- The logic errors are propagated to the primary outputs, and they are independent to the test vectors generated.
- The number of checkers inserted are restricted to have a low hardware overhead with maximum HT detection accuracy.

An automated low overhead online monitoring technique is developed, for HT detection at the gate-level netlist, and it has been applied to develop a secured system architecture directed towards design-for-trust. The effectiveness of the proposed scheme is experimentally validated, which shows close to 100% detection coverage with controlled design overhead of 10%. This technique avoids the functionality change of the original circuit due to the checker module, since the checker module is functionally free circuit. The proposed technique is resistant to cell replacement attack, since RE is not possible in the modified design and can be extended easily to secure the system designed.

9.5.3 Trojan Prevention

The emerging hardware threat in a large number of applications, from consumer electronics all the way to defense systems, is considered as a growing concern. These threats occur when the fab-less IC companies are outsourcing their design to external untrusted foundries for manufacturing process. IP piracy, illegal IC overbuilding, RE, and malicious circuit insertion known as hardware Trojan are the major hardware vulnerabilities [16,50]. The main goal of adversary in untrusted foundry is (a) to obtain the design functionality by RE or (b) to gain design knowledge that enables its manipulation. The adversary in the foundry may have access to the GDS-II file of the design and claim the IP ownership by means of RE. Due to these threats, the semiconductor industries lose several billion dollars every year [51]. Thus, an IC design engineer has a major challenge to produce a secured design such that the adversary may fail to obtain the circuit functionality through RE. Challenges of IC industries to regain the trust of end-users are discussed in [52]. From which design-for-security (DfS) techniques are considered to be authentic and hence chosen. The key-gates used for hardware encryption techniques are XOR/XNOR gates [50,53,54], multiplexers [53], and AND/OR gates [55]; in some encryption techniques, a combination of all the above gates are also used [56]. In EPIC [50], XOR/XNOR key-gates are inserted randomly into the design; when an invalid key is applied, the original logic is inverted by the key-gates to produce a wrong output. When a valid key is applied, these locking keys act as buffer and produce a correct logical output. However, the attacker can easily guess the valid keys to unlock the circuit with the basic encryption knowledge, and random insertion of key-gates does not always produce an incorrect output for an invalid keys. Hence, locations of key-gates play an important role in locking the digital circuits for security.

In order to enhance the logical locking technique by producing high output corruption, the key-gates are placed at the high fault impact nodes based on fault analysis approach [53,57]. The locations of inserting key-gates are selected upon three basic phenomena of IC testing principle: fault excitation, fault propagation, and fault masking. The basic idea is upon applying an incorrect key; that is, either a stuck-at-0 (s-a-0) or stuck-at-1 (s-a-1) is excited, and these faults are propagated towards the output to corrupt a maximum number of output bits. The nodes with maximum fault impact are the potential locations to insert key-gates. For an incorrect key, the designer will controllably corrupt the 50% of output bits by the fault excitation and propagation method [58]. The enhanced logic encryption technique is a primary input key-based hardware locking mechanism [59–62]. An average of 50.652% output corruption is produced when a wrong key is applied and the key size is 10 bits of the primary input, so it is not easy to unlock the circuit with the random attempts. These two characteristics of the proposed scheme strengthen the encryption scheme, so that the adversary has to put more effort to reveal the valid keys. Extra hardware is required to ensure high security of the design, which is practically reasonable for large circuits.

9.6 CONCLUSION

Secured IoT devices cannot be built unless the secured hardware is used as a strong foundation. If any HT is embedded in the IoT device, then it cannot be removed and it will destroy the entire smart environment throughout its lifetime. It is necessary to have a hardware security module embedded in the IoT devices to protect the entire setup including the piracy and security of data generating. Use of online monitoring checkers will be a countermeasure to HT during the runtime of IoT devices. These online checkers will detect the HT throughout the lifetime of the device, and once if the IC is found to be defective, it is easy to find the alternate solution for the user. However, prevention is better than cure, so HT prevention methods are most welcome countermeasure to protect the IP from design to end product stage. Logical obfuscation technique will provide an end-to-end security by functionally locking the IC design with a key stored in tamper-proof memory. This chapter describes the proper countermeasure against the most vulnerable hardware threat, HT in IoT devices. There is a possibility to attack the obfuscated module also, and hence, a strengthened locking mechanism is to be built to provide a trusted IoT devices to the end-users.

Bibliography

[1] Paddy Nixon, Gerard Lacey, Simon Dobson, et al. *Managing Interactions in Smart Environments: 1st International Workshop on Managing Interactions in Smart Environments (MANSE'99), Dublin, December 1999*. Springer Science & Business Media, 2012.

[2] D INFSO. 4networked enterprise & rfid infso g. 2 micro & nanosystems in co-operation with the rfid working group of the etp eposs. internet of things in 2020: A roadmap for the future [ol], 4.

[3] K. Sakamura. Computers everywhere: The future of ubiquitous computing and networks. In *MIC Japan/ITU/UNU WSIS Thematic Meeting, "Towards the Realization of the Ubiquitous Network Society"*, 2005.

[4] Force, Task United States. Defense Science Board. Task Force on High Performance Microchip Supply. *High performance microchip supply*, 2005.

[5] Lawrence L. Harada. Semiconductor technology and us national security. Technical report, Army War Coll Carlisle Barracks, Carlisle, PA, 2010.

[6] Stefan Mangard, François-Xavier Standaert, et al. Cryptographic hardware and embedded systems, ches 2010. In *12th International Workshop, Santa Barbara, CA*. Springer, 2010.

[7] Anju P. Johnson, Sikhar Patranabis, Rajat Subhra Chakraborty, and Debdeep Mukhopadhyay. Remote dynamic partial reconfiguration: A threat to internet-of-things and embedded security applications. *Microprocessors and Microsystems*, 52:131–144, 2017.

[8] CIPSEC. Enhancing critical infrastructure protection with innovative security framework (cipsec). H2020 European Project. Available online: www.cipsec.eu (accessed on 28 March 2017).

[9] Apostolos P. Fournaris, Lidia Pocero Fraile, and Odysseas Koufopavlou. Exploiting hardware vulnerabilities to attack embedded system devices: A survey of potent microarchitectural attacks. *Electronics*, 6(3):52, 2017.

[10] Carsten Maple. Security and privacy in the internet of things. *Journal of Cyber Policy*, 2(2):155–184, 2017.

[11] Soma Bandyopadhyay, Munmun Sengupta, Souvik Maiti, and Subhajit Dutta. A survey of middleware for internet of things. In *Recent Trends in Wireless and Mobile Networks*, pages 288–296. Springer, 2011.

[12] Bassam J. Mohd and Thaier Hayajneh. Lightweight block ciphers for iot: Energy optimization and survivability techniques. *IEEE Access*, 6:35966–35978, 2018.

[13] Bassam J. Mohd, Thaier Hayajneh, and Athanasios V. Vasilakos. A survey on lightweight block ciphers for low-resource devices: Comparative study and open issues. *Journal of Network and Computer Applications*, 58:73–93, 2015.

[14] Bassam Jamil Mohd, Thaier Hayajneh, Zaid Abu Khalaf, and Khalil Mustafa Ahmad Yousef. Modeling and optimization of the lightweight hight block cipher design with fpga implementation. *Security and Communication Networks*, 9(13):2200–2216, 2016.

[15] Bassam Jamil Mohd, Thaier Hayajneh, Khalil M. Ahmad Yousef, Zaid Abu Khalaf, and Md Zakirul Alam Bhuiyan. Hardware design and modeling of lightweight block ciphers for secure communications. *Future Generation Computer Systems*, 83:510–521, 2018.

[16] R. Sree Ranjani and M. Nirmala Devi. Malicious hardware detection and design for trust: An analysis. *Elektrotehniski Vestnik*, 84(1/2):7, 2017.

[17] Rajat Subhra Chakraborty, Seetharam Narasimhan, and Swarup Bhunia. Hardware trojan: Threats and emerging solutions. In *2009 IEEE International High Level Design Validation and Test Workshop*, pages 166–171. IEEE, 2009.

[18] Francis Wolff, Chris Papachristou, Swarup Bhunia, and Rajat S. Chakraborty. Towards trojan-free trusted ics: Problem analysis and

detection scheme. In *2008 Design, Automation and Test in Europe*, pages 1362–1365. IEEE, 2008.

[19] Masoud Rostami, Farinaz Koushanfar, Jeyavijayan Rajendran, and Ramesh Karri. Hardware security: Threat models and metrics. In *2013 IEEE/ACM International Conference on Computer-Aided Design (ICCAD)*, pages 819–823. IEEE, 2013.

[20] Adam Waksman and Simha Sethumadhavan. Silencing hardware backdoors. In *2011 IEEE Symposium on Security and Privacy*, pages 49–63. IEEE, 2011.

[21] Sally Adee. The hunt for the kill switch. *iEEE SpEctrum*, 45(5):34–39, 2008.

[22] DS Board. Defense science board (dsb) study on high performance microchip supply. Available online: www.acq.osd.mil/dsb/reports/ADA435563.pdf (accessed on 16 March 2015), 2005.

[23] Protection, Intellectual Property [VSIAlliance2000WhitePaper]. Intellectual Property Protection: Schemes, Alternatives and Discussion. 2000.

[24] Senate, US Inquiry into counterfeit electronic parts in the department of defense supply chain. S. Rpt,112–167, 2012.

[25] Crawford, Mark and Telesco, Teresa and Nelson, Christopher and Bolton, Jason and Bagin, Kyle and Botwin, Brad Defense industrial base assessment: Counterfeit electronics. *Washington, DC: US Department of Commerce Bureau of Industry and Security Office of Technology Evaluation*, 2010.

[26] Ujjwal Guin, Ke Huang, Daniel DiMase, John M. Carulli, Mohammad Tehranipoor, and Yiorgos Makris. Counterfeit integrated circuits: A rising threat in the global semiconductor supply chain. *Proceedings of the IEEE*, 102(8):1207–1228, 2014.

[27] Chipworks. Intel's 22-nm tri-gate transistors exposed. http://www.chipworks.com/blog/technologyblog/2012/04/23/intels-22-nm-tri-gate-transistors-exposed (accessed on 2012).

[28] Randy Torrance and Dick James. The state-of-the-art in semiconductor reverse engineering. In *Proceedings of the 48th Design Automation Conference*, pages 333–338, 2011.

[29] Bernstein, K. Integrity and reliability of integrated circuits (iris). https://www.darpa.mil/program/integrity-and-reliability-of-integrated-circuits, 2011

[30] Chipworks. Reverse engineering software. http://www.chipworks.com/en/technical-competitive-analysis/resources/reerse-engineering-software.

[31] Degate. http://www.degate.org/documentation/.

[32] Vivek Venugopalan and Cameron D. Patterson. Surveying the hardware trojan threat landscape for the internet-of-things. *Journal of Hardware and Systems Security*, 2(2):131–141, 2018.

[33] Dakshi Agrawal, Selcuk Baktir, Deniz Karakoyunlu, Pankaj Rohatgi, and Berk Sunar. Trojan detection using ic fingerprinting. In *2007 IEEE Symposium on Security and Privacy (SP'07)*, pages 296–310. IEEE, 2007.

[34] Yier Jin and Yiorgos Makris. Hardware trojan detection using path delay fingerprint. In *2008 IEEE International Workshop on Hardware-Oriented Security and Trust*, pages 51–57. IEEE, 2008.

[35] Miodrag Potkonjak, Ani Nahapetian, Michael Nelson, and Tammara Massey. Hardware trojan horse detection using gate-level characterization. In *2009 46th ACM/IEEE Design Automation Conference*, pages 688–693. IEEE, 2009.

[36] T. Saran, R. Sree Ranjani, and M. Nirmala Devi. A region based fingerprinting for hardware trojan detection and diagnosis. In *2017 4th International Conference on Signal Processing and Integrated Networks (SPIN)*, pages 166–172. IEEE, 2017.

[37] Sheng Wei, Jong Hoon Ahnn, and Miodrag Potkonjak. Energy attacks and defense techniques for wireless systems. In *Proceedings of the Sixth ACM Conference on Security and Privacy in Wireless and Mobile Networks*, pages 185–194, 2013.

[38] Sheng Wei, Saro Meguerdichian, and Miodrag Potkonjak. Gate-level characterization: Foundations and hardware security applications. In *Design Automation Conference*, pages 222–227. IEEE, 2010.

[39] Julien Francq and Florian Frick. Introduction to hardware trojan detection methods. In *2015 Design, Automation & Test in*

Europe Conference & Exhibition (DATE), pages 770–775. IEEE, 2015.

[40] Francis Wolff, Chris Papachristou, Swarup Bhunia, and Rajat S. Chakraborty. Towards trojan-free trusted ics: Problem analysis and detection scheme. In *2008 Design, Automation and Test in Europe*, pages 1362–1365. IEEE, 2008.

[41] Dongdong Du, Seetharam Narasimhan, Rajat Subhra Chakraborty, and Swarup Bhunia. Self-referencing: A scalable side-channel approach for hardware trojan detection. In *International Workshop on Cryptographic Hardware and Embedded Systems*, pages 173–187. Springer, 2010.

[42] Shekhar Borkar, Tanay Karnik, Siva Narendra, Jim Tschanz, Ali Keshavarzi, and Vivek De. Parameter variations and impact on circuits and microarchitecture. In *Proceedings of the 40th annual Design Automation Conference*, pages 338–342, 2003.

[43] Sheng Wei and Miodrag Potkonjak. Self-consistency and consistency-based detection and diagnosis of malicious circuitry. *IEEE Transactions on Very Large Scale Integration (VLSI) Systems*, 22(9):1845–1853, 2013.

[44] Sheng Wei, Saro Meguerdichian, and Miodrag Potkonjak. Malicious circuitry detection using thermal conditioning. *IEEE Transactions on Information Forensics and Security*, 6(3):1136–1145, 2011.

[45] Seetharam Narasimhan, Xinmu Wang, Dongdong Du, Rajat Subhra Chakraborty, and Swarup Bhunia. Tesr: A robust temporal self-referencing approach for hardware trojan detection. In *2011 IEEE International Symposium on Hardware-Oriented Security and Trust*, pages 71–74. IEEE, 2011.

[46] R. Sree Ranjani, P. K. Maneesh, and M. Nirmala Devi. Golden chip free ht detection and diagnosis using power signature analysis. In *Presented at the 7th IEEE International Workshop on Reli-ability Aware System Design and Test (RASDAT)*, 2016.

[47] R. Sree Ranjani and M. Nirmala Devi. Golden-chip free power metric based hardware trojan detection and diagnosis. *Far East Journal of Electronics and Communications*, 17:517–530, 2017.

[48] Rajat Subhra Chakraborty, Samuel Pagliarini, Jimson Mathew, Sree Ranjani Rajendran, and M. Nirmala Devi. A flexible online checking technique to enhance hardware trojan horse detectability by reliability analysis. *IEEE Transactions on Emerging Topics in Computing*, 5(2):260–270, 2017.

[49] R. Sree Ranjani. Online monitoring based design-for-trust technique to build a trusted hardware design. In *32nd International Conference on VLSI Design and 18th International Conference on Embedded Systems(PhD Forum)*, 2019.

[50] Jarrod A. Roy, Farinaz Koushanfar, and Igor L. Markov. Ending piracy of integrated circuits. *Computer*, 43(10):30–38, 2010.

[51] S. Jose. Innovation is at risk as semiconductor equipment and materials. *Semiconductor Equipment and Material Industry (SEMI)*, 2008.

[52] Mohammad Tehranipoor, Hassan Salmani, Xuehui Zhang, Michel Wang, Ramesh Karri, Jeyavijayan Rajendran, and Kurt Rosenfeld. Trustworthy hardware: Trojan detection and design-for-trust challenges. *Computer*, 44(7):66–74, 2010.

[53] Jeyavijayan Rajendran, Huan Zhang, Chi Zhang, Garrett S. Rose, Youngok Pino, Ozgur Sinanoglu, and Ramesh Karri. Fault analysis-based logic encryption. *IEEE Transactions on computers*, 64(2):410–424, 2013.

[54] Muhammad Yasin, Jeyavijayan J. V. Rajendran, Ozgur Sinanoglu, and Ramesh Karri. On improving the security of logic locking. *IEEE Transactions on Computer-Aided Design of Integrated Circuits and Systems*, 35(9):1411–1424, 2015.

[55] Sophie Dupuis, Papa-Sidi Ba, Giorgio Di Natale, Marie-Lise Flottes, and Bruno Rouzeyre. A novel hardware logic encryption technique for thwarting illegal overproduction and hardware trojans. In *2014 IEEE 20th International On-Line Testing Symposium (IOLTS)*, pages 49–54. IEEE, 2014.

[56] Yu-Wei Lee and Nur A. Touba. Improving logic obfuscation via logic cone analysis. In *2015 16th Latin-American Test Symposium (LATS)*, pages 1–6. IEEE, 2015.

[57] Jeyavijayan Rajendran, Youngok Pino, Ozgur Sinanoglu, and Ramesh Karri. Logic encryption: A fault analysis perspective. In *2012 Design, Automation & Test in Europe Conference & Exhibition (DATE)*, pages 953–958. IEEE, 2012.

[58] Nirmala M. Devi, Irene Susan Jacob, Sree R. Ranjani, and M. Jayakumar. Detection of malicious circuitry using transition probability based node reduction technique. *Telkomnika*, 16(2):573–579, 2018.

[59] R. Sree Ranjani and M. Nirmala Devi. A novel logical locking technique against key-guessing attacks. In *2018 8th International Symposium on Embedded Computing and System Design (ISED)*, pages 178–182. IEEE, 2018.

[60] R. Sree Ranjani and M. Nirmala Devi. Enhanced logical locking for a secured hardware ip against key-guessing attacks. In *International Symposium on VLSI Design and Test*, pages 186–197. Springer, 2018.

[61] R. Sree Ranjani and M. Nirmala Devi. Secured hardware design with locker-box against a key-guessing attacks. *Journal of Low Power Electronics*, 15(2):246–255, 2019.

[62] R. Sree Ranjani. Secured hardware design against trojans at gate-level netlist. In *32nd International Conference on VLSI Design and 18th International Conference on Embedded Systems(User Design Track)*, pages 186–197. Springer, 2018.

CHAPTER 10

Blockchain for Internet of Battlefield Things: A Performance and Feasibility Study

Abel O. Gomez Rivera and Deepak Tosh

University of Texas at El Paso

Jaime C. Acosta

CCDC Army Research Laboratory

CONTENTS

10.1 INTRODUCTION

Industrial consumers are catapulting the rapid evolution of Internet of Things (IoT), where the smart devices are capable of capturing information from their environment and transmit it to other devices and servers for enabling intelligence. In general, IoT devices are resource-constrained; that is, their capabilities and functions are limited. Due to their small-scale and limited functions, the available resources in IoT devices are not suitable for complex crypto-operations, thus undermining the security posture of the connected ecosystem. Commercial off-the-shelf (COTS) IoT devices have been frequently targeted for trivial cyber-attacks which aim to exploit well-known security vulnerabilities (e.g., unprotected telnet services). The Mirai botnet [1] is one such example of malware attack that has infected millions of IoT devices by targeting default authentication secret used in them. In particular, it targets the default authentication in the telnet service; the malware compromises IoT devices to create distributed denial of service (DDoS) attacks capable of disrupting Internet services like government operations or military missions. The rapid evolution of communication and data transmission of IoT enabled researchers to explore the (dis)advantages of implementing IoT in the Battlefield (IoBT) context. However, developing robust, lightweight, and novel security mechanisms to strengthen the security of IoBT in the first place remains challenging.

Traditionally, IoBT networks are centralized and vulnerable to a single point of failure attacks; that is, if the central server gets compromised, all data and operations in the network are affected [2]. Besides its security issues, centralized architecture also has performance challenges. The central server can process only a limited amount of information whenever its resources are at peak demand, and thus, the overall performance of the system gets constrained [3]. A distributed computing paradigm, namely, edge computing, can address the performance challenge of centralized IoBT networks. Edge computing can improve network latency and utilization of resources by distributing the tasks amount different devices at the edge of the network [4]. However, the distribution of tasks introduces novel challenges such as the security of multiple devices, data synchronization, and distributed communication. A decentralized platform, namely, Blockchain, can be used to eliminate the security issues involved in distributed IoBT infrastructures. Although blockchain technology was introduced through cryptocurrencies, its fundamental attributes can be successfully applied to networking and communication fields. Blockchain leverages a distributed network, where each peer maintains a local copy of an immutable distributed ledger that is only updated when the peers achieve common agreement through consensus [5]. Blockchain technology can be a useful tool in IoBT context because it can assure the integrity of all activities through validated transactions [6]. Since blockchain relies on cryptographic primitives, special considerations must be taken before its implementation in IoBT networks [7,8]. The state-of-art blockchain platform suffers from several performance challenges such as scalability issues in terms of both devices and transactions, and latency to validate transactions, and the network overhead is significant to maintain peer communication. The impacts of transaction throughput and network overhead in a blockchain-based tactical network are not well studied in the past literature. However, these attributes are particularly important to achieve trustworthy IoBT infrastructures that is dynamic and resource-constrained.

Military tactical networks, otherwise known as IoBT, consist of three major components, as illustrated in Figure 10.1: (a) heterogeneous devices, (b) battlefield terrain, and (c) mission functions. In this chapter, we analyze the impacts of transactions throughput and network overhead in a blockchain-based tactical network of heterogeneous devices on which the soldiers rely for situational awareness and make critical decisions for mission's success [9]. Soldiers in the battlefield depend on a reliable and secure data communication to succeed in the mission, as this can be disrupted by several factors like adverse location, a limited power of transmitter devices, passive devices of adversaries, dynamic topologies, and low bandwidth. IoBT infrastructure to have reliable communication in static and mobile networks, and blockchain technology can be used to strengthen the security of the battlefield network. Our contributions in this chapter are the following:

- Analyze the impacts of transaction throughput and network overhead in a blockchain-based tactical network.

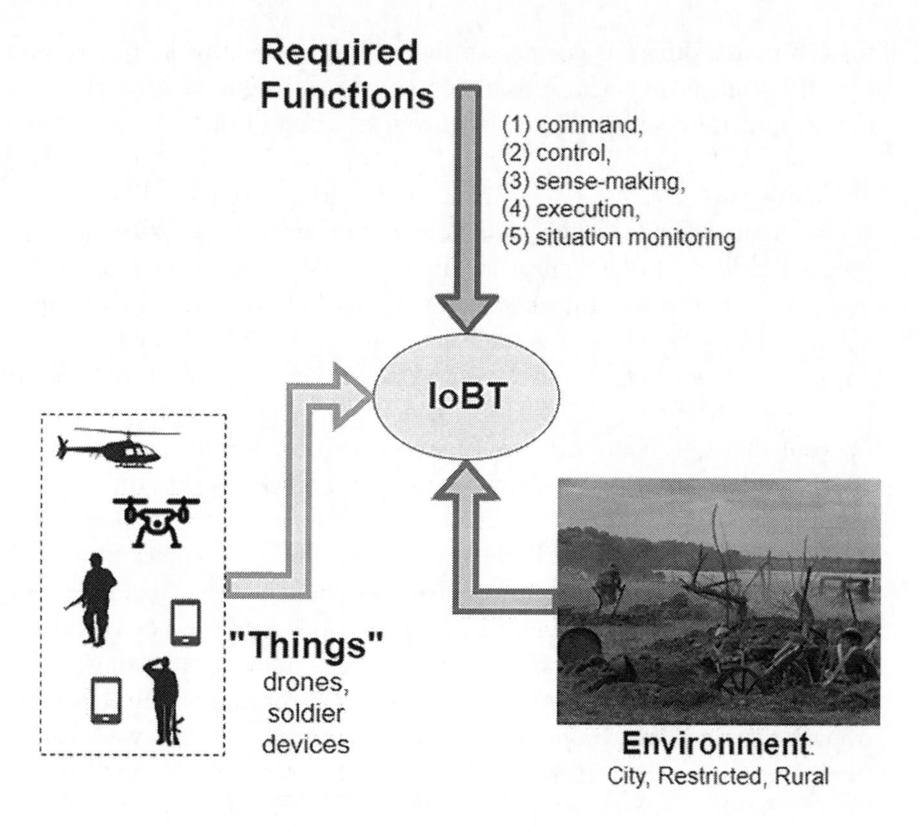

Figure 10.1 Elements of a typical IoBT environment.

- Feasibility of a permissioned blockchain variant, Hyperledger Sawtooth, for different types of IoBT networks.
- Report of results and insights on Hyperledger Sawtooth's performance under different configurations.

This chapter is organized as follows: we first introduce the IoBT concepts and challenges in making coordinated decisions in the battlefield. We also discuss the recent technological advancements along with some longstanding challenges by reviewing related works in the domain of IoBT. By understanding the advantages and disadvantages of previously proposed models, we offer insights on how the distributed ledger technology can address several IoBT challenges. Then, the (dis)advantages of different types of blockchains along with their characteristics, feasibility, and integration challenges in IoBT are presented. In order to demonstrate feasibility of blockchain in IoBT environment, we report our simulation results and insights by conducting several emulations with IoBT relevant and diverse network topologies with varying configurations. Finally, the chapter concludes with a summary of the essential contributions and plans for future research directions.

10.2 CHALLENGES OF IoBT: AN OVERVIEW

The necessity of real-time monitoring of physical processes has led to rapidly evolving IoT devices. Such devices or "things" have become more intelligent with a degree of autonomy. The benefits of the miniaturization and communication capabilities of these devices are being exploited in tactical environments. The mission progress, troop health monitoring, terrain information, and many other tasks in military missions can be achieved through the help of IoT technology. To offer such solution where IoBT would survive and operate securely, several challenges, such as secure and reliable communication, must be addressed. A battlefield or tactical environment, as illustrated in Figure 10.2, consists of different components that spans from the human capital to the technology that soldier carry on their person and the "smart" recognition devices. The broad adoption of COTS IoT devices increases the attack surface of the integrated platform to disrupt the normal activities (e.g., command and control) and introduce anomalies. A comprehensive and successful adoption of IoT technology in IoBT will require to comply with five essential [10] functions necessary to achieve specific mission

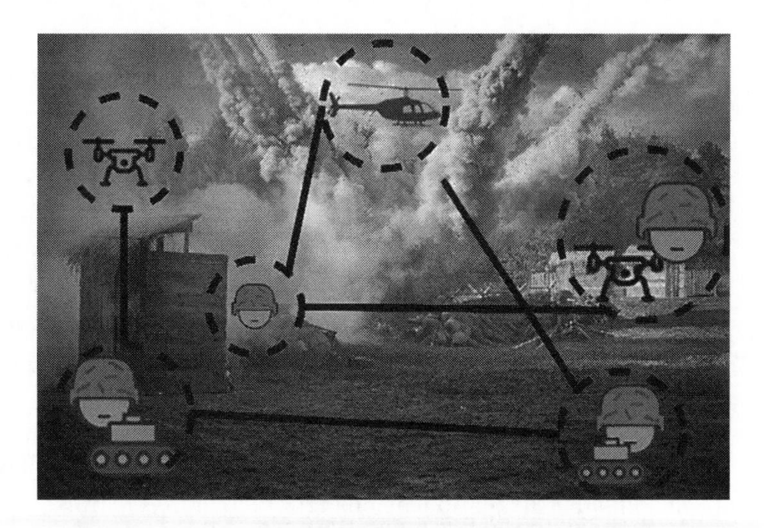

Figure 10.2 Sample network topology in tactical scenario.

goals, (a) command, (b) control, (c) sense-making, (d) execution, and (e) situation monitoring.

A significant security feature that IoBT networks must have is resilience in terms of information and processing. IoBT networks are vulnerable to several cyber-attacks such as denial of service (DoS), spoofing, and man-in-the-middle (MITM). Although there exist numerous [10,11] security issues when considering heterogeneous elements in a battlefield, problems such assniffing network traffic, jamming communication mediums, and compromising node location are often targeted due to their easy exposure to adversaries. Before we dig deeper into blockchain-based solutions, it is important to explore some of the fundamental challenges and requirements of IoBT networks.

10.2.1 Security and Trustworthiness

In general, battlefields consist of interconnected heterogeneous devices that communicate through diverse mediums to provide up-to-date intelligence of the inaccessible terrain or location of adversaries. Soldiers and central command rely on intelligence data provided by IoT devices to make critical and timely decisions that can potentially impact the mission objective. Real-time and reliable access to the intelligence data is a significant need in IoBT applications; the loss or corruption of such data could be catastrophic, resulting in loss of lives and waste of resources. Since information flow is crucial for the mission goal, attackers commonly target to disrupt the communication mediums by overloading the network or interfering with transmitter devices. Several attacks such as DoS, data and device spoofing, and others [12] target the availability of soldiers to access intelligence information.

Standard attacks that target information flow can be carried out in in many ways, e.g., exploiting security vulnerabilities in communication protocols such as TCP/IP, interfering procedure of information handling, and introducing rogue devices and corrupting data flow in the IoBT network. DoS attacks can be mitigated through intrusion detection systems (IDS) that continuously monitor systems and networks to detect anomalies, and another way to eliminate DoS attacks is through firewalls, which are used to examine all inbound traffic. However, due to the ad-hoc demands of IoBT applications and the limited resources in IoBT devices, the state-of-art solutions, such as IDS and firewalls, are not efficient for IoBT applications [13]. Thus, decentralized security through redundancy in the IoBT environment is very important in order to avoid various security attacks and enable a trustworthy platform for conducting tactical operations reliably.

10.2.2 Communication

Robust and reliable communication infrastructure such as Long-Term Evolution (LTE), Wi-Fi, and Ethernet are available to the consumer industry, and in general, many protocols and applications assume a stable connection on which they can transmit information. From the military viewpoint, communication infrastructure such as Wi-Fi and LTE is useful in short-range missions that take place in adequate environments that allow the unrestricted transmission of information. For extended missions that usually take place on uneven terrain and cover wide areas, technologies such as mobile subscriber equipment (MSE), satellites, long-range radios, and others [14] are used to transmit information.

State-of-art communication technology use in military networks may provide few tactical advantages to resourceful attackers that can compromise the communication between soldiers, devices, and central command. Previous work [15–17] has studied different communications protocols that have the capabilities to provide a robust, secure, and reliable infrastructure to IoBT. A logical approach to address the true mobility requirement of IoBT is to implement ad-hoc networks that allow for continuous communication and flexible topology. Ad-hoc networks do not rely on existing infrastructure and partnered with secure communication protocols; they can function as required by IoBT. However, the efforts in previous work do not deeply address the challenges and trade-offs of ad-hoc networks in an IoBT context.

10.2.3 Node Location

In most cases, a network setup will directly affect network traffic, consumption, throughput, and more. Additionally, the locations of nodes and routers on a network will directly affect the security of a system. When a router is the central point of communication between all nodes, a particular target is identified for attackers. Redundantly connected systems, which at their core are decentralized, allow for the mitigation of threats such as data corruption, data loss, and service interruptions. These issues may have independent solutions, although they may be costly. Take, for example, [18], where analysis of different attack management platforms for DoS attacks is presented. Although the solutions are many, they require a memory-intensive solution that may not be viable in all environments. For communication security, defenses such as those describe in [19] involve adding more nodes into a network that may be used as an early warning system. Again, node availability may be a factor.

Centralized networks suffer from attack vectors that make it difficult always to mount a perfect defense. Rather than combining many different solutions, we propose using blockchain, a highly customizable platform that contains many native security traits. Blockchain is a decentralized, trust-less platform that creates, distributes, and records information with all participating nodes. In this platform, there is no central authority, and a copy of live information is distributed to all. When the system is attacked, no data is lost or easily corrupted, and remaining nodes can continue working in the system.

10.3 RELATED WORKS

The exponential growth of IoT technology enables businesses and home users the ability to collect, monitor, and analyze devices across the world. More recently, the military is looking at ways to adopt this technology in order to perform more efficiently and securely on the battlefield. However, such adoption has several roadblocks. Besides the technical requirements of communication, availability, trust, and security, the battlefield environment has infrastructure, organizational, and planning requirements. Suri [20] describes six barriers that limit the free adoption of IoT technology and discusses two major challenges that must be addressed: (a) there exist an active adversary that interferes with the network and (b) limited resources in terms of communication and network infrastructure. Tortonesi [21] addresses the communication and interoperability challenges of IoBT and proposes a new paradigm that prioritizes information dissemination. The platform, namely, SPF (Sieve, Process, and Forward), integrates IoT data filtering and analytic functions in military networks. Overall, IoT technology and IoBT reliant on centralized communication paradigms that heavily depend on management and control devices. This introduces significant security vulnerabilities, such as a single point of failure. A natural step to address the issues of IoBT is a flexible platform that can adapt to a dynamic topology.

Mobile ad-hoc networks (MANETs) provide a platform that does not rely on central devices to control the operational flow of a system. Self-organizing MANETs combine wireless communication and node mobility to provide a network that does not rely on a fixed infrastructure such as the Internet. MANETs rely on the capabilities of the devices to control and command; in many cases, such devices are resource-constrained IoT devices with limited functions. The limitation of IoT devices can impact the overall performance of the network in different ways, e.g., limiting the network bandwidth. State-of-art IoBT networks are drastically scaling in terms of devices and transactions. The exponential increase of IoBT in terms of devices introduces security vulnerabilities and performance challenges to the already limited bandwidth networks. The implementation of MANETs in IoBT networks opens the following vulnerabilities [22]:

- *Wireless Communication*: Unlike wired networks, attackers can easily eavesdrop on an active interface. If the communication transmitted over the interface is not properly secure, the attacker can obtain unauthorized access. In a decentralized network, any node can transmit or receive information, increasing the possibility of attackers as they can target weak nodes, unlike centralized architectures on which the role of each node is defined.

- *Mobility of Nodes*: The dynamic nature of MANET network allows nodes to move freely in and out of the network. MANETs need to ensure the integrity and trustfulness of nodes through the duration of the mission. Without a central authority, nodes in the network need to agree through consensus protocols, what nodes are part of the network. Malicious users can disrupt the communication trough jammer devices preventing node to reconnect.

- *Honest Nodes*: Since communication heavily depends on the nodes of the networks, the cooperative nature and honesty of each node are crucial for the proper operation of the network. In general, centralized architectures use Public Key Infrastructures (PKI) to verify honest nodes; a decentralized PKI is still an open research question.

- *Wide Surface Attack Area*: The lack of standardized infrastructure and the ad-hoc nature of MANETs open a wide area for attackers, which an attacker can exploit to attack the network from different places.

- *Low-End Devices*: In general, low-end devices do not have the capabilities to implement robust security protocols, limiting the use of state-of-art crypto solutions. Commonly centralized architectures delegate high-security task to resourceful nodes. In a decentralized network, every node needs to ensure the integrity and protection of the data.

Farooq and Zhu [23] discuss a device-to-device (D2D) architecture and propose a model that characterizes the connectivity of devices in IoBT. Their model provides a framework to tune the physical parameters (e.g., number of combat units, the transmission power of devices) of the network to achieve optimal dissemination of intelligence information. Although the challenges to implement a pure peer-to-peer (P2P) network are not discussed, Farooq and Zhu introduce the possibility to use D2D or P2P networks in IoBT. Self-sovereignty in IoBT devices demands new paradigms and protocols that define rules to disseminate confidential information and to create robust access policies. A P2P-based network provides better capacity to implement robust security protocols because the network does not depend on central authorities to determine security protocols and access rules. P2P networks are reliant in the security and processing capacity of the devices in the network; a logical approach to implement P2P-based networks is to use a modular, flexible, and decentralized platforms such as blockchain. However, IoBT devices, in general, are resource-constrained, and the feasibility and challenges of blockchain in IoBT environment have not been broadly studied. Sudhan and Nene [24] propose a blockchain-based framework of network-enabled military operations that ensure data availability and data integrity for military networks. The feasibility of the framework is analyzed in three case studies: (a) data communication in battlefield management system (BMS), (b) logistics support for the armed forces, and (c) smart contracts in ammunition management. Sudhan and Nene demonstrate how the features of blockchain technology can address the challenges and issues of battlefield environment (e.g., smart contracts to track the supply chain of ammunition). A multi-layered architecture of blockchain-enabled IoBT, illustrated in Figure 10.3, aims to improve the trustworthiness of actions or commands. The architecture has three layers: (a) battlefield sensing layer which consists networking-capable sensors used to transmit intelligence data, (b) network layer with a unique purpose to capture transactions generated in the sensing layer, and (c) consensus and service layer, which uses consensus mechanism to agree on valid transactions

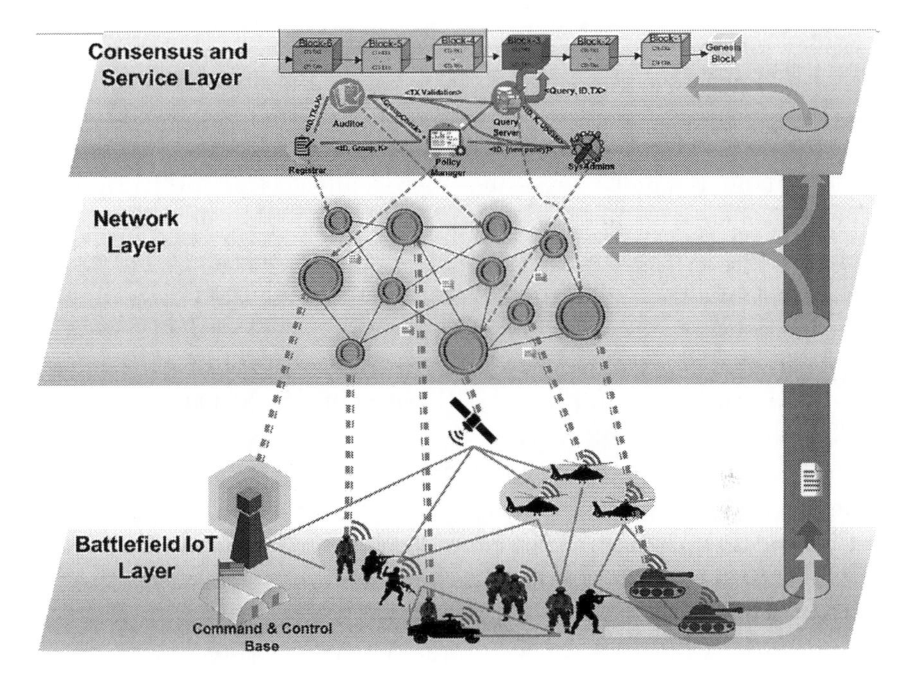

Figure 10.3 Blockchain-enabled IoBT architecture [8].

through a distributed ledger. Besides the proposed architecture, Tosh [8] discusses possible research challenges (e.g., interoperability, participants selection, number of nodes) that can impact the performance and security of IoBT.

A critical component missing in previous blockchain-based IoBT architectures is the empirical evaluation of the proposed blockchain-based network. Due to the critical nature of the information in IoBT network, real-time access, low latency, and high availability are essential attributes that blockchain platforms must meet. In general, IoBT networks are dynamic, extremely mobile, and with multiple dispersed devices, and due to these conditions, off-the-self blockchain platforms are not often suitable for IoBT. From the variety of blockchain-related works, only a few analyze the performance of such distributed platforms in low-bandwidth systems. Weston [25] evaluates the Ethereum platform, which is a blockchain technology used for digital currency, in a disrupted, intermittent, and low bandwidth (DIL) network. Their experiments consist of simulated environments that replicate realistic scenarios of IoBT networks, all nodes in the network submit one transaction every five seconds with a payload of 500 bytes. Their results suggest that blockchain technology requires stable-connected networks to operate correctly, and the Ethereum platform particularly introduces a high overhead in terms of TCP traffic.

Previous research has proposed centralized and decentralized (blockchain) solutions to address the challenges of IoBT. Because of the dynamic nature and tactical space of IoBT, centralized solutions are not suitable. Opposite to centralized architectures, decentralized architectures like blockchain have the potentials to address IoBT challenges, but more studies to determine the feasibility of blockchain-based IoBT networks are needed. Weston [25] provides insightful results in terms of resource utilization and latency. However, their evaluation focus on a public blockchain platform that, in general, has performance issues such as scalability of nodes and limited throughput. In [4], we proposed an edge-based blockchain platform, which aims to improve the performance of centralized platforms. The proposed edge-based blockchain platform addressed the scalability of blockchain in terms of transactions. In an edge-based architecture, the tasks of a central server are distributed between various blockchain edge nodes. The edge nodes process the data transactions made by IoT sensors through the Hyperledger Sawtooth platform. Our edge-based blockchain platform demonstrated the potentials of blockchain and edge computing to enable a robust and flexible platform.

In this chapter, we expand our previous analysis [26] of the feasibility of a private blockchain, namely, Hyperledger Sawtooth in IoBT context. Our previous study of the impacts of transactions throughput and network overhead in a blockchain-based tactical network focused on different testing parameters with realistic environments. We showed that Hyperledger Sawtooth though configurable is only feasible when enough resources are available. To expand our previous study of Hyperledger Sawtooth in IoBT environments, this chapter introduces an edge-based blockchain topology on which IoBT devices move the processing of information to resourceful devices located at the edge of the network. Before introducing our simulated environment, we discuss blockchain variants and how their distinguishing attributes can be used in IoBT.

10.4 BLOCKCHAIN USE CASES FOR TACTICAL NETWORKS

State-of-art blockchain platforms (Figure 10.4) can be adapted to find a model that is resilient, effective, and computationally feasible. Such blockchain-based model can address the standard requirements

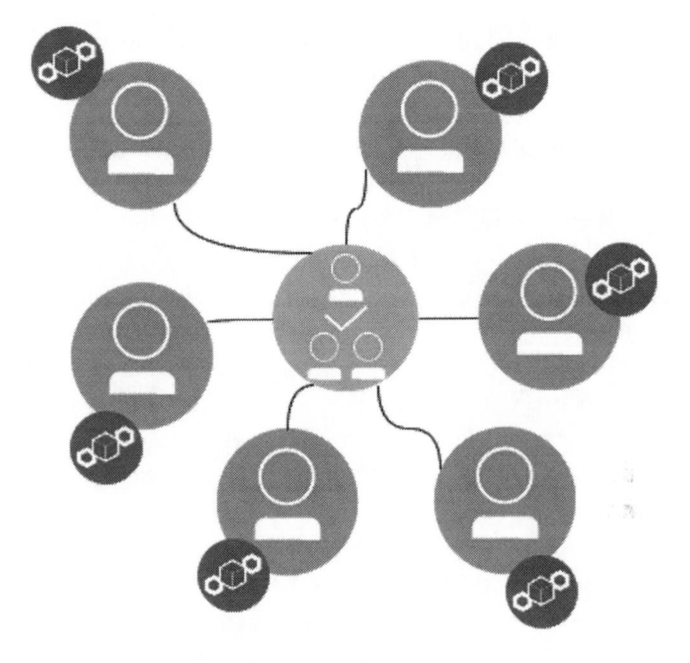

Figure 10.4 State-of-art blockchain system.

(e.g., high availability) of battlefields, but do not address all security concerns of the nodes such as information dissemination and overall transaction performance. In general, transactions in blockchain platforms are group into blocks and are disseminated through communication protocols, e.g., gossip protocol to other nodes. Nodes keep a local copy of all transactions in a distributed ledger [27]. A significant feature of blockchain is data availability on which all users can receive and publish blocks. Over time as the network grows, publishing blocks to every node increase network traffic, and this creates congestion and overhead in the communication protocols. Such overhead has a direct impact on the overall performance, feasibility, and operability of the system. We discuss the feasibility of two types of blockchain platforms in IoBT. Both types have relevant security features (e.g., consensus protocol) that focus on node security and data dissemination.

10.4.1 Permission-Less Variants

Access policies and role-based privileges determine what users can publish transactions and blocks in a blockchain platform. In a permissionless chain (otherwise known as public chain), every node in the network has public readability of the distributed ledger;that is, every node can publish, read, and access transactions or blocks. A public chain has two distinguishing features. First, there are no restrictions to join the network; any device with the proper services (e.g., transaction processor) can become a node in the network. The second is the public readability; that is, transactions recorded in the distributed ledger are available to all nodes and at all levels. Public readability is one of the major concerns of public chains. Rogue nodes can have the same access to data as honest nodes. Besides the security issues of public chains, permission-less variants have efficiency trade-offs [28] that can impact the overall performance in IoBT context.

Permission-less blockchain may not be suitable for battlefields. Tactical networks need to restrict who joins the network, and this creates a necessity to implement role-based access, which is not available in public chains. Although MANETs can be adapted to work over public chains, because they allow crypto-solutions such as encryption at the network level, their performance deficiencies [28] combined with the complexity of open chain platforms may introduce significant overheads to IoBT environments. For our purposes, we will incorporate security at the blockchain level, with a permissioned variant.

10.4.2 Permissioned Variants

Permissioned blockchain, also called private chains, restricts access to the network through role-based access controls. The public readability of distributed ledger is eliminated;that is, full transparency is not required. Role-based access control is achieved through management identity protocols [29] that are deployed on top of a blockchain platform. Besides the protections in place to access data, permissioned blockchain provides additional performance benefits (e.g., network latency) through relaxing components such as consensus protocols. Private chains can assume that all nodes are trusted, and thus, relaxing consensus protocols allows for faster processing of transactions. Performance modeling of permissioned blockchains such as Hyperledger Fabric and Hyperledger Sawtooth has been proposed [30–32] in the past. The previous performance models studied the feasibility of private blockchain in constrained-resources networks and identified the most common challenges and constraints of such platforms. We consider that permissioned platforms are suitable for IoBT when paired with military protocols and procedures.

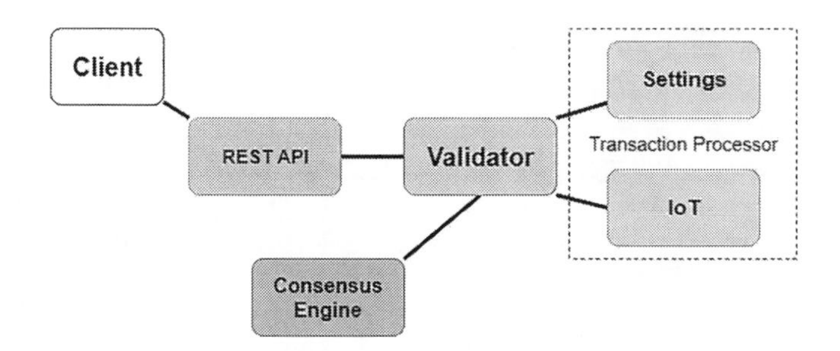

Figure 10.5 Components of a Hyperledger Sawtooth node.

From the multiple permissioned platforms available (e.g., Hyperledger Fabric, Multichain), we consider Hyperledger Sawtooth [33] due to its distinguishing features such as role-based identity manager and modular framework. State-of-art Sawtooth node, illustrated in Figure 10.5, consists of three major components: (a) consensus protocol called proof of elapsed time (POET), (b) REST API, and (c) transaction processor. Having selected Sawtooth, we aim to study the feasibility of the platform in IoBT networks. Military battlefields can be set up under different constraints and resources, making flexible blockchain platforms critical. We measure the feasibility of Hyperledger Sawtooth through a wide range of tactical environments under different testing parameters that stress network configuration in terms of overhead and delays.

10.5 EVALUATION

In this section, we analyze the feasibility of Hyperledger Sawtooth for a sample military battlefield network through simulations on the Common Open Research Emulator (CORE), a software used to simulate various types of networks [34]. Evaluations are based on a series of tests that measure how well Sawtooth would perform under different parameter changes. Each simulation presents a different topology with a combination of parameters, changing bandwidth, and transaction rate.

10.5.1 Setup

Due to the importance of a network setup in military environments, we test using four topologies: (a) fully connected, (b) mesh, (c) tree, and (d) edge-based (Figure 10.6). Each setup can be directly related to a particular military scenario, from command center setups to ad-hoc battlefield networks.

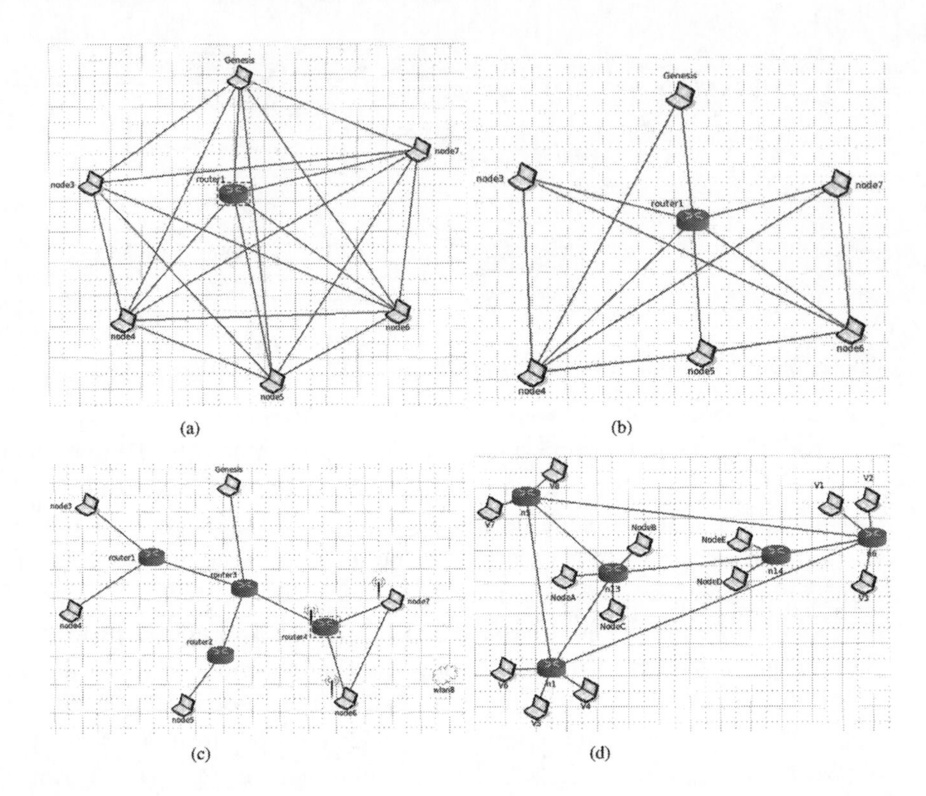

Figure 10.6 Sample IoBT relevant topologies for evaluation. (a) Full connectivity, (b) mesh topology, (c) tree topology and (d) edge-based topology.

The fully connected, mesh, and tree topologies consider a six-node tactical mission that was prepared with the intent of successful data reconnaissance on a battlefield using resource-constrained devices. The edge-based topology considers a transmitter–receiver setup on which constrained-resource devices offload blockchain-related operations to resourceful nodes. The edge-based topology consists of eight validator nodes and five transmitter nodes. The blockchain environment is set up and configured to work as follows. Sawtooth's blockchain environment consists of a multi-layer approach, where each layer communicates via a REST API. The topmost layer is that of collection nodes or participants of the chain. The information they collect is sent to a validator node that lives in the second layer. Validators verify that the information sent by the previous layer is in the proper order and stamped by the sender's key. Finally, once a user-specified number of transactions is collected, a block for the chain is formed. A block in a blockchain platform can be described as a collection of records that are stamped and approved a final time before being added to the permanent ledger. Any validator can perform this action. For our simulation, client nodes could be interpreted as the nodes collecting field intelligence and actively sending image metadata to the blockchain for processing. The nodes take images, and then, metadata is extracted using a Python script. This metadata is of use to mobile command stations connected to the blockchain nodes, but only interpret information, not provide. Each simulation occurs for one hour, under three different bandwidth options: 1 Gbps, 100 Mbps, and 64 Kbps. A traffic capture was taken for each test, where metric data was separated into two categories: (a) network maintenance packets and (b) image data packets.

For every bandwidth option and topology pairing, a total of ten tests were conducted. Each test changed the rate at which transactions were sent to a validator node in the network. These rates included 1, 10, 25, 50, and 100 transactions every two and five seconds, yielding a total of 120 tests for analysis.

10.5.2 Substituting Unlimited Term

CORE's term for an unrestricted bandwidth is known as unlimited. However, the performance of this term is completely dependent on machine hardware. To factor out this dependency, we compare the unlimited bandwidth to that of 1 Gbps, the highest numerical value available in CORE. Table 10.1 shows the number of transactions sent in a 10-minute period with a rate of 50 transactions every two seconds for the three topologies, under the unlimited and 1 Gbps bandwidth options.

As seen in the table, there is some deviation between the unlimited option and 1 Gbps. For every topology, the 1 Gbps was able to successfully deliver 20% to 30% more data packets and thus more transactions. This can be attributed to the routing and communication protocols for unlimited and 1 Gbps. Where the 1 Gbps bandwidth has built-in packets

TABLE 10.1 Data Packet Comparison for "Unlimited" and 1 Gbps Bandwidths

	Full	**Mesh**	**Tree**
Unlimited	93169	77555	55689
1 Gbps	114477	103679	84012

delays and optimized routing, the unlimited does not. For this reason, the testing set we will use in this chapter includes only the three bandwidths specified above.

10.5.3 Network Maintenance

Packet captures for each simulation are split into two categories: (a) maintenance and (b) data. Network maintenance refers to those packets which help maintain constant communication and operability for the blockchain. Packet types include ARP, MDNS, and TCP acknowledgments. Our evaluation demonstrates that for any topology and transaction rate under 1 Gbps and 100 Mbps bandwidth allotment, maintenance packets constitute an approximate 49%–51% of the total capture. Figure 10.7 shows the average number of maintenance packets observed in three topologies, full, mesh, and tree. Figure 10.8 shows the average number of maintenance packets observed in the edge-based topology. Note that the remaining percentage, being data packets, is reflected in Figures 10.9 and 10.10 discussed in Section 10.5.4.

There exists a small performance difference between two topologies, mesh and full. Using the 1 Gbps bandwidth, the mesh topology, in combination with a two- second transaction rate, consistently uses the highest number of packets. In comparison, the full topology paired with a 100 Mbps bandwidth and two-second transaction rate produced more packets. There are two main reasons for this. In the case of the mesh topology, the combination of bandwidth and node placement made for an efficient system, where a high level of transactions was sent and validated. When observing the tree topology, the number of packets rose due to a buffer constraint contained in Sawtooth's validation scheme. We observe that the network must resend packets when transactions overpass the buffer capacity. In comparison with mesh, full, and tree topologies, the edge-based topology (Figure 10.8) decreases the overall number of packages. The main reason for this is due to the roles of nodes, in an edge-based topology, only validator nodes must maintain active communication. The edge-based topology shows an increase of packets at the five-second rate. We observe this increase because the validator nodes receive transactions in a broader window; this triggers Sawtooth's heartbeat protocol.

The major outliers for these simulations are those that use the 64 Kbps bandwidth. The first point is that the amount of transactions sent is significantly lower, which leads to an assumption that network maintenance packets would also decrease. However, the maintenance

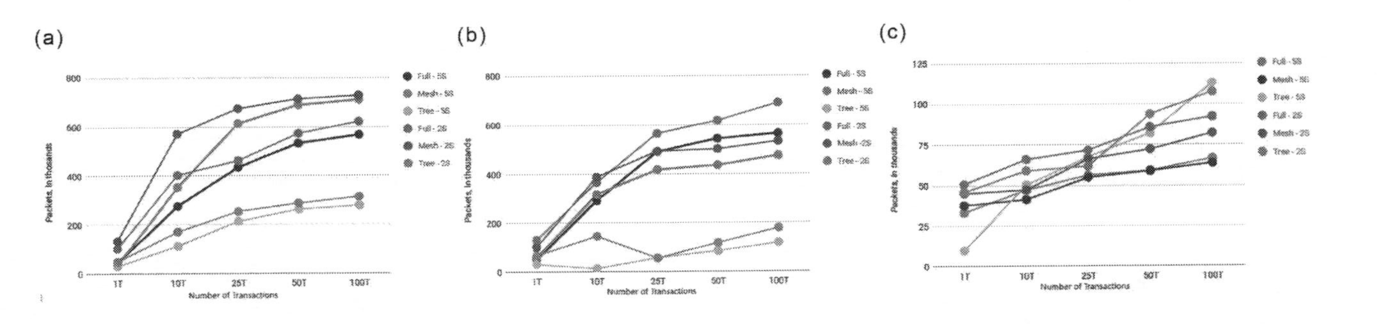

Figure 10.7 Maintenance packets for (a) 1 Gbps, (b) 100 Mbps, and (c) 64 Kbps simulations. Full – 5S is an abbreviated format for full topology – "x" transactions every five seconds. 1T is an abbreviated format for one transaction.

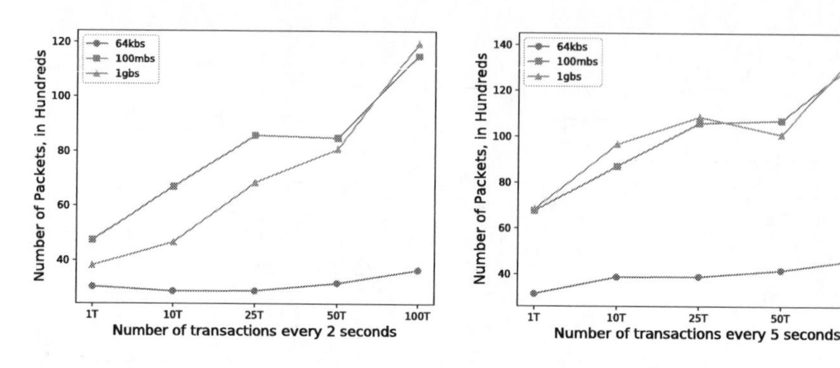

Figure 10.8 Maintenance packets for edge-based simulations.

percentages rose to an average range of 54%–70% due to the bottleneck created by the low bandwidth. Transactions are lost as the buffer cannot process in time, and thus, the network becomes congested with re-transmission of acknowledgments, Sawtooth's native heartbeat function, and others. Figure 10.7 demonstrates the number of packets needed to maintain network functionality under a 64 Kbps bandwidth setup.

10.5.4 Data Throughput

In this section, we look for packets that relate specifically to image metadata transmission or propagation of the blocks in the chain.

10.5.4.1 Topology

Node placement and connectivity in these simulations play a factor. Overall, the full and mesh topologies functioned at very similar rates and performance peaks. Having 1 Gbps bandwidth allows the mesh topology to perform better than the full topology, as seen in Figure 10.9. Full topology is followed by the tree topology, the weakest performer. The edge-based topology when the 1 Gbps is available depends on the transaction rate when working under a five-second rate; the edge-based overpasses the other topologies, but at the two-second rate, it performs similar to the full and mesh topologies.

When 100 Mbps bandwidth was available, the mesh, full, and edge-based topologies had overlapping results. Figure 10.9 shows how for tree, mesh, and full simulations that send "x" amount of transactions every two seconds, mesh performs better at certain rates like 10 transactions and 100 transactions, versus the full topology at all other points. The simulations that send "x" amount of transactions every five seconds show

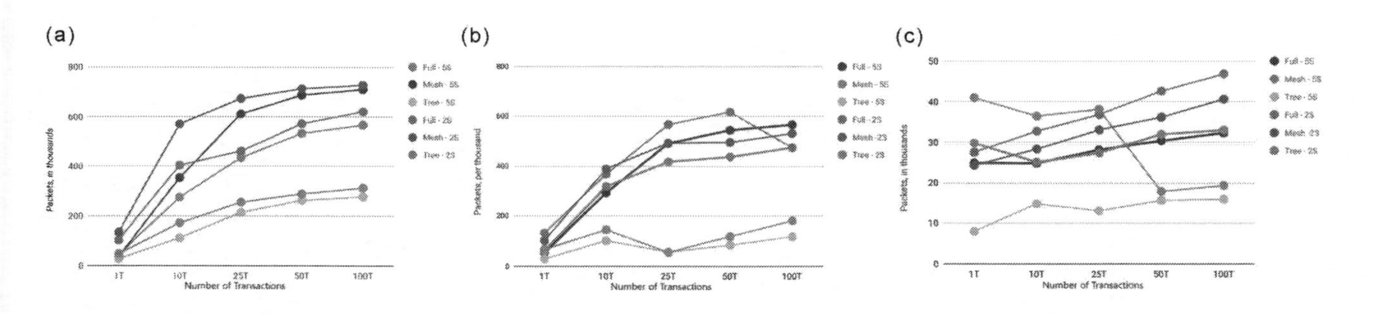

Figure 10.9 Data packets for (a) 1 Gbps, (b) 100 Mbps, and (c) 64 Kbps simulations. Full - 5S is an abbreviated format for full topology - "x" transactions every five seconds. 1T is an abbreviated format for 1 transaction.

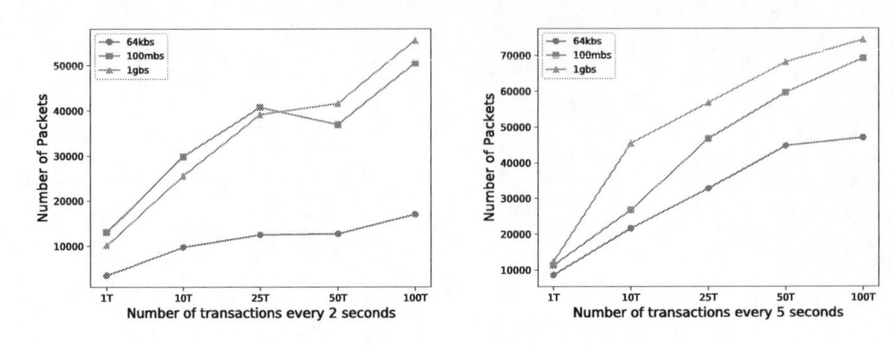

Figure 10.10 Data packets for edge-based simulations.

the mesh topology as the most effective choice, with a full topology trailing closely behind. For all tests under the 1 Gbps and 100 Mbps bandwidth, the tree topology consistently had the lowest rates of data throughput. However, it is not valid for the 64 Kbps captures. Referring to Figure 10.9, the tree topology had the highest starting point for simulations that sent transactions every two seconds and, in a more common result, the lowest starting point with simulations that sent transactions every five seconds.

Figure 10.10 shows the performance of the edge-based simulation sending "x" amount of transactions every two and five seconds. The edge-based topology has similar behavior in all three bandwidths, with the 64 Kbps at the lowest in both transaction rates. At two-second transaction rate, the edge-based simulations show that the available bandwidth significantly impacts the topology. For all four topologies, as the rate increased, the throughput decreased significantly.

10.5.4.2 Bandwidth

There are three significant points of focus that the data reflects in Figure 10.9. First, the highest amount of data throughput is observed under the 1 Gbps bandwidth, the highest value being around 700K data packets at 100 transactions every two seconds. For the edge-based topology (Figure 10.10), the highest amount of data throughput is also observed under the 1 Gbps bandwidth at around 70K data packets. Next, for 1 Gbps and 100 Mbps bandwidths, the throughput rose steadily until the 100th transaction marker. At that point, the tree topology under 100 Mbps had a sharp decline. This coincides perfectly with the network maintenance seen in the previous subsection, as more network failures caused this topology to underperform. The edge-based simulation, as observed in the network maintenance subsection, shows the overall lower number of packets, and we observe the same behavior in the number of data packets with a steady increase through all transaction rates.

Finally, the 64 Kbps captures are highly variable for simulations of tree, mesh, and full. In some instances, one transaction every five seconds was able to outperform ten transactions every two seconds, as noted by mesh and tree in Figure 10.9 for 64 Kbps. We can attribute this to a combination of both node instability and network processing failures due to increasing congestion of transaction validation. Additionally, the number of transactions is significantly smaller when compared to the two previous bandwidths. Regarding the edge-based simulation, the 64 Kbps bandwidth is not highly variables as the other three simulations, but it still underperforms. We attribute the more stable behavior of the 64 Kbps bandwidth to the nature of edge-based networks. Edge nodes only communicate to a subset of nodes of the network.

10.5.5 Successfully Validated Transactions

After detailing how the topologies and different bandwidths interact, we must also denote the success rate (i.e., the percentage of successfully validated transactions). This section uses ten-minute tests for all topologies under two bandwidth parameters with an increasing transaction rate that ranges from 1 to 10,000, except for the edge-based topology that uses all three bandwidths. To find the success rate of each combination, we take the number of successfully validated transactions and divide by the total transactions sent between all nodes. We will denote it as a percentage.

10.5.5.1 100 Mbps Bandwidth

Consider Figure 10.11. For the full topology, we start with a 100% success rate, then a small decline as the transaction rate increases. This increase and decrease of success are consistent, up until 1,000 transactions. After this, we observe a stagnant success percentage, followed by a sharp decline. At 7,500 transactions or more, this topology ceases being effective with a final success percentage of 38%.

The mesh topology shows a very different rate of decline when compared to the full topology. We start at 100%, then begin a steady decline as the transactions increase. At 250 transactions, a small increase in success rate is seen, making it the highest value before a steady decline

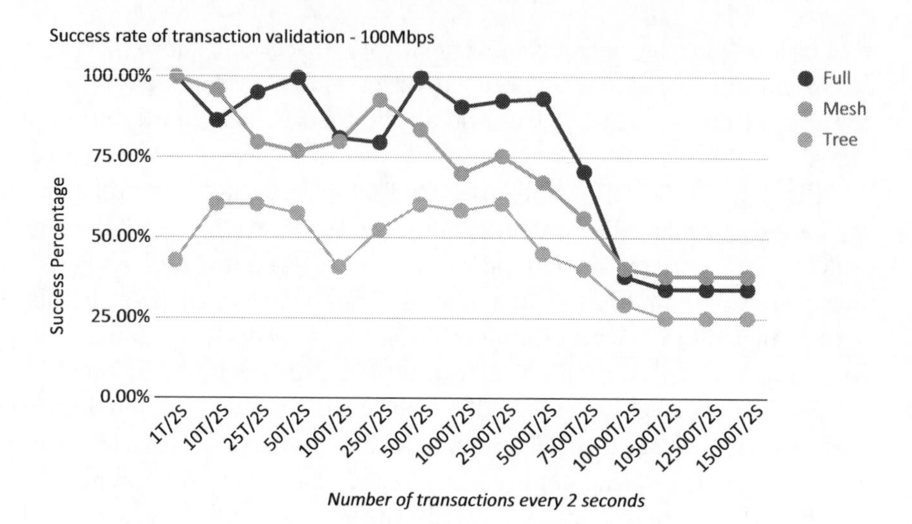

Figure 10.11 Success rate of 100 Mbps. Full is an abbreviated format for full topology. 1T/2S is an abbreviated format for 1 transaction every two seconds.

of success as transaction rates increase. We end the simulations at a success rate of 40%, just above the full topology.

The tree topology is consistently the lowest performing of the three. The highest success percentage observed was 60%, with transactions in the range of 10 and 250, excluding 100 transactions. This topology always underperforms when we compare it to the full and mesh, ending the simulations at a 25% success rate.

10.5.5.2 64 Kbps Bandwidth

Figure 10.12 highlights that when the bandwidth is decreased to 64 Kbps, the success rate never climbs above 65%. The rate of decline is steadier in these tests with an astonishing result: the tree topology is able to have the highest performance marker at the end of the set of tests. Although the tree topology is highly variable, it is able to end at a higher success rate due to the combination of network routing, packet delays, and topology setup. As much of the routing redundancy is reduced in the tree topology and the network packet bottleneck is reduced, the Sawtooth validator queue can take more incoming transactions rather than rejecting them.

For the full topology, the highest success rate is approximately 60% at a transaction rate of 10 per second and 64 Kbps bandwidth. The 60%

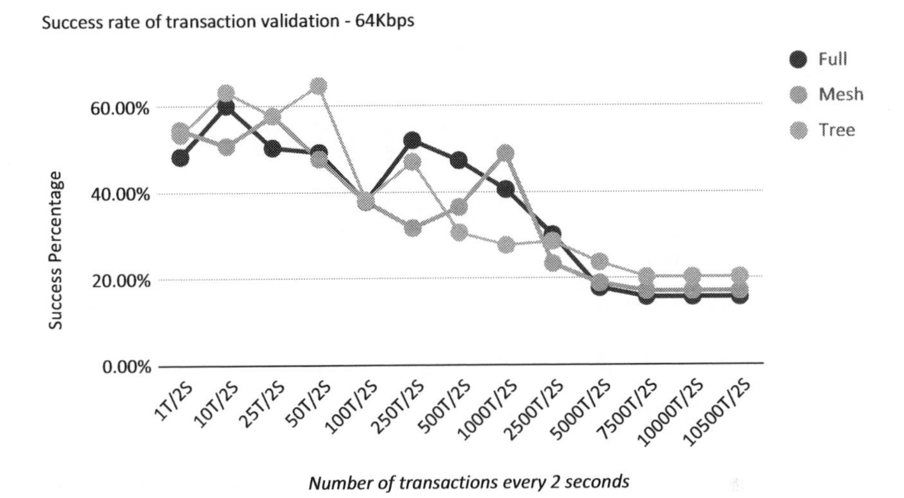

Figure 10.12 Success rate of 64 Kbps. Full is an abbreviated format for full topology. 1T/2S is an abbreviated format for 1 transaction every two seconds.

success rate is also observed at 7,500 transactions when 100 Mbps bandwidth is available. The mesh topology is very much similar to full, showing the highest rate of success at 58% with 64 Kbps bandwidth and a 25 transaction rate. This value is observed for mesh again at 7,500 transactions and 100 Mbps of bandwidth. Between full and mesh, the data shows that the full topology had the highest numbers of success data points overall.

Though volatile, the tree topology managed to end the set of tests with the highest success marker. However, the performance of this topology is still lower when compared to the full setup. As it does better than all other topologies before 100 transactions at 64 Kbps bandwidth, it is a viable option for a very small subset. An additional point is that the tree success rate remained somewhat consistent between both bandwidth values. This means that if more bandwidth is available, the tree topology will run at a capped performance peak, with the decay of success extending further out as the bandwidth grows.

10.5.6 Edge-Based Simulation

Figure 10.13 highlights the percentage of successful transactions for the edge-based simulations under all three bandwidths and both transactions rate. As observed in previous sections, the 64 Kbps bandwidth

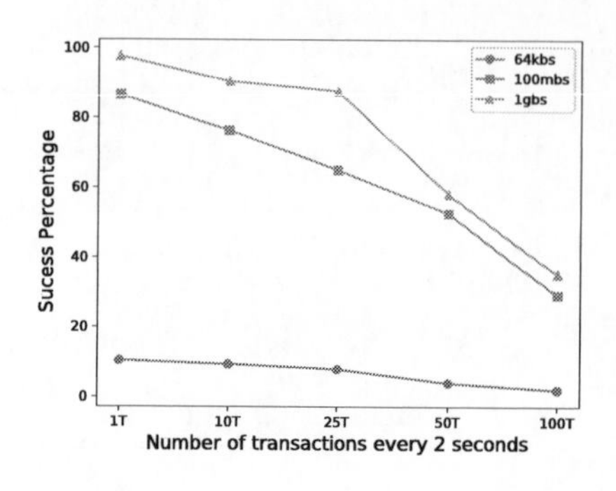

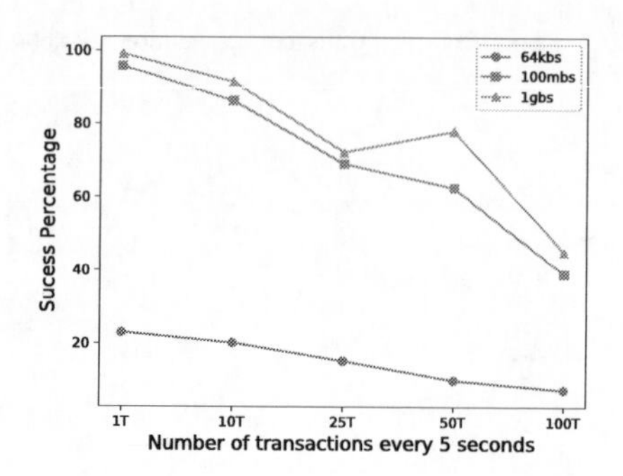

Figure 10.13 Success rate for edge-based simulations.

underperforms by a significant difference. Overall, the edge-based topology shows a similar behavior than the other three topologies. At a low transaction rate, the topology achieves a high percentage of confirmed transactions, but as the transaction rate is increased, the throughput performance decrease. Regarding the 1 Gbps and 100 Mbps, the edge-based topology has a consistent performance with 1 Gbps, always over-performing the 100 Mbps.

It is worth noting that neither of the charts shows tests reaching a complete lack of transaction processing. The leading cause of non-validation is due to a queue maintained by all validators. When the queue is full, the transaction is dropped and therefore lost. The logic then follows that while the queues may always be full, there exists a small percentage that is validated, as noted by the high transaction rates in both the 100 Mbps and 64 Kbps charts.

10.5.7 Findings

With the data available, we can conclude several things. First, is that a decentralized network approach (full, mesh, and edge-based) are the most effective when it comes to a blockchain setup. We are able to sustain a more efficient system on these platforms. When it comes to the most ideal setup, we consider two things: (a) data throughput and (b) success of validation. Figure 10.9 showed that the mesh topology was able to produce the highest amount of data packets over one hour; however, the success rate of the mesh topology was lower overall than that of the full topology for both 100 Mbps and 64 Kbps bandwidth. Overall, the edge-based is the more stable setup with the lowest overhead to the network. It can achieve a high percentage of confirming transactions with a low number of data packets. The ideal setup then becomes a full topology with the highest success rate demonstrated at 1 and 5,000 transaction(s) every two seconds when 100 Mbps bandwidth is available or a full topology between 1 and 250 transactions every two seconds when 64 Kbps bandwidth is present, or an edge-based topology under a 1Gbps and between 1 and 25 transactions every two seconds.

Second, we are able to conclude that the Sawtooth, though configurable, is feasible when higher resources are available. While we are able to perform tests under small time constraints, the amount of storage and bandwidth needed overtime is too significant when we consider the low-resource devices that a soldier may carry. Furthermore, these tests approximate small military operations. When scaled to a bigger size, there is a risk that the blockchain platform may not provide a 100% operational guarantee to maintain the ledger due to resource constraints.

10.5.8 Other Factors

Other testing parameters can play a significant role in the metric data produced. In our simulations, the nodes are stationary. Nodes that are able to travel in and out of range of a wireless network may significantly alter the data, as Sawtooth's native response is to hold transactions in a buffer. When the buffer is full, transactions cease to be stored. When a node reconnects, it begins not only sending all of the stored transactions but also receiving information from others. Nodes and the network resources might not be able to handle an increased amount of transactions if multiple nodes were to disconnect and reconnect.

One crucial factor to note is that the impact node quantities have on a network. Distribution of information will take longer as more nodes and routing protocols are established, slowing down the transaction rate desired and overall performance. Additionally, the POET consensus method is one of two different methods that may be utilized with Sawtooth. In this configuration, we do not observe any form of stale blocks, as the rules specify a block must be formed once the minimum number of validated transactions is achieved. This cleans up network propagation, whereas other consensus methods may not be as efficient.

10.6 CONCLUSION

Enabling a secure yet decentralized network for IoBT applications requires adoption of novel solutions that are scalable and robust against adversarial attacks. It is known that maintaining centralized networks in battlefield is difficult and costly, while failure and safety issues are of major concerns. Centralized network infrastructure is exposed to many security vulnerabilities, such as a single point of failures and DoS, as it can give attackers a predictable target, and depending on the level of security used, it can be easy to compromise. A blockchain-integrated platform for IoBT can be a potential choice to address many security challenges of IoBT networks and possibly help in designing a decentralized security solution. However, there are not enough studies in place to determine the performance of current state-of-art blockchains in IoBT context. Therefore, we integrated a permissioned blockchain, namely, Hyperledger Sawtooth, with a distributed IoBT environment to achieve necessary security needs and evaluated the network constraints on the blockchain under various topologies with different parameters (e.g., rate of transactions). The blockchain-based IoBT network is decentralized and can meet the security requirements (e.g., trusted communication channels) established by the military. Although simulations show that this particular variant is not efficient enough for battlefields, because it implements a buffer queue that as the rate increases, it becomes a bottleneck for the network decreasing the throughput performance, they show promise for other similar blockchain implementations.

In the future, we plan to expand our experiments in following three areas: (a) *Blockchain platform survey* – Hyperledger Sawtooth is only one of the many permissioned blockchain platforms that exist. We plan to develop a survey of different blockchain platforms that will be used to identify the strengths and weaknesses of each platform, and what are the unique characteristics that could be used for IoBT. We plan to analyze and test the attributes of each blockchain platform in terms of effectiveness and readiness for IoBT context. (b) *MANET analysis* – Future experiments will analyze the challenges and issues of MANETs in permissioned blockchains. Through the security attributes of MANETs (e.g., security at the network layer), a blockchain platform could potentially relax the robust security of communication protocols, meaning that devices will have more available resources that can be dedicated to the processing of transactions and blocks. (c) *Network Partition* – Military operations can take place in large areas with uneven terrain, on which military personnel sub-divide to cover as much terrain as possible. Communication between sub-teams and the central command is crucial for the success of the mission and life of soldiers. However, reliable communication is not always possible.

Sub-teams should be capable of maintaining a sustainable local communication while disconnected from the central command. Furthermore, whenever a connection is available, the local sub-network must be capable of updating its state with the main network. Future experiments will analyze the impact of partitions on the network, and the effects to validator nodes that are constantly disconnected from the main network.

Bibliography

[1] C. Kolias, G. Kambourakis, A. Stavrou, and J. Voas, "DDoS in the IoT: Mirai and other botnets," *IEEE Computer*, vol. 50, no. 7, pp. 80–84, 2017.

[2] M.-H. Maras, "Internet of Things: security and privacy implications," *International Data Privacy Law*, vol. 5, no. 2, pp. 99–104, May 2015. [Online]. Available: https://academic.oup.com/idpl/article/5/2/99/645234.

[3] A. Dey, K. Stuart, and M. E. Tolentino, "Characterizing the impact of topology on iot stream processing," in *2018 IEEE 4th World Forum on Internet of Things (WF-IoT)*, Feb 2018, pp. 505–510.

[4] A. O. G. Rivera, D. K. Tosh, and L. Njilla, "Scalable blockchain implementation for edge-based internet of things platform," in *MILCOM 2019 - 2019 IEEE Military Communications Conference (MILCOM)*, 2019, pp. 1–6.

[5] D. K. Tosh, S. Shetty, P. Foytik, C. Kamhoua, and L. Njilla, "Cloud-pos: A proof-of-stake consensus design for blockchain integrated cloud," in *IEEE International Conference on Cloud Computing (CLOUD)*. IEEE, 2018.

[6] M. Crosby, "BlockChain Technology: Beyond Bitcoin," *Applied Innovation Review (AIR)*, vol. 2, pp. 6–10, 2016.

[7] N. Kshetri, "Can blockchain strengthen the internet of things?" *IT Professional*, vol. 19, no. 4, pp. 68–72, 2017.

[8] D. K. Tosh, S. Shetty, P. Foytik, L. Njilla, and C. A. Kamhoua, "Blockchain-Empowered Secure Internet-of-Battlefield Things (IoBT) Architecture," in *IEEE Military Communications Conference (MILCOM)*, Oct 2018, pp. 593–598.

[9] N. Suri, M. Tortonesi, J. Michaelis, P. Budulas, G. Benincasa, S. Russell, C. Stefanelli, and R. Winkler, "Analyzing the Applicability of Internet of Things to the Battlefield Environment," May 2016.

[10] A. Kott and D. S. Alberts, "How do you command an army of intelligent things?" *Computer*, vol. 50, no. 12, pp. 96–100, Dec 2017.

[11] T. Abdelzaher, N. Ayanian, T. Basar, S. Diggavi, J. Diesner, D. Ganesan, R. Govindan, S. Jha, T. Lepoint, B. Marlin, K. Nahrstedt, D. Nicol, R. Rajkumar, S. Russell, S. Seshia, F. Sha, P. Shenoy, M. Srivastava, G. Sukhatme, A. Swami, P. Tabuada, D. Towsley, N. Vaidya, and V. Veeravalli, "Toward an internet of battlefield things: A resilience perspective," *Computer*, vol. 51, no. 11, pp. 24–36, Nov 2018.

[12] G. Padmavathi and D. Shanmugapriya, "A survey of attacks, security mechanisms, and challenges in wireless sensor networks," *International Journal of Computer Science and Information Security*, vol. 4, no. 1, 2009.

[13] R. Sepe, "Denial of Service Deterrence," *SANS Institute Information Security Reading Room* [white paper], p. 25, 2015.

[14] P. Sass, "Communications networks for the force xxi digitized battlefield," *Springer US Mobile Networks and Applications*, vol. 4, pp. 139–155, 1999.

[15] A. E. Cohen, G. G. Jiang, D. A. Heide, V. Pellegrini, and N. Suri, "Radio frequency iot sensors in military operations in a smart city," in *MILCOM 2018 - 2018 IEEE Military Communications Conference (MILCOM)*, Oct 2018, pp. 763–767.

[16] Courtesy, "Four future trends In tactical network modernization,"[Online]. Available: https://www.army.mil/article/216031/four_future_trends_in_tactical_network_modernization, last accessed: May 31, 2019.

[17] A. Castiglione, K. R. Choo, M. Nappi, and S. Ricciardi, "Context aware ubiquitous biometrics in edge of military things," *IEEE Cloud Computing*, vol. 4, no. 6, pp. 16–20, Nov 2017.

[18] T. Peng, C. Leckie, and K. Ramamohanarao, "Survey of network-based defense mechanisms countering the DoS and DDoS problems," *ACM Computing Surveys*, vol. 39, no. 1, pp. 3–es, Apr 2007. [Online]. Available: http://portal.acm.org/citation.cfm?doid=1216370.1216373.

[19] S. Misra, S. K. Dhurandher, A. Rayankula, and D. Agrawal, "Using honeynodes for defense against jamming attacks in wireless infrastructure-based networks," *Computers & Electrical Engineering*, vol. 36, no. 2, pp. 367–382, Mar 2010. [Online]. Available: http://www.sciencedirect.com/science/article/pii/S0045790609000536.

[20] N. Suri, M. Tortonesi, J. Michaelis, P. Budulas, G. Benincasa, S. Russell, C. Stefanelli, and R. Winkler, "Analyzing the applicability of internet of things to the battlefield environment," in *2016 International Conference on Military Communications and Information Systems (ICMCIS)*, May 2016, pp. 1–8.

[21] M. Tortonesi, A. Morelli, M. Govoni, J. Michaelis, N. Suri, C. Stefanelli, and S. Russell, "Leveraging internet of things within the military network environment — challenges and solutions," in *2016 IEEE 3rd World Forum on Internet of Things (WF-IoT)*, Dec 2016, pp. 111–116.

[22] A. Pathan, *Security of Self-Organizing Networks: MANET, WSN, WMN, VANET.* CRC Press, 2016. [Online]. Available: https://books.google.com/books?id=ZtBnZoijaDcC.

[23] M. J. Farooq and Q. Zhu, "Secure and reconfigurable network design for critical information dissemination in the internet of battlefield things (iobt)," in *2017 15th International Symposium on Modeling and Optimization in Mobile, Ad Hoc, and Wireless Networks (WiOpt)*, May 2017, pp. 1–8.

[24] A. Sudhan and M. J. Nene, "Employability of blockchain technology in defence applications," in *2017 International Conference on Intelligent Sustainable Systems (ICISS)*, Dec 2017, pp. 630–637.

[25] N. Weston, J. Willard, and P. Wang, "Performance of blockchain technology on DoD tactical networks," in *Disruptive Technologies*

in Information Sciences II, M. Blowers, R. D. Hall, and V. R. Dasari, Eds., vol. 11013, International Society for Optics and Photonics. SPIE, 2019, pp. 109–119. [Online]. Available: https://doi.org/10.1117/12.2520541.

[26] E. D. Buenrostro, A. O. G. Rivera, D. Tosh, J. C. Acosta, and L. Njilla, "Evaluating usability of permissioned blockchain for internet-of-battlefield things security," in *MILCOM 2019 - 2019 IEEE Military Communications Conference (MILCOM)*, 2019, pp. 841–846.

[27] T. T. A. Dinh, R. Liu, M. Zhang, G. Chen, B. C. Ooi, and J. Wang, "Untangling Blockchain: A Data Processing View of Blockchain Systems," *arXiv:1708.05665 [cs]*, Aug. 2017, arXiv: 1708.05665. [Online]. Available: http://arxiv.org/abs/1708.05665.

[28] Z. Zheng, S. Xie, H. Dai, X. Chen, and H. Wang, "An Overview of Blockchain Technology: Architecture, Consensus, and Future Trends," in *2017 IEEE International Congress on Big Data (BigData Congress)*. IEEE, June 2017, pp. 557–564, event-place: Honolulu, HI, USA. [Online]. Available: http://ieeexplore.ieee.org/document/8029379/.

[29] "Hyperledger indy," [Online]. Available: https://www.hyperledger.org/projects/hyperledger-indy, last accessed: May 31, 2019.

[30] B. Ampel, M. Patton, and H. Chen, "Performance modeling of hyperledger sawtooth blockchain," in *2019 IEEE International Conference on Intelligence and Security Informatics (ISI)*, July 2019, pp. 59–61.

[31] A. Baliga, N. Solanki, S. Verekar, A. Pednekar, P. Kamat, and S. Chatterjee, "Performance characterization of hyperledger fabric," in *2018 Crypto Valley Conference on Blockchain Technology (CVCBT)*, June 2018, pp. 65–74.

[32] Z. Shi, H. Zhou, Y. Hu, S. Jayachander, C. de Laat, and Z. Zhao, "Operating permissioned blockchain in clouds: A performance study of hyperledger sawtooth," in *2019 18th International Symposium on Parallel and Distributed Computing (ISPDC)*, June 2019, pp. 50–57.

[33] "Introduction — Sawtooth v1.1.4 documentation," [Online]. Available: https://sawtooth.hyperledger.org/docs/core/releases/latest/introduction.html#, last accessed: May 31, 2019.

[34] J. Ahrenholz, "CORE Documentation," [Online]. Available: https://www.nrl.navy.mil/itd/ncs/products/core, last accessed: May 31, 2019.

CHAPTER 11

Internet of Things in 5G Cellular Networks: Radio Resource Perspective

Ajay Pratap

Indian Institute of Technology (BHU) Varanasi, India

Sajal K. Das

Missouri University of Science and Technology Rolla, MO, USA

CONTENTS

The fifth-generation (5G) of mobile communication systems is expected to become an all-encompassing solution to fundamentally every broadband wireless communication need of the next decade. Advanced modulation and coding schemes facilitate 5G to fulfill the extensive demand of the Internet of Things (IoT)-enabled cellular networks. The use of advanced modulation and coding schemes allow cellular 5G networks to fulfill maximum IoT resource demand without the degradation of quality. In this work, consider advanced modulation schemes supported with different codec standards to find their impact on cell capacity, which further improves the performances of IoT-enabled 5G networks. First, we made capacity estimation using these advanced modulation schemes supported with different codec standards. Later, we estimate service blocking performances by applying $M/M/m/m$ queueing model. Through the experiment, we observed that the advanced modulation schemes supported with lower codec enhance the system capacity. Provisioning of higher cell capacity by having advanced modulation schemes with lower codec further improves the task execution performances in IoT-enabled 5G networks.

11.1 INTRODUCTION

As per the International Data Corporation (IDC) report, the 5G services will drive 70% of companies to spend $1.2 billion on connectivity management solutions [1]. The new paradigm of Internet of Things (IoT) business model needs new performance criteria such as coverage of wireless communication, ultra-low latency, ultra-reliability, throughput, and massive connectivity for ultra-dense deployed IoT devices. Moreover, a number of connected devices are expected more than 100 billion by 2050 compared to 30 billion of 2020, as shown in Figure 11.1. To fulfill these demands, the evolving 5G cellular model is expected to furnish a new interface for the future of IoT devices. The deployment of 5G is at its early stage, which intents at access point improvement, higher frequency, new radio access technology along with new modulation, and coding schemes [5,9,11,27]. Moreover, a new generic radical change needs to appear in the IoT-enabled 5G cellular model with architectural and business perspective. People expect a heterogeneous 5G cellular network, such as shown in Figure 11.2 to satisfy the IoT users' demand and maximize the revenue of different cellular operators altogether. The heterogeneous 5G networks will have a multi-tier model consisting of fog access point (FAP) [21,23], small cell access point (SAP) [22,26], and device-to-device (D2D)

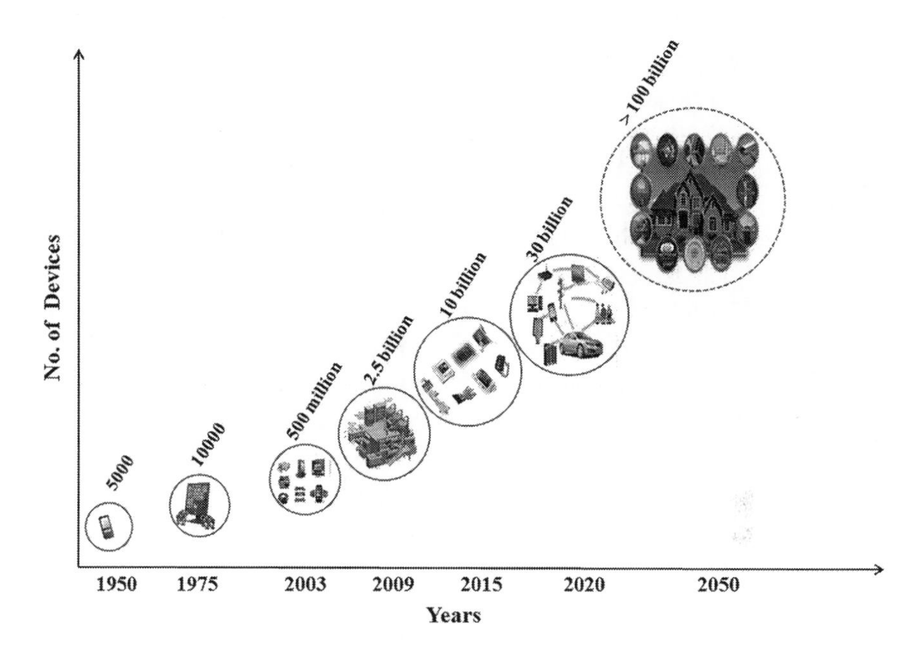

Figure 11.1 Expected number of IoT devices by 2050.

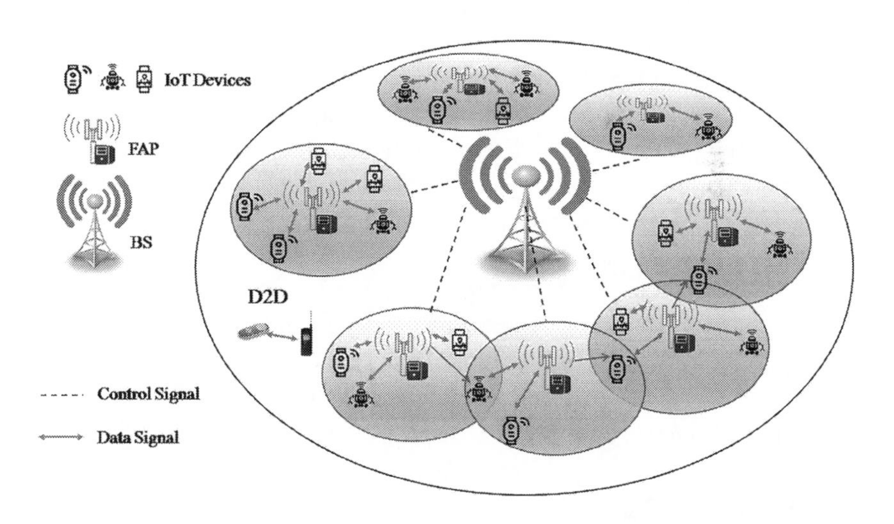

Figure 11.2 IoT-enabled 5G networks.

communication-enabled smart IoT devices [13,24,25,27,28] working altogether underlying cellular model.

Several new technologies have already emerged for IoT devices, and many of them are already well employed in other scenarios such as Bluetooth 5.0 Low Energy [7], IEEE 802.15.4 [17], LoRa [29], Sigfox [2],

WiFi HaLow [3], and Narrow-band IoT [4]. Disregarding, these existing technologies are not sophisticated enough or do not show up to fulfill all requirements to be used on an enormous scale [4]. However, the upcoming 5G technology will not completely replace these current technologies; instead, it will complement these technologies in the future applications and services [16]. Thus, there is a challenge to deal with the coordination and coexistence between upcoming and existing technologies for next-generation IoT-enabled wireless networks.

The latest trend of multiple applications such as live voice call (e.g., Skype, WhatsApp, Facebook), instant image uploading, and video downloading/streaming with the increase of IoT devices (such as mobile phones, laptop, and tablets) has reached exponential increase in the cellular data demand. As the total available spectrum in conventional sub 6 GHz bands is already limited, the utilization of new adaptive modulation and coding schemes are needed in order to generate efficient applications for IoT-enabled 5G cellular architecture. Advanced modulation and coding schemes facilitate IoT devices to get efficient services in 5G cellular model. Use of advanced modulation and coding schemes allow 5G networks to fulfill maximum IoT users' demand without degradation in quality of services. In this work, we consider the advanced modulation schemes supported with different codec standards at the narrow-band IoT to find their impact on 5G cell capacity with further impact over blocking performance in different applications.

11.2 NARROW BAND AND CODEC SCHEME FOR IoT-ENABLED 5G

In this section, we discuss the applicability of narrow band, modulation, and codec schemes for IoT-enabled 5G cellular networks. Narrow-band IoT model speaks to a noteworthy advancement towards the association between cellular technology and IoTs as described in the following:

11.2.1 Narrow-Band IoT

Narrow band has a versatile, adaptive, and scalable frequency band from 1.4 MHz to 20 MHz. Radio resource data transmission is separated into equivalent sub-directs of 180 kHz in recurrence space and a transmission time interval (TTI) of 1 ms in time domain. A radio resource in narrow band consists of 180 kHz (12 consecutive sub-carriers, each of which 15 kHz) and 0.5 ms [31]. The frame length is 10 ms. A TTI is comprised of two-time slots of 0.5 ms each. Each slot is further divided into six and seven orthogonal frequency-division multiple access (OFDMA) symbols in the extended and the normal cyclic prefix (CP), respectively. Hence, a radio resource in frequency/time domain of across one sub-channel in frequency domain and one-time slot in time domain considered as physical resource block (PRB). The smallest resource unit that can be assigned to an IoT for data transmission is known as PRB as shown in the Figure 11.3.

Based on the sub-carriers spacing of 15 kHz, OFDM symbol duration can be calculated as $1/15 = 66.67\mu s$, fast Fourier transform (FFT) size is set as 2,048, and the sampling rate f_s is set as $\Delta f \times N_{FFT} = 33.72$ MHz and the sampling interval T_s as $1/f_s$. In order to minimize the latency constraint of IoT-enabled network, there is a need to reduce the TTI spacing at 30 kHz [10]. Thus, the OFDM symbol duration T_{OFDM} be 33.33 μs FFT size N_{FFT} to become 1024, while sampling rate f_s

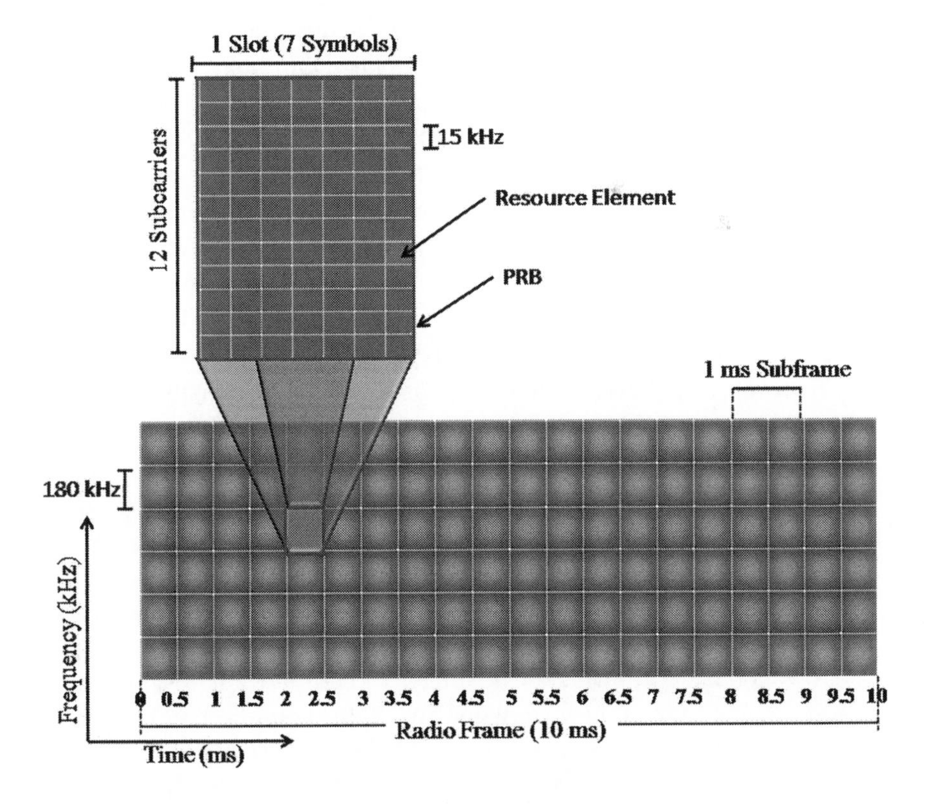

Figure 11.3 Narrow-band IoT carrier.

is kept at 30.72 MHz. The frame duration of 10 ms can further be divided into 40 subframes. Each subframe duration T_{sf} be 0.25 ms. Moreover, each subframe could contain 6 or 7 symbols. At the lower offered load in IoT-enabled 5G networks, 0.25 ms TTI is an attractive choice to obtain lower latency due to lower control overhead [15]. There is a possible direction to different TTI sizes for the future IoT-enabled 5G networks.

11.2.2 Adaptive Modulation and Codec Schemes

Investigating the suitable modulation and codec schemes for next-generation IoT-enabled 5G network is crucial to accomplish task within certain deadline. Due to latency constraint, time synchronization and phase coherence for different tasks become challenging research problem. Moreover, due to short packet size of different tasks, a large coding cannot be suitable such as for convention data networks. However, the overhead required to maintain the synchronization and phase coherency becomes significantly large while using the conventional coherent modulation schemes [8]. Particularly, this overhead can be reduced by exploiting the non-coherent modulation/demodulation and codec schemes.

In cellular networks, uplink information is modulated to only one carrier, adjusting the phase or amplitude of the carrier or both. The modulation methods are Quadrature Phase Shift Keying (QPSK), 16 QAM (Quadrature Amplitude Modulation), and 64 QAM. QPSK and 16 QAM are available for all the devices, whereas 64 QAM is dedicated to devices whose demand are higher, i.e., up to 300 Mbps in cellular model. QPSK allows good transmitter dexterity when functioning at full transmission power. The device uses lower transmitter power when operating at 16 QAM or 64 QAM. Binary phase shift keying (BPSK) has been assigned for control channels.

Adaptive multi-rate (AMR) or AMR-narrow-band (AMR-NB) codec is added into 3GPP Release 98 for enabling codec rate harmony; 8 kHz sampling rate of AMR provides 300 Hz–34 Hz audio bandwidth. AMR-wideband (AMR-WB) was introduced in 3GPP Release 5 as an extension of AMR. The range of AMR-WB and AMR-NB is shown in Table 11.1. AMR-WB data rates vary between 1.75 kbps and 23.85 kbps. The use of sampling rate of 16 kHz provides 50–7,000 Hz audio bandwidth which results in better voice quality. The 12.65 kbps rate of AMR-WB is similar to 12.65 kbps of AMR. The bit rates of 1.8 kbps and 1.75 kbps are utilized for silence insertion descriptor (SID) transmission.

TABLE 11.1 AMR Codec Radio Bandwidths

AMR-NB (kbps)	AMR-WB (kbps)
12.2	23.85
10.2	19.85
7.95	18.25
7.4	15.85
6.7	14.25
5.9	12.65
5.15	8.85
4.75	6.6
1.8	1.75

To enhance the inseparability of IoT devices concerning energy and latency constraint, AMR-WB can be a good module to incorporate in [12,20,30]. An IoT device for performing codec rate adaptation accounts for storing instructions and processing circuit to communicate in cellular model. The instruction to select a particular codec rate is generated by macro-base station (MBS). Moreover, MBS helps IoT devices in order to select the appropriate FAP to get their computation load offloaded [23,25]. The instruction consists of getting a bit rate suggestion for an FAP, selecting an appropriate AMR-WB and sending the task to respective computational capability enabled server, i.e., FAP. In the case of handover of task from one FAP to another FAP, IoT device receives an instruction to re-select an appropriate AMR-WB and offload the task to another FAP with help of MBS.

11.3 CAPACITY ESTIMATION OF 5G NETWORKS

We have assumed that IoT devices generate different tasks and MBS offloads them to the respective FAP for computation. We have assumed that all tasks are critical delay sensitive, and they get blocked in the case of unavailability of radio resources, i.e., PRBs. Moreover, depending on the availability of PRBs, MBS schedules two kinds of tasks: (a) newly arrived tasks and (b) already existing tasks in the system. The basic idea of task offloading is to keep track of guaranteed bit rate (GBR) of different tasks as given in the following equation:

$$\sum_{i=1}^{K} GBR_i + GBR_{new} \leq R_{max}, \tag{11.1}$$

where K is the total number of existing tasks in the system, and R_{max} is an average uplink cell throughput. GBR_{new} represents the GBR of newly arrived tasks. In the following section, we have shown the task offloading procedure via resource allocation mechanism.

11.3.1 Resource Allocation Procedure

Based on available PRBs, the task offloading criterion can be expressed as follows:

$$\sum_{i=1}^{K} N_i + N_{new} \leq N_{tot} \tag{11.2}$$

where N_i is the assigned number of PRBs, N_{new} is the requesting task's PRB demand, and N_{tot} is the total number of PRBs in the system. Figure 11.4 shows the resource allocation and task execution procedure.

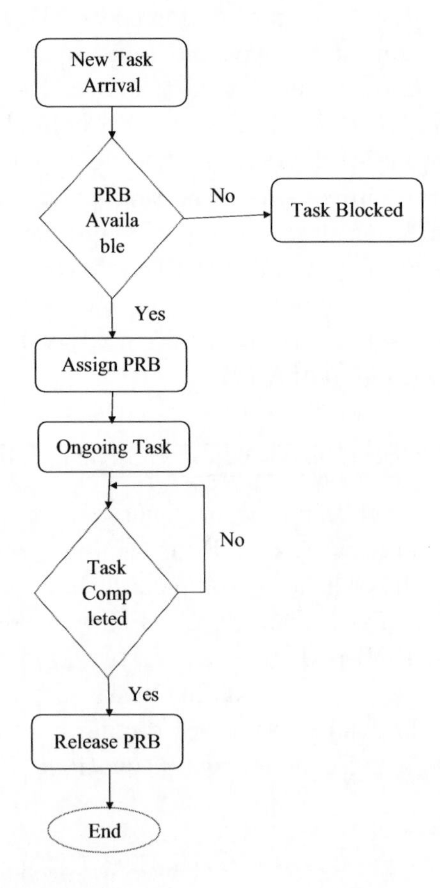

Figure 11.4 Proposed algorithm.

Whenever new task arrives in the network, MBS first checks total available resources. If no resource is available, task gets blocked; otherwise, task is assigned with valid resource. Upon the completion of task, allocated resources get released and the process ends.

To estimate the total number of PRBs in the system, we need to evaluate the total capacity of cellular 5G network. The theoretical capacity analysis is done in the following subsection.

11.3.2 Theoretical Capacity Analysis of Cellular 5G Networks

Let TF_{used} and TF_{total} represent the number of used PRBs and total available PRBs in the system. The total number of available PRBs for a service can be calculated as follows:

$$TF_{total} = (\text{No. of TTIs}) \times (\text{No. of PRBs per TTI}). \tag{11.3}$$

We have assumed that IoTs and MBS have already been synchronized with each other before sending any task. However, synchronization between IoTs and MBS can be done using the technique given in [14]. The number of used PRBs by newly arrived tasks can be written as follows:

$$\begin{aligned} TF_{used} &= \sum_{k=1}^{K} \{N_{Ack,k}(1 + R_{ave_k} \times \varepsilon_k) + N_{sche,k}\} \\ &\quad + \sum_{l=1}^{L} N_{NonA,l}(1 + R_{ave_l} \times \varepsilon_l) + \sum_{z=1}^{Z} N_{grant,z} \end{aligned} \tag{11.4}$$

where number of active and non-active IoTs are represented as K and L, respectively. Number of required TTI-PRB to transmit a task of an IoT k in active status is denoted as $N_{Ack,k}$. Let $N_{NonA,l}$ be the number of required TTI-PRB of an IoT l to transmit a silence indicator (SID) packet under discontinuous transmission (DTX) status. Numbers of average re-transmissions for IoT under non-active state and active state are denoted as R_{ave_l} and R_{ave_k}, respectively. Active IoTs are those who currently connected with MBS (i.e., radio resources are allocated to those IoTs for transmission of task to FAP). Non-active IoTs are in idle mode who receives paging and broadcast information from the MBS. The $N_{sche,k}$ represents the number of TTI-PRB grid needed by the kth active IoT to send the scheduling request in order to transmit task, and once it accepted, MBS generates grant message and that will need $N_{grant,k}$ number of TTI-PRB. The number of accepted IoTs can be less than or equal to requested IoTs based on the availability of resources at MBS, so $Z \leq K$ IoTs. Ratios of resources for re-transmissions to resources for initial transmissions of an IoT in the active and non-active states are given as ε_k and ε_l. Hence, it is quite obvious to write the following equation:

$$\frac{TF_{used}}{TF_{total}} \leq 1 \tag{11.5}$$

Using above equations (11.3) and (11.5), the total number of supported IoTs in cellular 5G network can be estimated. To simplify the estimation, we considered the some assumptions: (a) let v be the average activity factor for all IoTs, (b) let average re-transmission for active and not active states for all IoTs be the same; i.e., R_{aveR}, (c) $\varepsilon_k = \varepsilon_l = 1$, (d) $N_{Ack,k} = N_{NonA,l} = N_{ave}$, and (e) the ratio between the arrival of task in active and non-active states is given as p. Let N_{sup} be the total number IoTs that can be supported by an MBS. The values of these variables can be modified based on a particular application. Moreover, the next-generation wireless network will consist of D2D and machine-to-machine (M2M) connection modes while sharing data among themselves without sole dependency over MBS [19]. Thus, we can say these devices in a non-active stage with respect to MBS, and accordingly, the value of p can be modified. If v is the task activity factor (derived in the next subsection), then we can rewrite the equation (11.4) as follows:

$$TF_{used} = N_{sup}[(1 + R_{ave_k}) \times N_{ave} + N_{sche} + N_{grant}] \times ((1 - p) \times v + p). \tag{11.6}$$

Using equations (11.5) and (11.6) we can calculate N_{sup} as follows:

$$N_{sup} \leq \frac{TF_{total}}{((1 - p) \times v + p)[(1 + R_{ave_k}) \times N_{ave} + N_{sche} + N_{grant}]}. \tag{11.7}$$

We can formulate the number of PRBs required for one task as follows:

$$N_0 = ((1 - p) \times v + p)[(1 + R_{ave_k}) \times N_{ave} + N_{sche} + N_{grant}]. \tag{11.8}$$

Task's packet size for different AMR-WB codecs are derived in the following subsection.

11.3.3 Traffic Model of Tasks

We have considered two state activity model for traffic scenario as shown in Figure 11.5.

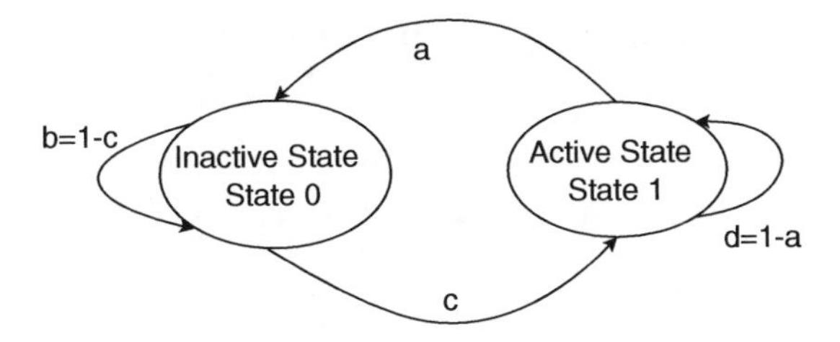

Figure 11.5 Two-state model.

In this model, the probability of transition from state 1 to state 0 is equal to a. The probability of transition from state 0 to state 1 is c. Let, T be the encoder frame duration and R be task encoder frame rate defined as $R = 1/T$. Further, let P_1 and P_0 are defined as the probability of an IoT for being in state one and state zero, respectively. Thus, we can write the steady-state equations as follows:

$$P_0 = \frac{a}{a+c}, P_1 = \frac{c}{a+c}. \tag{11.9}$$

The Task Activity Factor (TAF) v can be written as follows:

$$v = P_1 = \frac{c}{a+c}. \tag{11.10}$$

During active period, a task encoder generates 253 bits payload per task packet in every 20 ms, corresponding to AMR-WB 12.65 kbps. Additional 64 bits for headers of RTP/UDP/IP/PDCP/RLC with header compression will be included in each task packet. During off period, a 120 bits SID packet including header generated in once every 160 ms. Figure 11.6 represents different task packet size on the basis of AMR-WB codecs. The task packet size can be written as follows

$$Task\ packet\ size \text{ (in bits)} = (b_{AMR-WB}) \times 20 + 64 \tag{11.11}$$

where b_{AMR-WB} is the bit rate supported by the AMR-WB codec.

11.3.4 Estimation of Task Blocking Performances Based on M/M/m/m Model

On the basis of available PRBs, a task can be either accepted or rejected. If available PRBs are sufficient to maintain the connection, then only task can be accepted. The number of PRBs required by any task can be obtained by equation (11.8). Here, it is assumed to be equivalent to

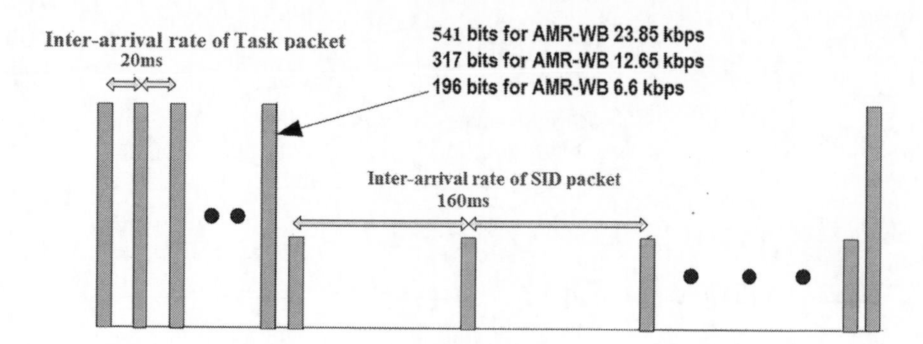

Figure 11.6 Traffic model.

allocate a server using $M/M/m/m$ model to a task if and only if total available PRBs in the system are more than N_0. The total available PRBs in the system can be obtained from equation (11.3). The number of IoTs that can get services from MBS depends upon the cell capacity. If the number of IoTs goes beyond the cell's capacity, then transfer tasks get dropped, or new tasks get blocked. Thus, this phenomenon can be modeled by using $M/M/m/m$, where the first M indicates the nature of task arrival process, and the second M shows the nature of probability distribution of the service time. In this model, we consider a maximum of m numbers of tasks can be served by m number of PRBs. Tasks are generated according to the Poisson process with rate λ defined by the expression $\lambda = \lambda_{new} + \lambda_{trans}$ where λ_{new} is the new task arrival rate (TAR), and λ_{trans} is the transfer TAR. The service time is exponentially distributed with mean $1/\mu$. The service rate is defined as follows:

$$1/\mu = \frac{\lambda_{new}}{\lambda_{new} + \lambda_{trans}} \times \frac{1}{\mu_{new}} + \frac{\lambda_{trans}}{\lambda_{new} + \lambda_{trans}} \times \frac{1}{\mu_{trans}} \tag{11.12}$$

where μ_{new} and μ_{trans} are the service rates for new and transfer tasks, respectively.

Figure 11.7 shows the discrete-time Markov chain for the $M/M/m/m$ model. Task blocking occurs if system is occupied with m tasks; thus, the system is in the state of utilizing all m PRBs. Solving the global balance equation for $M/M/m/m$ model by utilizing all m PRBs is given as follows [6]:

$$P_m = \frac{(\frac{\lambda}{\mu})^m/m!}{\sum_{n=0}^{m}(\frac{\lambda}{\mu})^n/n!}, \tag{11.13}$$

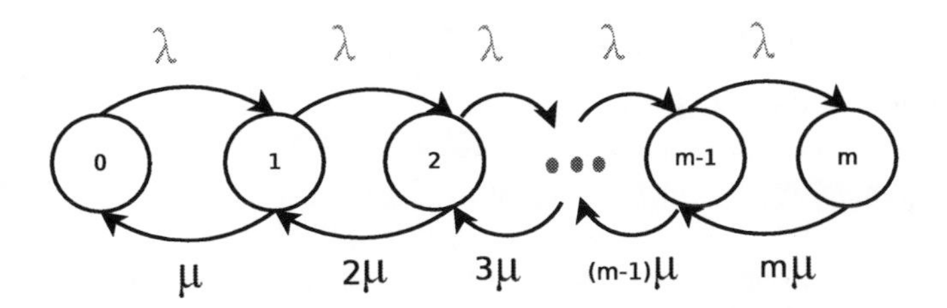

Figure 11.7 Discrete-time Markov chain for the $M/M/m/m$ system.

where P_m represents the blocking probability of the system. Let P_b and P_d denote the new task blocking probability and transfer task dropping probability, respectively. For the sake of simplicity, we have assumed $P_b = P_d = P_m$ in our model.

11.4 PERFORMANCE STUDY

In this section, we first explain the parameters used for simulation environment considering the different advanced modulation and codec schemes supported in cellular 5G networks followed by the obtained results and discussion.

11.4.1 Performance Values

We have considered the bandwidth values from 1.4 MHz to 20 MHz. We have considered QPSK 1/3, 16 QAM 3/4, and 64 QAM 5/6 modulation schemes. AMR-WB 6.6 kbps, AMR-WB 12.65 kbps, and AMR-WB 23.85 kbps codec standards are considered. We considered the homogeneous system in which the new task arrival is Poisson and independent with the TAR λ per cell site. The offered load is uniform throughout the network and within the cell site. We assume that N_{sche} and N_{grant} messages can be monitored by the control signals [18] such as physical uplink control channel (PUCCH), physical downlink control channel (PDCCH), and physical channel hybrid ARQ indicator channel (PHICH). We can further write the value of N_{ave} as follows.

$$N_{ave} = \left\lceil \frac{Task\ packet\ size\ (bits)}{Payload\ size\ (bits)} \right\rceil. \tag{11.14}$$

The task termination rate (i.e., μ) is 0.005 s^{-1}. Here, we consider that $\lambda_{trans} = 0.01$ task/s, whereas λ_{new} varies from 0.5 tasks/s to 15 tasks/s. The simulation environment is shown in Table 11.2.

TABLE 11.2 Parameter Values

Parameters (Units)	Details
$1/\mu$ (s)	200
λ_{new} (tasks/s)	0.5–5, 0.01
λ_{trans} (tasks/s)	12, 0.1–2.0
Modulation schemes	QPSK 1/3, 16 QAM 3/4, 64 QAM 5/6
b_{AMR-WB} (kbps)	23.85, 12.65, 6.6
System bandwidth (MHz)	1.4, 3, 5, 10, 15, 20
R_{avgR} %	40
v %	50
p	0.125

11.4.2 Obtained Results and Discussions

Figure 11.8 shows the task blocking probabilities at 1.4 MHz using QPSK 1/3 and 16 QAM 3/4. Here, we considered three AMR-WB codec standards. The number of tasks that can be supported in a cell (i.e., the cell capacity) in AMR-WB 23.85 kbps is less than that of AMR-WB 12.65 kbps and AMR-WB 6.6 kbps. The task blocking probability in AMR-WB 23.85 kbps is more than that of AMR-WB 12.65 kbps and AMR-WB 6.6 kbps. The tasks blocking probabilities at 1 task/s and 15 tasks/s are 0.876928 and 0.991675, respectively, for AMR-WB 23.85 kbps using QPSK 1/3 MCS. Thus, we observed that the task blocking probability is increasing with the increase in new TAR. If TAR is 1 task/s and modulation and codec schemes are QPSK 1/3 with AMR-WB 23.85, QPSK 1/3 with AMR-WB 12.65, (QPSK 1/3 with AMR-WB 6.6 or 16 QAM 3/4 with AMR-WB 23.85) and (16 QAM 3/4 with

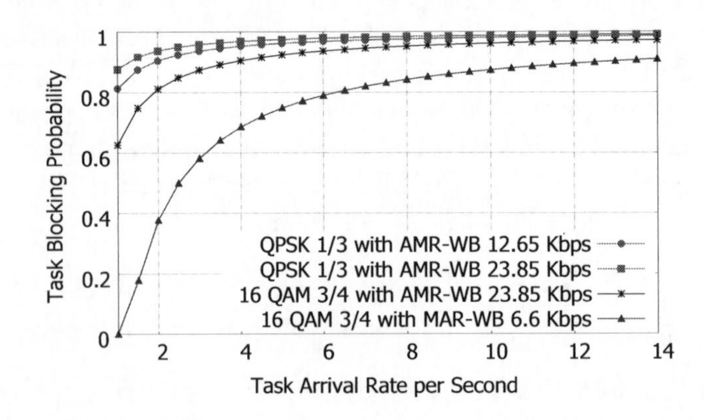

Figure 11.8 Task blocking probabilities at 1.4 MHz.

AMR-WB 12.65 or AMR-WB 6.6), then task blocking probabilities become 0.876928, 0.813011, 0.626676, and 0.260737, respectively. So, task blocking probability decreases when higher modulation and lower codec schemes are applied.

Figure 11.9 shows the task blocking probabilities for 3 MHz using QPSK 1/3 and 16 QAM 3/4. The task blocking probabilities using QPSK 1/3 with AMR-WB 12.65 kbps at 5 tasks/s and 15 tasks/s are 0.905294 and 0.968365, respectively. So, it is concluded that the task blocking probability increases with the increase in TAR. On the other hand, task blocking probability for QPSK 1/3 with AMR-WB 6.6 kbps and 16 QAM 3/4 with AMR-WB 23.85 kbps are the same because the total number of IoTs supported (i.e., cell capacity) in both of these cases are the same. If we consider new TAR as 1 call per second, modulation and codec schemes as QPSK 1/3 with AMR-WB 23.85, QPSK 1/3 with AMR-WB 12.65, (QPSK 1/3 with AMR-WB 6.6 or 16 QAM 3/4 with AMR-WB 23.85), and (16 QAM 3/4 with AMR-WB 12.65 or AMR-WB 6.6), then task blocking probabilities are obtained as 0.690317, 0.533953, 0.095590, and 0.000000, respectively.

Figure 11.10 shows the task blocking probabilities for 10 MHz bandwidth. The task blocking probabilities for QPSK 1/3 with AMR-WB 6.6 kbps and 16 QAM 3/4 with AMR-WB 23.85 kbps at 3 tasks/s and 15 tasks/s are 0.069736 and 0.809872, respectively. On the other hand, if we consider new TAR as 3 tasks/s modulation and codec schemes as QPSK 1/3 with AMR-WB 23.85, QPSK 1/3 with AMR-WB 12.65, (QPSK 1/3 with AMR-WB 6.6 or 16 QAM 3/4 with AMR-WB 23.85), and (16 QAM 3/4 with AMR-WB 12.65 or AMR-WB 6.6), then task

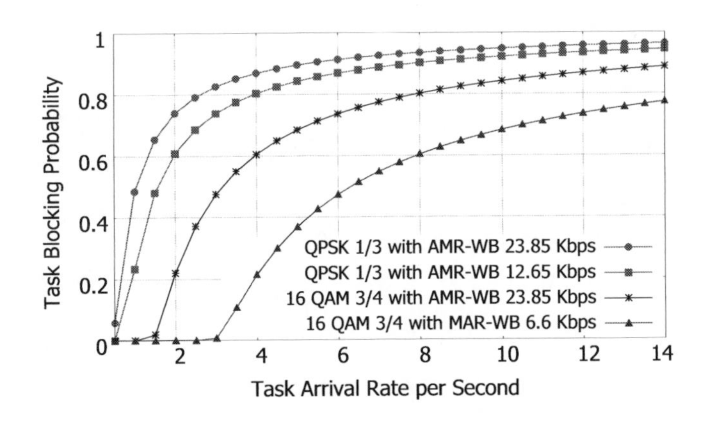

Figure 11.9 Task blocking probabilities at 3 MHz.

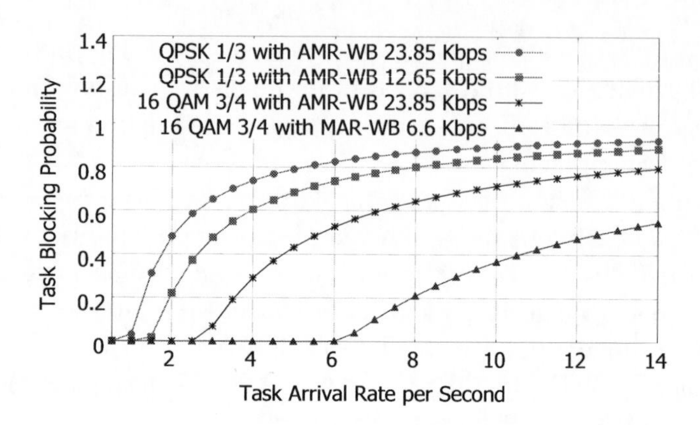

Figure 11.10 Task blocking probabilities at 10 MHz.

blocking probabilities are 0.652046, 0.475243, 0.069763, and 0.000000, respectively. Thus, we can conclude, with the use of proper modulation and coded scheme, total capacity of network can be increased.

Figure 11.11 shows the task blocking probabilities for 15 MHz bandwidth. If TAR is 9 tasks/s, modulation and codec schemes are QPSK 1/3 with AMR-WB 23.85, QPSK 1/3 with AMR-WB 12.65, (QPSK 1/3 with AMR-WB 6.6 or 16 QAM 3/4 with AMR-WB 23.85), and (16 QAM 3/4 with AMR-WB 12.65 or AMR-WB 6.6), then task blocking probabilities are 0.825312, 0.736048, 0.527134, and 0.000542, respectively. The task blocking probabilities for 16 QAM 3/4 with AMR-WB 12.65 kbps or AMR-WB 6.6 kbps at 8.5 tasks/s and 15 tasks/s are 0.0 and 0.366331, respectively. Thus, it is observed that the task blocking probability increases with the increase in TAR.

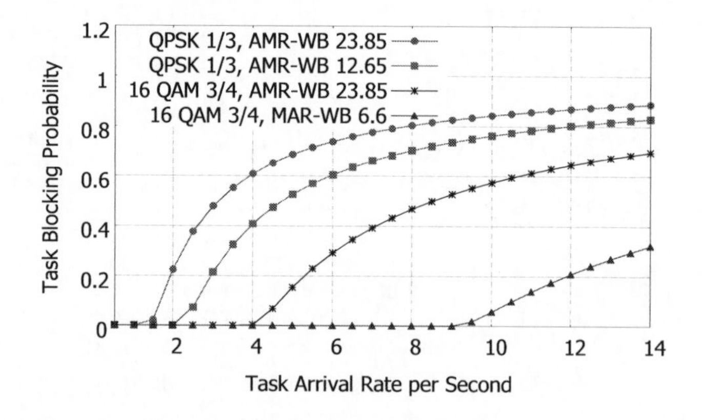

Figure 11.11 Task blocking probabilities at 15 MHz.

Figure 11.12 shows the task blocking probabilities for 20 MHz bandwidth. The TAR for 16 QAM 3/4 with AMR-WB 12.65 kbps or AMR-WB 6.6 kbps do not much influence the task blocking probability at 20 MHz. If we consider TAR as 12 tasks/s modulation and codec schemes as QPSK 1/3 with AMR-WB 23.85, QPSK 1/3 with AMR-WB 12.65, (QPSK 1/3 with AMR-WB 6.6 or 16 QAM 3/4 with AMR-WB 23.85) and (16 QAM 3/4 with AMR-WB 12.65 or AMR-WB 6.6), then task blocking probabilities are obtained as 0.825234, 0.736202, 0.472155, and 0.000172, respectively. So, task blocking probability decreases when higher modulation and lower codec schemes are applied.

Figures 11.13 and 11.14 show a comparison of task blocking probabilities for different bandwidth levels acceptable for cellular

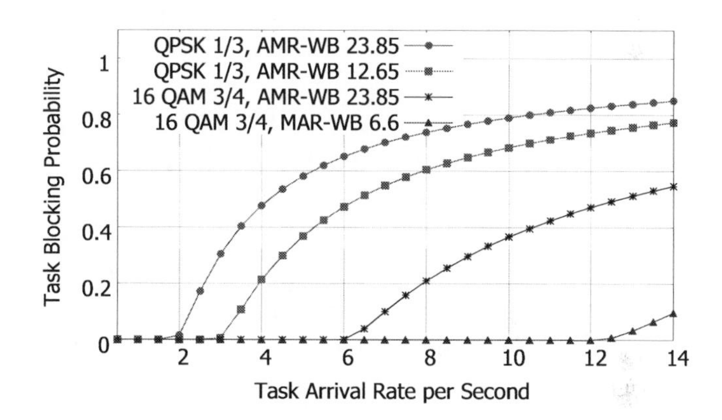

Figure 11.12 Task blocking probabilities at 20 MHz.

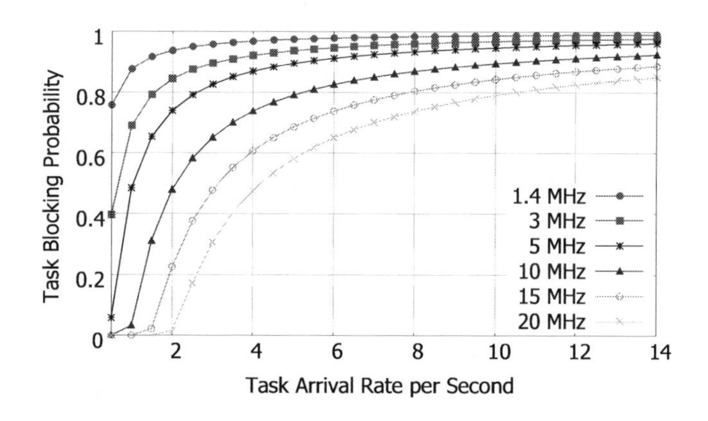

Figure 11.13 Task blocking probabilities at QPSK 1/3, AMR-WB 23.85 kbps

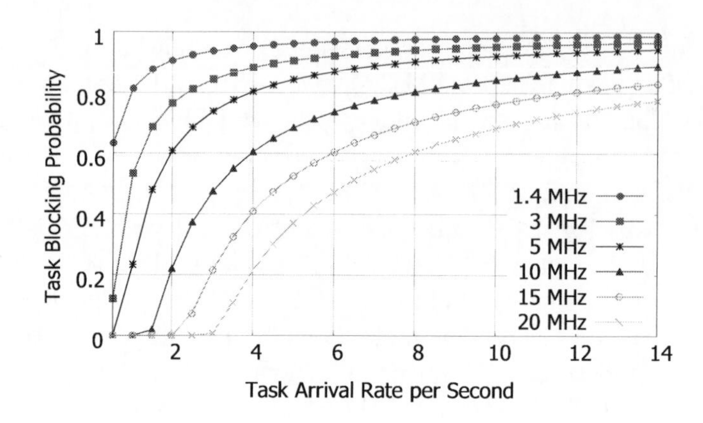

Figure 11.14 Task blocking probabilities at QPSK 1/3, AMR-WB 12.65 kbps.

5G network. From Figures 11.13 and 11.14, we found that task blocking probability of QPSK 1/3 with AMR-WB 23.85 kbps is more than task blocking probability of QPSK 1/3 with AMR-WB 12.65 kbps. The task blocking probability at 1.4 MHz is more than the task blocking probability at 3 MHz, 5 MHz, 10 MHz, 15 MHz, and 20 MHz. This is because with increase of bandwidth results in higher capacity of system to support the more number of tasks. Thus, blocking probability decreases with respect to increase in the system capacity by increasing the bandwidth levels available in the 5G system.

In Table 11.3, we have shown task blocking probabilities at different bandwidth, modulation, and coding schemes supported in the narrowband IoT model. The values of λ_{trans} and λ_{new} are set at 12 tasks/s and 2 task/s, respectively. From the result, we can see, with an increase in bandwidth, task blocking probability decreases. The reason is that with an increase of bandwidth, the total capacity of cellular 5G networks increases, consequently a lower task blocking probability obtained.

11.5 CONCLUSION

In this chapter, we have estimated the number of tasks that can be supported by a cell at different bandwidth levels available for cellular 5G network using advanced modulation schemes supported with different codec standards. It is observed that the use of advanced modulation schemes and lower AMR-WB codec enhances the cell capacity in cellular 5G networks. This further decreased the blocking probability in cellular

TABLE 11.3 Task Blocking Probability at Different Bandwidths

Modulation and coding scheme	1.4 MHz	3 MHz	5 MHz	10 MHz	15 MHz	20 MHz
QPSK 1/3 with AMR-WB 23.85	0.99108	0.97751	0.96251	0.92503	0.88755	0.8506
QPSK 1/3 with AMR-WB 12.65	0.98643	0.96608	0.94359	0.88683	0.83007	0.77368
QPSK 1/3 with AMR-WB 6.6	0.97287	0.93217	0.88683	0.79616	0.69551	0.54708
16 QAM 3/4 with AMR-WB 6.6	0.94574	0.86434	0.77368	0.54708	0.32075	0.09645

5G networks providing different services. Also, we found that blocking probability decreased with use of higher bandwidth levels due to increase of system capacity. From the results, we conclude that the use of advanced modulation schemes with lower AMR-WB codec reduced the blocking probability to a large extent in cellular 5G by providing different services. In the future, this scheme can be justified on more realistic scenario of 5G technologies. The proposed scheme can be applied to other upcoming advanced technologies such as millimeter wave structure-based multi-tier heterogeneous networks by changing the respective set of network's constraints.

Bibliography

[1] I-Scoop, 5G and IoT: the mobile broadband future of IoT. https://www.i-scoop.eu/internet-of-things-guide/5g-iot/. Accessed: 2020-03-17.

[2] Sigfox, Radio technology keypoints. https://www.sigfox.com/en/sigfox-iot-radio-technology. Accessed: 2020-03-17.

[3] Wi-Fi Alliance® introduces low power, long range Wi-Fi HaLow™. https://www.wi-fi.org/news-events/newsroom/wi-fi-alliance-introduces-low-power-long-range-wi-fi-halow. Accessed: 2020-03-17.

[4] Mosa Ali Abu-Rgheff. *5G Enabling Technologies: Narrowband Internet of Things and Smart Cities*, pages 151–188. IEEE, 2019.

[5] Mamta Agiwal, Navrati Saxena, and Abhishek Roy. Towards connected living: 5G enabled internet of things (IoT). *IETE Technical Review*, 36(2):190–202, 2019.

[6] Dimitri P. Bertsekas, Robert G. Gallager, and Pierre Humblet. *Data networks*, volume 2. Prentice-Hall International, Upper Saddle River, NJ, 1992.

[7] SIG Bluetooth. Bluetooth core specification version 5.0.(2016), 2016.

[8] Qiwang Chen, Lin Wang, Pingping Chen, and Guanrong Chen. Optimization of component elements in integrated coding systems for green communications: A survey. *IEEE Communications Surveys & Tutorials*, 21(3):2977–2999, 2019.

[9] Ivo B. F. de Almeida, Luciano L. Mendes, Joel J. P. C. Rodrigues, and Mauro A. A. da Cruz. 5G Waveforms for IoT Applications. *IEEE Communications Surveys & Tutorials*, 21(3):2554–2567, 2019.

[10] Peng Guan, Xi Zhang, Guangmei Ren, Tingjian Tian, Anass Benjebbour, Yuya Saito, and Yoshihisa Kishiyama. Ultra-low latency for 5G-a lab trial. *arXiv preprint arXiv:1610.04362*, 2016.

[11] Samer Henry, Ahmed Alsohaily, and Elvino S. Sousa. 5G is Real: Evaluating the Compliance of the 3GPP 5G New Radio System With the ITU IMT-2020 Requirements. *IEEE Access*, 8:42828–42840, 2020.

[12] Sree Ram Kodali, Mohan Rao Thota, and Manish G. Vemulapalli. Adaptability in EVS Codec to Improve Power Efficiency, December 27 2018. US Patent App. 15/941,148.

[13] Arthur M. Langer. *Analysis and Design of Next-Generation Software Architectures: 5G, IoT, Blockchain, and Quantum Computing.* Switzerland, AG: Springer Nature, 2020.

[14] Gilsoo Lee, Walid Saad, and Mehdi Bennis. An online optimization framework for distributed fog network formation with minimal latency. *IEEE Transactions on Wireless Communications*, 18(4):2244–2258, 2019.

[15] Pei-Kai Liao, Tao Chen, and Jiann-Ching Guey. Flexible frame structure for ofdm systems, January 10 2019. US Patent App. 16/127,533.

[16] Bao-Shuh Paul Lin, Yi-Bing Lin, Li-Ping Tung, and Fuchun Joseph Lin. Exploring the Next Generation of the Internet of Things in the 5G Era. *5G-Enabled Internet of Things*, page 67, 2019.

[17] Gang Lu, Bhaskar Krishnamachari, and Cauligi S. Raghavendra. Performance evaluation of the IEEE 802.15. 4 MAC for low-rate low-power wireless networks. In *IEEE International Conference on Performance, Computing, and Communications, 2004*, pages 701–706. IEEE, 2004.

[18] Ajay R. Mishra. *Fundamentals of Network Planning and Optimisation 2G/3G/4G: Evolution to 5G.* Hoboken, NJ, USA: John Wiley & Sons, 2018.

[19] Shahid Mumtaz, Anwer Al-Dulaimi, Valerio Frascolla, Syed Ali Hassan, and Octavia A. Dobre. Guest editorial special issue on 5G and beyond—mobile technologies and applications for IoT. *IEEE Internet of Things Journal*, 6(1):203–206, 2019.

[20] Fanny Parzysz and Yvon Gourhant. Drastic energy reduction with gDTX in low cost 5G networks. *IEEE Access*, 6:58171–58181, 2018.

[21] Ajay Pratap, Ragini Gupta, Venkata Sriram Siddhardh Nadendla, and Sajal K. Das. On Maximizing Task Throughput in IoT-Enabled 5G Networks Under Latency and Bandwidth Constraints. In *2019 IEEE International Conference on Smart Computing (SMARTCOMP)*, pages 217–224, June 2019.

[22] Ajay Pratap, Rajiv Misra, and Sajal K. Das. Maximizing Fairness for Resource Allocation in Heterogeneous 5G Networks. *IEEE Transactions on Mobile Computing*, pages 1–16, 2019.

[23] Ajay Pratap, Federico Concone, Venkata Sriram Siddhardh Nadendla, and Sajal K. Das. Three-dimensional matching based resource provisioning for the design of low-latency heterogeneous iot networks. In *Proceedings of the 22nd International ACM Conference on Modeling, Analysis and Simulation of Wireless and Mobile Systems*, pages 79–86, 2019.

[24] Ajay Pratap and Rajiv Misra. Firefly inspired improved distributed proximity algorithm for D2D communication. In *2015 IEEE International Parallel and Distributed Processing Symposium Workshop*, pages 323–328. IEEE, 2015.

[25] Ajay Pratap, Shivani Singh, Shaswat Satapathy, and Sajal K. Das. Maximizing joint data rate and resource efficiency in d2d-iot enabled multi-tier networks. In *2019 IEEE 44th Conference on Local Computer Networks (LCN)*, pages 177–184. IEEE, 2019.

[26] Ajay Pratap, Rishabh Singhal, Rajiv Misra, and Sajal K. Das. Distributed randomized k-clustering based PCID assignment for ultra-dense femtocellular networks. *IEEE Transactions on Parallel and Distributed Systems*, 29(6):1247–1260, 2018.

[27] Kinza Shafique, Bilal A. Khawaja, Farah Sabir, Sameer Qazi, and Muhammad Mustaqim. Internet of Things (IoT) for next-generation smart systems: A review of current challenges, future trends and

prospects for emerging 5G-IoT scenarios. *IEEE Access*, 8:23022–23040, 2020.

[28] Rishabh Singhal, Ajay Pratap, Rajiv Misra, and Sajal K. Das. Factor Graph-based Message Passing Technique for Distributed Resource Allocation in 5G Networks. In *2019 11th International Conference on Communication Systems & Networks (COMSNETS)*, pages 359–366. IEEE, 2019.

[29] Lorenzo Vangelista, Andrea Zanella, and Michele Zorzi. Long-range IoT technologies: The dawn of LoRa^{TM}. In *Future Access Enablers of Ubiquitous and Intelligent Infrastructures*, pages 51–58, Switzerland, AG: Springer, 2015.

[30] Chih-Hsiang Wu. Device and method of performing a codec rate adaptation in a wireless communication system, September 12 2019. US Patent App. 16/421,487.

[31] Tongyang Xu, Christos Masouros, and Izzat Darwazeh. Waveform and space precoding for next generation downlink narrowband IoT. *IEEE Internet of Things Journal*, 6(3):5097–5107, 2019.

CHAPTER 12

An SDN-IoT – Based Framework for Future Smart Cities: Addressing Perspective

Uttam Ghosh

Vanderbilt University

Pushpita Chatterjee and Sachin Shetty

Old Dominion University

Raja Datta

Indian Institute of Technology Kharagpur

CONTENTS

In *this chapter*, a software-defined network (SDN)-based framework for future smart cities has been proposed and discussed. It also comprises a distributed addressing scheme to facilitate the allocation of addresses to devices in the smart city dynamically. The framework is dynamic, and modules can be added and omitted by a centralized controlling unit without disturbing the other components of the framework, and other modules may be updated accordingly. In the proposed addressing scheme, a new Internet of Things (IoT) device will receive an IP address from one of their existing neighboring peer devices. This allows devices in the city to act as a proxy and generate a set of unique IP addresses from their own IP addresses, which can then be assigned to new (joining) devices, hence reducing addressing overhead and latency, as well as avoiding the need to send broadcast messages during the address allocation process. Thus, it achieves considerable bandwidth and energy savings for the IoT devices.

12.1 INTRODUCTION

It has been estimated that approximately 65% of the world's population will eventually live in cities by the 2040 [1]. There has been a trend of making cities smarter, for example, by leveraging existing and emerging technologies such as Internet of Things (IoT). The latter can be broadly defined to be a (heterogeneous) network of a broad range of physical Internet-connected devices, such as smart vehicles, smart home appliances, and other devices with embedded software or hardware (e.g., sensors) that can be used to connect, sense/collect, and disseminate/exchange large volume of data. This also allows us to offer advanced services that can be used to improve the quality of service (QoS) delivery and life.

The increasing trend of smart cities is partly due to the lower of technological and cost barriers in deploying communication networks (e.g., wireless and 5G) in a broad range of settings, such as residential and commercial buildings, utility networks, transportation networks, and those in the critical infrastructure sectors [2,3]. In such settings, it is clear that data plays a key role, for example, in informing decision and strategy making and formulating. Such data can be collected by the broad range of IoT devices and networks, and can be compiled and analyzed to achieve improved service delivery in healthcare, manufacturing, utility, supply chain, and many other services. However, there exist a number of challenges in dealing with such data, due to the volume, variety, velocity, and veracity (also commonly referred to as the four V's of big data). For example, the management and performance optimization of IoT–based smart cities and programmability of things can be extremely complex, and also the inter-connectivity can introduce security implications. Therefore, how to ensure that the underpinning communication infrastructure in the smart city is scalable, reliable, secure, and efficient can be challenging, both operationally and research-wise.

Emerging software-defined networking (SDN) decouples the control plane and data plane, and subsequently, it enables the control plane to become directly programmable and the underlying infrastructure to be abstracted for the applications and the network services. SDN controller, also called network operating system (NOS), is logically centralized and responsible for controlling, managing, and dynamically configuring the devices at the data plane of the network. It is effective in taking decisions for the routing, QoS, and load balancing dynamically. It is easy to add new network functionalities through application programs due to the programmability feature of SDN controller. Moreover, SDN enhances the network performance by providing security and the network virtualization features. SDN controller is capable of monitering all the nodes and their traffic and eliminating the attacker node from the network on-fly by writing effective flow rules on the switches at data plane [4].

Motivation: Each device in the infrastructure should have a unique address by which it can be identified. This unique address enables unicast communication and routing between devices in the infrastructure. However, as more IoT devices are introduced in the smart city, the demand for these unique addresses increases rapidly. Manual configuration of IoT devices in most of the cases is inapplicable and error prone due to large size of the network. Further, centralized Dynamic Host Configuration Protocol (DHCP) [5] is not a suitable solution as the sever has to maintain configuration information of all the nodes in the network.

Duplicate address detection (DAD) mechanism [6] can be used to resolve address conflict in the smart city. In DAD, a joining node chooses a tentative IP address randomly and verifies the whether this address is available for use or not. In order to verify the uniqueness of the address, the joining node floods a duplicate address probe (DAP) message throughout the smart city and starts a timer to receive address conflict notice (ACN) message from the network. If no ACN message is received, then the joining node concludes that the tentative address is free to use and configures itself with the address permanently. It has to run the DAD process again in case the joining node receives a ACN message from the network. The addressing overhead for DAD mechanism is very high as it needs to flood a message throughout the network. Further, the broadcast storm problem [7] can be seen in DAD. Figure 12.1 shows the DAD mechanism where a new node tries to join the network.

Contribution: It can be seen from the above discussion that there is a need to design a distributed addressing scheme to efficiently handle the ever increasing requirement in SDN-IoT–based smart city networks. Further, the addressing scheme should assign unique IP addresses to the devices of the network for the correct routing and unicast communications. Furthermore, the scheme needs to be scalable and should not degrade its performance with respect to addressing overhead when the network size is very large like a smart city. This chapter has two significant contributions:

- First, an SDN-based IoT framework for a smart city architecture
- Second, a distributed addressing scheme to efficiently assign a unique IPv6 address to each device in the proposed smart city framework.

With this chapter, readers can have a more thorough understanding architectures of SDN, IoT, and SDN-IoT–based smart cities. It further proposes an IPv6 addressing mechanism to allocate unique address to each IoT devices in a SDN-IoT–based smart city.

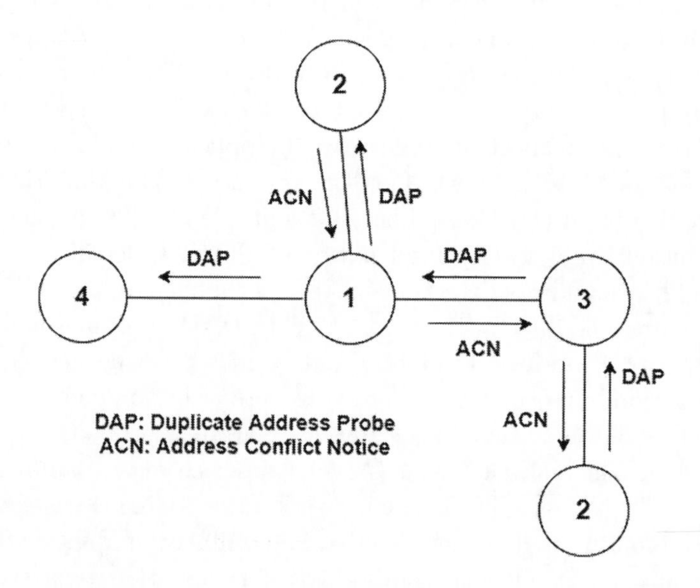

Figure 12.1 Duplicate Address Detection (DAD) mechanism.

Chapter Organization: The rest of the chapter is organized as follows: Section 12.2 presents a background of SDN, IoT, and IPv6 addressing. Section12.3 discusses state-of-the-art literature on SDN-IoT–based networks and also address allocation techniques in various wireless networks. The proposed framework for SDN-IoT–based smart city with an addressing scheme is presented in Section12.4. Finally, Section 12.5 concludes the chapter.

12.2 BACKGROUND

In this section, we give an overview of basic preliminary concepts of SDN, IoT, and IPv6 addressing.

12.2.1 An Overview of SDN

This section presents an overview of SDN architecture and its working principles. It also presents the need of SDN and how SDN is different as compared to the traditional networking. Figure 12.2 presents the major elements, planes (layers), and interfaces between layers of SDN architecture. It has three planes: data plane, control plane, and application plane.

Data Plane: The first plane in SDN architecture is the data plane (also known as infrastructure plane) that consists of hosts and traffic

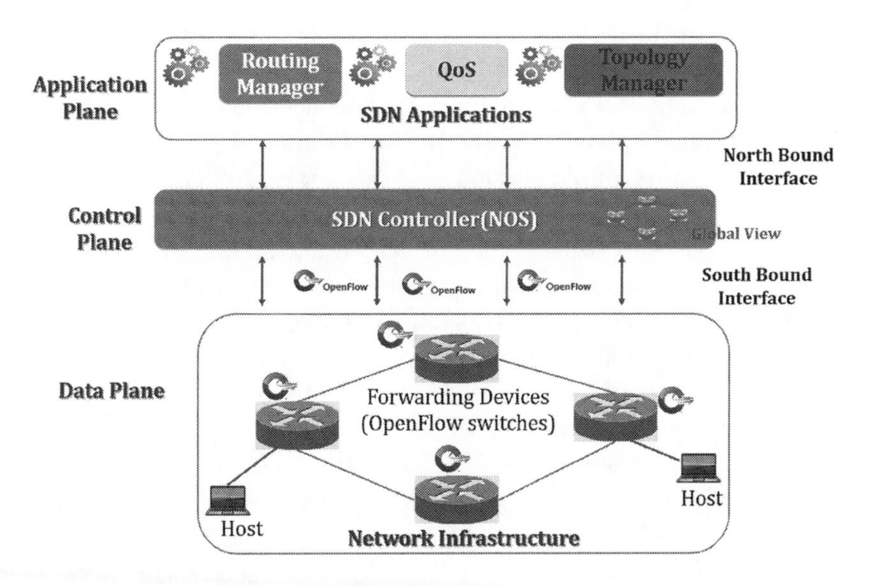

Figure 12.2 A layering architecture of SDN.

forwarding devices. These traffic forwarding devices are known as OpenFlow (OF) switches. These switches are called dump switches and able to forward the data from source host to destination host only after receiving the instructions (flow rules) from the SDN control layer.

Control Plane: The second plane in SDN architecture is the control plane that may comprise an SDN controller or a set of SDN controllers. SDN controller (also called network operating system (NOS)) is a logical entity (software programs) which is programmable. It is logically centralized. Hence, it can track the network topology (global view of the network) and the statistics of the network traffic periodically. Further, SDN controller is responsible for controlling, managing, and dynamically configuring the devices at the data plane of the network. It efficiently provides routing, QoS, and security, and also balances the load in the network.

Application Plane: The third and final plane is the application plane in SDN architecture. This plane runs application programs and uses application programming interface (API) to control the network resources with the SDN controller. These application programs periodically collect information from SDN controller and provide services (e.g., routing, QoS, and load balancing). This plane also provides a programming interface to the network administrator for developing applications according to the requirements of the network. For instance, an application can be built to monitor all the devices and their traffics periodically for detecting the misbehaving devices in the network.

The northbound API defines the connection between application plane and control plane, whereas the southbound API defines the connection between control plane and date plane. OF protocol has been widely used as the southbound API. The SDN controller uses OF protocol to send the flow rules to the OF switches in data plane. OF protocol uses secure socket layer (SSL) and transmission control protocol (TCP) for providing secure communication and reliable delivery of data between the controller and OF switches, respectively.

The working principle of SDN is presented in Figure 12.3. A device H1 (source) sends the packets of a flow to another device H2 (destination) through OF switches S3-S2-S1 in an SDN-based network [4]. Here, the SDN controller detects topology of the network using link layer discovery protocol (LLDP) as shown in Figure 12.4. Thus, it knows the global topology of the network and responsible for the routing between the devices.

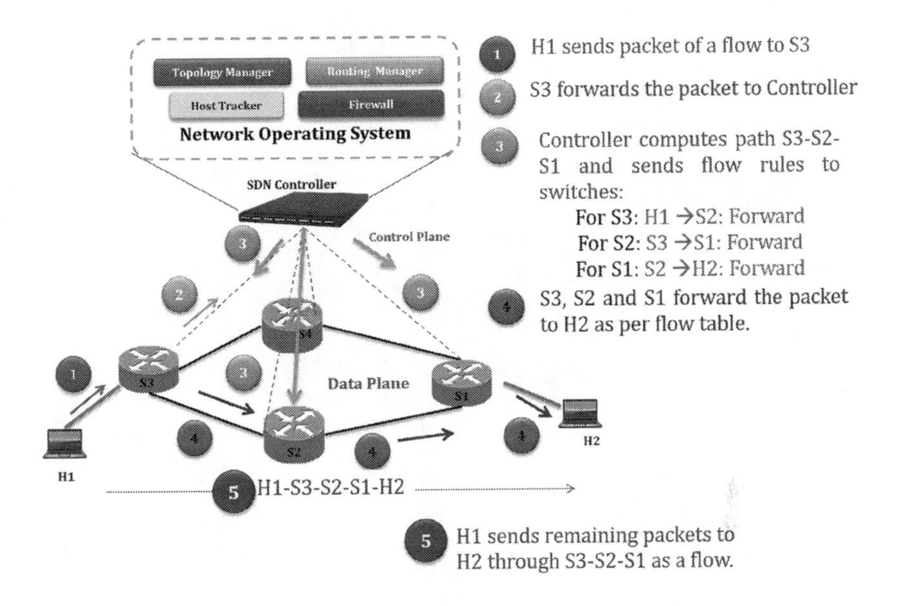

Figure 12.3 Working principles of SDN.

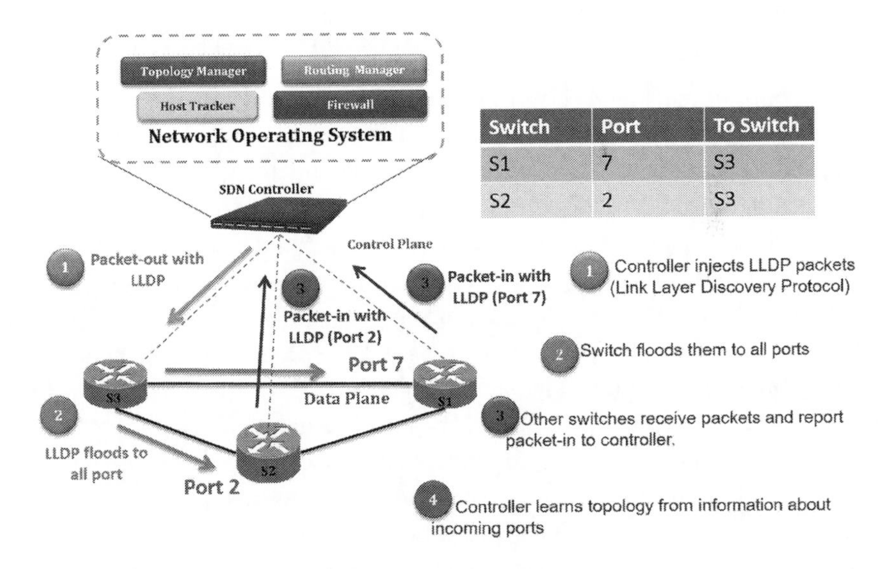

Figure 12.4 Topology detection using LLDP.

12.2.2 An Overview of IoT and Smart Cities

The Internet of Things (IoT): An IoT is a heterogeneous network of physical objects (things) that are embedded with electronics, sensors, software, actuators, RFID tags, and other technologies for connecting

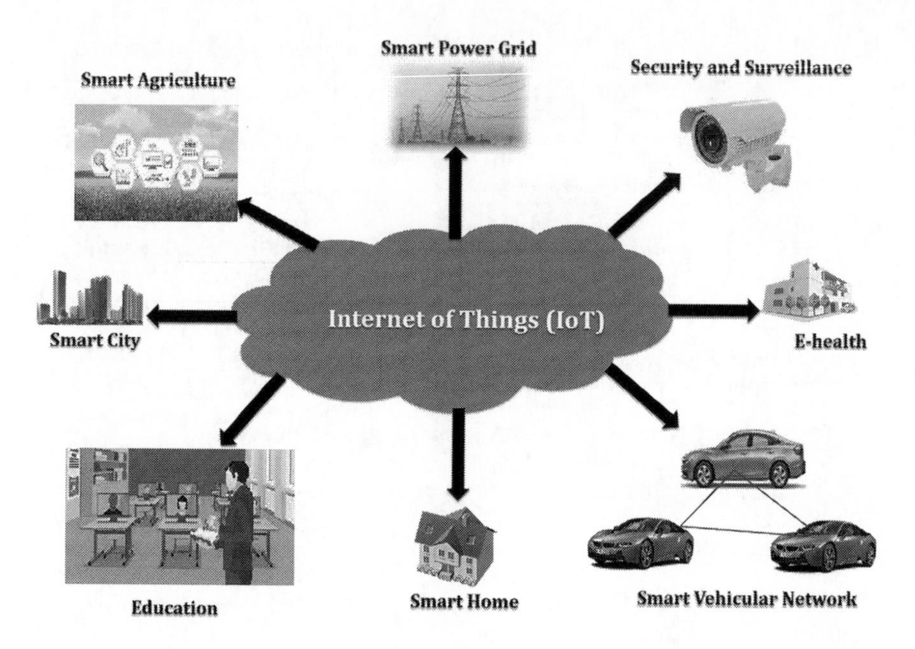

Figure 12.5 Internet of Things (IoT).

and communicating a large amount of data with other devices and networks over the Internet to offer a new class of services at anytime, anywhere, and for anyone (Figure 12.5). It can form a large network by combining wired networks and different types of wireless networks such as wireless sensor networks (WSNs), Zigbee, WiFi, mobile ad-hoc networks (MANETs), and radio frequency identification (RFID). IoT can be applied to make the physical infrastructures more smart, secure and reliable, and fully automated systems. These physical infrastructures include buildings (homes, schools, offices, factories, etc.), utility networks (gas, electricity, water, etc.), healthcare systems, transportation vehicles (cars, rails, planes, etc.), transportation networks (roads, railways, airports, harbors, etc.), and information technology networks. IoT collects, stores, and exchanges a large volume of heterogeneous data from various types of networks and provides critical services in smart homes and buildings, healthcare systems, transportation networks, utility networks, industrial control and monitoring systems, and so on [4,8–10].

Figure 12.6 shows the layering architecture of IoT. It comprises three main layers: sensing layer, network layer, and application layer. The sensing layer, also known as a perception layer, consists of physical

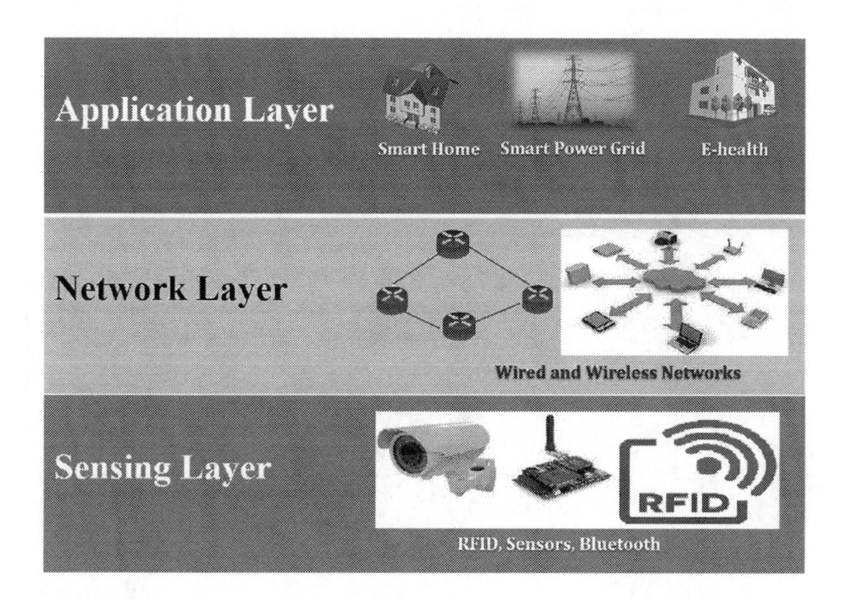

Figure 12.6 The three-layered architecture of IoT.

objects and sensing devices. This layer is responsible for sensing and collecting the data from the physical objects. Network layer bridges between sensing layer and application layer. It carries the data collected from the physical objects through sensors. The network can be wireless or wired network for the transmission. Thus, network layer is responsible for connecting the smart things, network devices ,and networks to each other, and also for transmitting the data from physical objects to the gateway of the network. Application layer is responsible for providing the services to the users based on their demands and applications. The applications of IoT can be smart homes and buildings, smart grids, smart health, smart cities, etc.

Smart City: A smart city is an urban area that uses different types of IoT devices to collect, process, and analyze the data for monitoring and managing traffic and transportation systems, utilities, power grids, waste management, water supply networks, schools, libraries, hospitals, security and surveillance systems, and other community services. It helps city officials to interact directly with both community and city infrastructure, and also to monitor and manage the city resources efficiently and smartly [11–14]. The main components of a smart city are depicted in Figure 12.7.

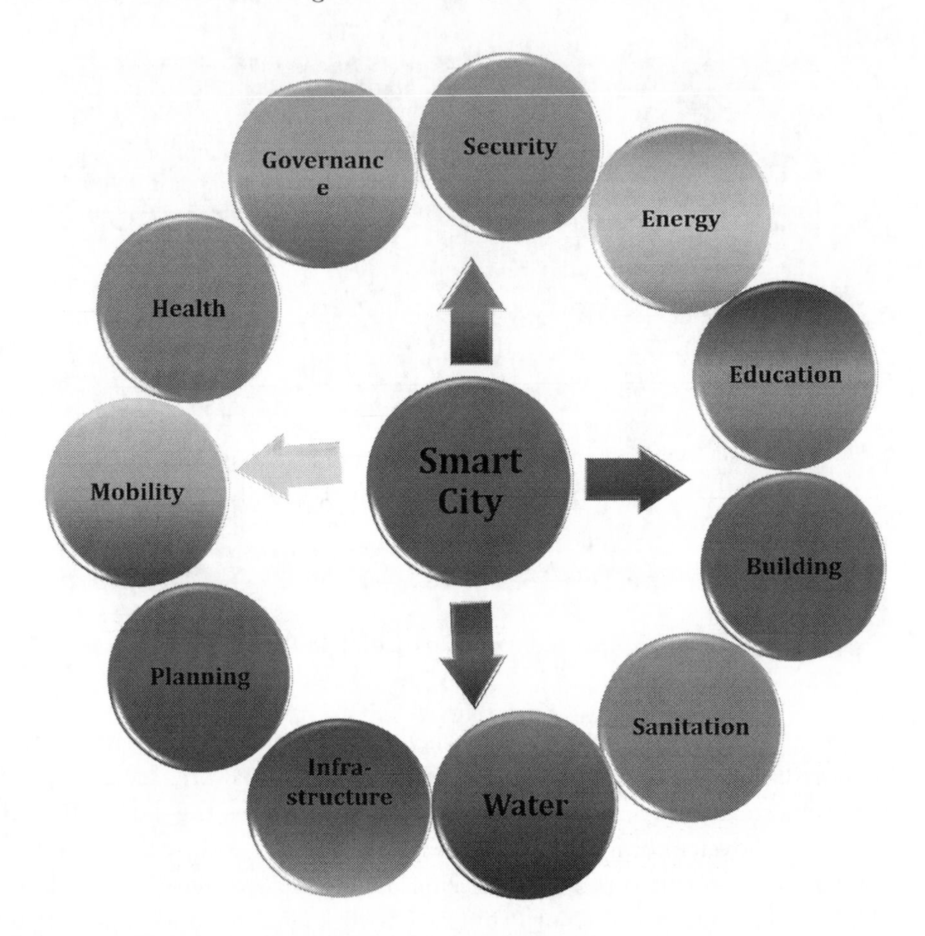

Figure 12.7 Overview of Smart city components.

12.2.3 An Overview of IPv6 Addressing

Internet Protocol version 4 (IPv4) is the most widely deployed IP used to connect devices to the Internet. IPv4 addresses are 32-bit long and can be used to assign a total of 2^{32} devices (over 4 billion devices) uniquely. However, with the growth of the Internet and IoT, it can be expected that the number of IPv4 addresses may eventually run out as each device that connects to the Internet and IoT requires an IP address. A new IP addressing system Internet Protocol version 6 (IPv6) is being deployed to fulfill the need for more IP addresses. An IPv6 addresses are 128-bit long and can be used to assign a total of $3.4 * 10^{38}$ devices uniquely. Further, it supports auto-configuration and provides better QoS, mobility, and security as compared to IPv4. Figures 12.8 and 12.9 present the headers of IP version 4 and IP version 6 respectively.

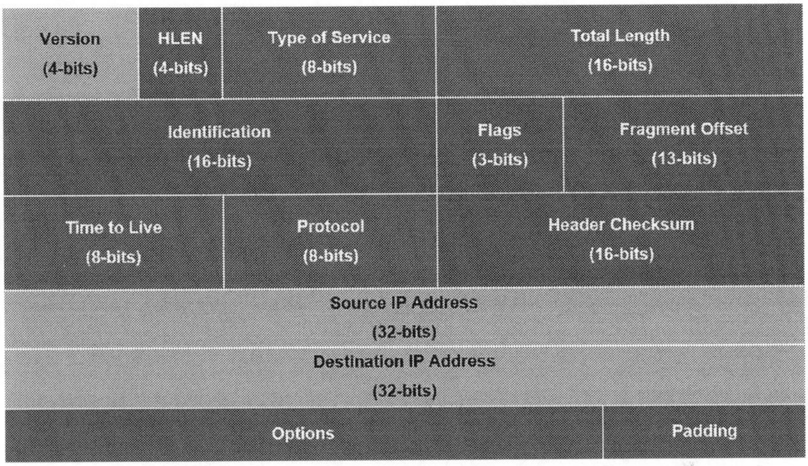

Version (4-bits)
HLEN (4-bits)
Type of Service (8-bits)
Total Length (16-bits)
Identification (16-bits)
Flags (3-bits)
Fragment Offset (13-bits)
Time to Live (8-bits)
Protocol (8-bits)
Header Checksum (16-bits)
Source IP Address (32-bits)
Destination IP Address (32-bits)
Options
Padding

Figure 12.8 IP version 4 (IPv4) header format.

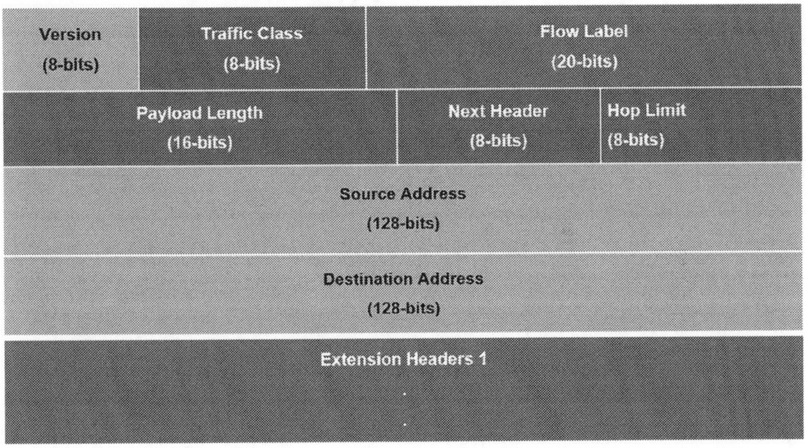

Version (8-bits)
Traffic Class (8-bits)
Flow Label (20-bits)
Payload Length (16-bits)
Next Header (8-bits)
Hop Limit (8-bits)
Source Address (128-bits)
Destination Address (128-bits)
Extension Headers 1
.
.

Figure 12.9 IP version 6 (IPv6) header format.

IPv6 Address Representation: An IPv6 address is represented as eight groups of four hexadecimal digits where each group represents 16 bits. These groups are separated by colons (:). An example of an IPv6 address is

2031:0000:130f:0000:0000:09c0:876a:130b

Leading zeroes in a group are optional and can be omitted. One or more consecutive groups containing zeros can be replaced by double colons (::), but only once per address. Therefore, the example address can be written as

2031:0:130f::9c0:876a:130b

IPv6 Header Format: The header format of IPv6 is shown in Figure 12.9. Here, the fields of IPv6 header have been discussed briefly:

Version: This field indicates the version of Internet Protocol (IP) which contains bit sequence 0110.

Traffic Class: This field presents the class or priority of IPv6 traffic as it is similar to service field in IPv4 header. The router discards the least priority packets if congestion occurs in the network.

Flow Label: Source node uses flow label field to label the packets belonging to the same flow in order to request special handling (e.g., QoS or real-time service) by intermediate IPv6 routers. It also specifies the lifetime of the flow.

Payload Length: This field indicates the total size of the payload including extension headers (if any) and upper layer data.

Next Header: This filed is used to indicate the type of extension header (if any) immediately following the IPv6 header. It also specifies the upper-layer protocols (UDP, TCP) in some cases.

Hop Limit: This field is same as time-to-live (TTL) field in the IPv4 header. It specifies the maximum number of routers an IPv6 packet can travel. The value of the hop limit gets decremented by one by each router that forwards the packet. The router discards the packet if the value of the hop limit reaches to 0. This field prevents the packet from circulating indefinitely in the network.

Source Address: This field specifies the IPv6 address of the original source of the packet.

Destination Address: This field indicates the IPv6 address of the final destination. In order to correctly route the packet, the intermediate routers use destination address of the packet.

Extension Header: This field have been introduced to allow the incorporation and usage of several options whenever is needed. The size of the IPv6 main header is 40 bytes long. Next header field of IPv6 main header points to the first extension header, and the first extension header points to the second extension header and so on.

12.3 RELATED WORKS

A number of different approaches have been explored in the literature, including the use of SDN. For example, there have been attempts to integrate SDN and IoT technologies into the heterogeneous communication infrastructure in smart cities [15–18], by say utilizing SDN to manage and determine the correctness of network operations at runtime. This is because we can leverage the globalized view and the programmability features available in the SDN controller to control, configure, monitor and detect faults, and mitigate abnormal operation(s) in the underpinning infrastructure, hence allowing us to achieve efficiency and reliability.

Mavani et al. have performed several works on secure addressing and privacy preserving methodologies for IoT and mobile environment paradigm [19–21]. In IoT, billions of devices can be addressed using IPv6 addressing scheme. Attackers can spoof addresses from unsecure wireless communication channels and advertise them as a legitimate device. Malicious users can track activity of these devices by spoofing IPv6 addresses to mitigate this type of attacks by hiding the IPv6 address from attacker. They have proposed a secure privacy preserving method[19], which changes the IPv6 address of each device periodically and pseudorandomly in order to hide its identity. They analyzed the method using Cooja simulator to show that the method does not inflict much overhead for random changing of address and reconfiguration. In [20,21], they investigated the use of secure addressing and privacy mechanisms for IPv6 over low-power wireless personal area network (6LoWPAN) and designed a method to provide resilience against address spoofing and better reconfiguration time from attack disruption. They showed the efficacy of their proposal by time complexity analysis and simulation with benchmark data, but overall, this does not pose much overhead to provide resilience against address spoofing.

Brilli et al. proposed a secure privacy aware two-layer addressing scheme for 6LoWPAN wireless network in order to improve security and privacy along with reducing the chance of spoofing by hiding the traceability of the user [22]. With a minimal overhead and using standard 6LoWPAN messages, security and privacy have been ensured in an energy constrained environment. Wang et al. proposed a long-thin and tree-based topology in addressing-based routing optimization in vehicular scenarios (AROV) [23] to provide unique address to sensor nodes in 6LoWPAN WSNs using a concept of super node for multi-hop sensor nodes serves as address initiator for its all neighbor nodes. They have shown it mitigates address failure and also gives performance in routing by reducing latency. The authors also proposed location aware addressing for 6LoWPAN WSNs [24]. In this addressing scheme without using DAD, a node can obtain a globally unique address. The address initialization is performed zone-wise where zones are independent of the one in another; therefore, this parallel and address initialization took less time. Wang et al. further proposed stateful address configuration mechanism for wireless body area networks [25]. The uniqueness of the address is maintained without DAD. Automatic reclamation of unused or released address has been carried out without any extra overhead. Using simulation, they have shown the efficacy of performance by reducing the address configuration delay and cost. For heterogeneous wireless network, a dynamic IP address assignment architecture [26] has been proposed by Khair et al. The addressing mechanism introduced security and service reliability with a reduced Opex. However, this scheme does not perform well in heterogeneous heavy traffic scenarios as it incurs significant overhead. Li et al. presented address configuration algorithm for network merging in ad-hoc network scenario [27]. By restricting the new address generation, only duplicate addresses during merging their scheme significantly improve the network performance.

In [28], an IP-based vehicular content-centric networking framework has been proposed by Wang et al., by employing the unicast address-centric paradigm to achieve content acquisition. They avoid the broadcast centric communication. Using the unicast communication, they have shown it substantially reduces the content acquisition cost and gives better performance in success rate content acquisition.

In [29], El-Shekeil et al. investigated several conflict scenarios of using private IP for enterprise network. They formulated the problem to minimize the routing table sizes as NP-hard. They devised effective heuristics formulation in order to solve the problem. To prove the efficacy of the same, they provided empirical result which showed significant reduce in the number of subnet entries and the routing table sizes.

A MANET is a collection of mobile nodes with a dynamic self-configured network. It has no fixed and pre-established infrastructure without any centralized administrations or base stations. It can be integrated with IoT to implement smart cities. Therefore, IP addressing is very important and challenging issue for a MANET as it is an infrastructure-less and highly dynamic network. In light of this, Ghosh et al. proposed IPv6-based and IPv4-based secure distributed dynamic address allocation protocols [30–35]. In these protocols, the new node gets an IP address from its neighbors acting as proxies. The new node becomes proxy once it receives an IP from the network. Further, these protocols can handle the network events such as network partitioning and merging without using complex DAD mechanisms.

Akhtar et al. proposed a congestion avoidance algorithm [36] for IoT-MANET which used bandwidth as the main component to find the optimal route. By getting feedback about the residual bandwidth of network path each channel aware routing scheme (BARS) that can avoid congestion by monitoring residual bandwidth capacity in network paths they significantly improve network parameters like of latency, end-to-end delay and packet delivery ratio for both static and dynamic network topologies. A secure SDN-based framework has been proposed for content centric application has been devised by Ghosh et al. In [37], secure multi-path routing protocol has been designed which significantly improves the network performance. This work is pretty much feasible to incorporate for futuristic smart cities. Ghosh et al. proposed a SDN-based secure framework for smart energy delivery system [38] or smart cities, which addressed a number of fault injections and controller failure scenarios as well. In [39], Alnumay et al. designed and developed a trust-based system for securing IoT applications using a predictive model of ARMA/GARCH (1,1), which significantly improve network functionalities in smart city scenarios.

12.4 THE PROPOSED SDN-IoT–BASED SMART CITY FRAMEWORK

Here, we propose our SDN-IoT–based smart city framework, which is configured, controlled, and managed by a global control center as shown in Figure 12.10. The proposed framework supports heterogeneous networks and contains different types of networks including Zigbee, MANETs, sensor networks, and Bluetooth.

We also present a SDN-IoT–based layered smart city framework in Figure 12.11. Our proposed architecture has three layers, described as follows. The first layer is the infrastructure layer, which consists of the following two sublayers: first, the IoT devices sublayer and, second, the forwarding devices sublayer. The IoT devices sublayer contains different types of wireless devices (e.g., Zigbee, sensors, and Bluetooth) to create different types of IoT application domains. These wireless devices collect large volume of data from the networks and send them to the global control center for further processing. The IoT device sublayer also contains actuators to receive control commands from the global control center and execute them. The forwarding devices sublayer consists of Openflow (OF) gateways, which facilitate the forwarding of control and data packets to the global control center.

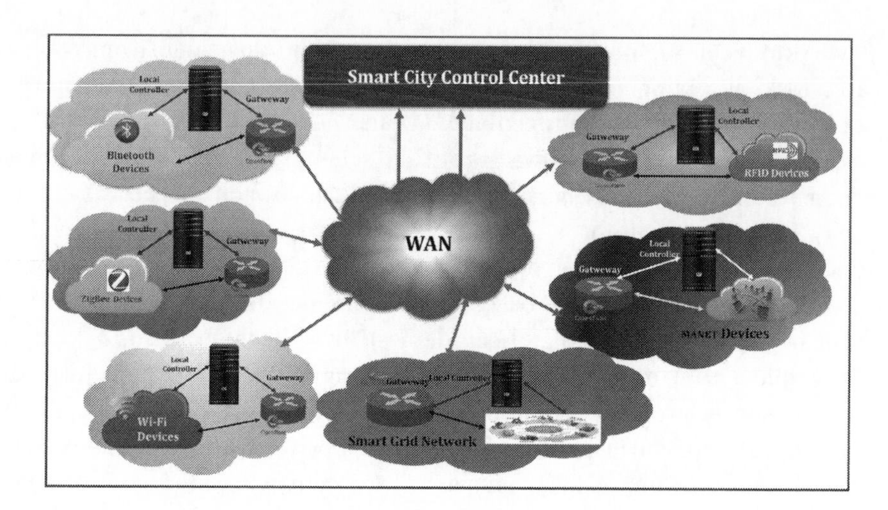

Figure 12.10 An SDN-IoT–based smart city framework.

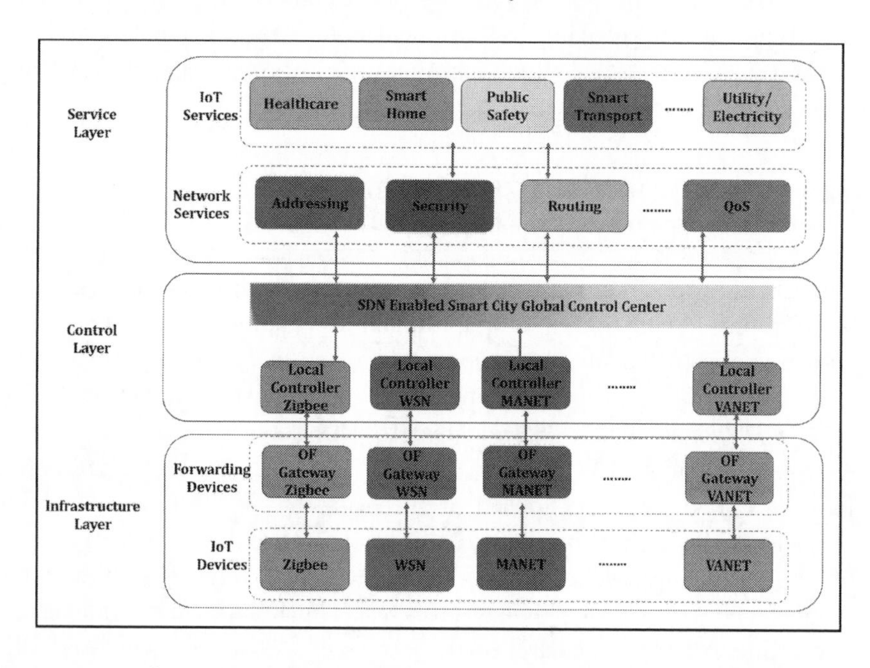

Figure 12.11 An SDN-IoT–based layered smart city framework.

The control layer contains a global SDN controller and a number of local SDN controllers. The global SDN controller is mainly responsible for controlling and monitoring communications between global control center to IoT application domains and an application domain to other application domains, and the local SDN controller controls and monitors the communication between devices inside an application domain. The application layer provides IoT services (e.g., smart homes, smart grids, and smart transportation) using SDN controllers. It further provides network services such as routing, security, and QoS in the city.

12.4.1 The Proposed Addressing Scheme

Here, we discuss our proposed IPv6 addressing scheme that is designed to provide unique addresses to IoT devices in the infrastructure. Using the proposed addressing scheme, unique IP addresses can be generated from the IP address of an existing device in the city (network), which can then be provided to new/joining IoT devices. In other words, without the need to broadcast any message over the entire city, any new/joining IoT device can acquire an IP address from its peers/neighboring devices. This concept is adopted from [40].

Here, we discuss the algorithm given in Function ip-generation that generates unique IPv6 addresses for new IoT devices joining the network. As discussed, an IPv6 address comprises eight (8) groups of four (4) hexadecimal (*HEX*) digits, which are separated by colons (e.g., 2031:0000:130f:0000:0000:09c0:876a:130b). The IPv6 address is logically divided into two parts: a 64-bit network prefix and a 64-bit interface identifier. For ease of presentation, we express the address in 16-byte dotted decimal (*DEC*) format: $(b_{15}.b_{14}.b_{13}.b_{12}.b_{11}.b_{10}.b_9.b_8.b_7.b_6.b_5.b_4.b_3.b_2.b_1.b_0)_{DEC}$ wherein $b_{15}.b_{14}.b_{13}.b_{12}.b_{11}.b_{10}.b_9.b_8$ and $b_7.b_6.b_5.b_4.b_3.b_2.b_1.b_0$ are the network prefix (which is fixed for a network domain) and the device identifier, respectively.

We assume that the global SDN controller runs an addressing application to configure all the local SDN controllers in the many different IoT application domains. Each local SDN controller also runs the proposed addressing application to configure any SDN and IoT devices in its domain. We further assume that a local SDN controller is configured with an IP address, say CEDF:0CB8:8BA3:8A2E::0001, by the global SDN controller. In our context, CEDF:0CB8:8BA3:8A2E is the network domain, and 0000:0000:0000:0001 is the identifier of the local SDN controller. The local SDN controller can assign the network prefix CEDF:0CB8:8BA3:8A2E and the device identifiers ranging from 1.0.0.0.0.0.0.1 to 255.0.0.0.0.0.0.1 and from 0.0.0.0.0.0.0.2 to 0.0.0.0.0.0.0.255 to IoT devices in the domain.

In our example, the IoT device that has host identifier 0.0.0.0.0.0.0.2 and a proxy with host identifier 0.0.0.0.0.0.0.255 can allocate

getmyip $\leftarrow (b_{15}.b_{14}.b_{13}.b_{12}.b_{11}.b_{10}.b_9.b_8.b_7.b_6.b_5.b_4.b_3.b_2.b_1.b_0)_{DEC}$;
Set *static count* $\leftarrow 0$, *count*1 $\leftarrow 1, j \leftarrow 0, i \leftarrow 0$;
count $\leftarrow$ (*count* + 1); $j \leftarrow$ *count*;
if $b_7 == 0$ *and* $b_0 == 1$ **then**
    ▷ local SDN controller
    if $j \leq 255$ **then**
        $IP_N \leftarrow b_{15}.b_{14}.b_{13}.b_{12}.b_{11}.b_{10}.b_9.b_8.j.b_6.b_5.b_4.b_3.b_2.b_1.b_0$;
    end
    else
        *count*1 $\leftarrow$ (*count*1 + 1); $i \leftarrow$ *count*1;
        if $i \leq 255$ **then**
            $IP_N \leftarrow b_{15}.b_{14}.b_{13}.b_{12}.b_{11}.b_{10}.b_9.b_8.b_7.b_6.b_5.b_4.b_3.b_2.b_1.i$;
        end
    end
end
else
    ▷ Other IoT devices acting as proxies
    if $j \leq 255$ **then**
        if $b_7 == 0$ *and* $b_0 \neq 1$ **then**
            $IP_N \leftarrow b_{15}.b_{14}.b_{13}.b_{12}.b_{11}.b_{10}.b_9.b_8.j.b_6.b_5.b_4.b_3.b_2.b_1.b_0$;
        else if $b_7 \neq 0$ *and* $b_6 == 0$ **then**
            $IP_N \leftarrow b_{15}.b_{14}.b_{13}.b_{12}.b_{11}.b_{10}.b_9.b_8.b_7.j.b_5.b_4.b_3.b_2.b_1.b_0$;
        else if $b_6 \neq 0$ *and* $b_5 == 0$ **then**
            $IP_N \leftarrow b_{15}.b_{14}.b_{13}.b_{12}.b_{11}.b_{10}.b_9.b_8.b_7.b_6.j.b_4.b_3.b_2.b_1.b_0$;
        else if $b_5 \neq 0$ *and* $b_4 == 0$ **then**
            $IP_N \leftarrow b_{15}.b_{14}.b_{13}.b_{12}.b_{11}.b_{10}.b_9.b_8.b_7.b_6.b_5.j.b_3.b_2.b_1.b_0$;
        else if $b_4 \neq 0$ *and* $b_3 == 0$ **then**
            $IP_N \leftarrow b_{15}.b_{14}.b_{13}.b_{12}.b_{11}.b_{10}.b_9.b_8.b_7.b_6.b_5.b_4.j.b_2.b_1.b_0$;
        else if $b_3 \neq 0$ *and* $b_2 == 0$ **then**
            $IP_N \leftarrow b_{15}.b_{14}.b_{13}.b_{12}.b_{11}.b_{10}.b_9.b_8.b_7.b_6.b_5.b_4.b_3.j.b_1.b_0$;
        else if $b_2 \neq 0$ *and* $b_1 == 0$ **then**
            if $b_2 == 255$ *and* $b_0 == 255$ *and* $j == 255$ **then**
                $b_0 = 254$;
                $IP_N \leftarrow b_{15}.b_{14}.b_{13}.b_{12}.b_{11}.b_{10}.b_9.b_8.b_7.b_6.b_5.b_4.b_3.b_2.j.b_0$;
            else
                $IP_N \leftarrow b_{15}.b_{14}.b_{13}.b_{12}.b_{11}.b_{10}.b_9.b_8.b_7.b_6.b_5.b_4.b_3.b_2.j.b_0$;
            end
        end
    end
end
return $(IP_N)_{HEX}$;

Algorithm 1: ip-generation().

addresses from 0.0.0.0.0.0.1.2 to 0.0.0.0.0.0.255.2 and addresses from 0.0.0.0.0.0.1.255 to 0.0.0.0.0.0.255.255 in the dotted decimal format (*DEC*), respectively. Therefore, one can easily see that a node with host identifier 0.255.255.255.255.255.255.255 can assign addresses in the range between 1.255.255.255.255.255.255.255 and 255.255.255.255.255.255.255.254, with a network prefix of CEDF:0CB8:8BA3:8A2E.

Figure 12.12 describes a simple example of how a peer or neighboring IoT device can allocate unique address (i.e., acting as a *proxy*), where

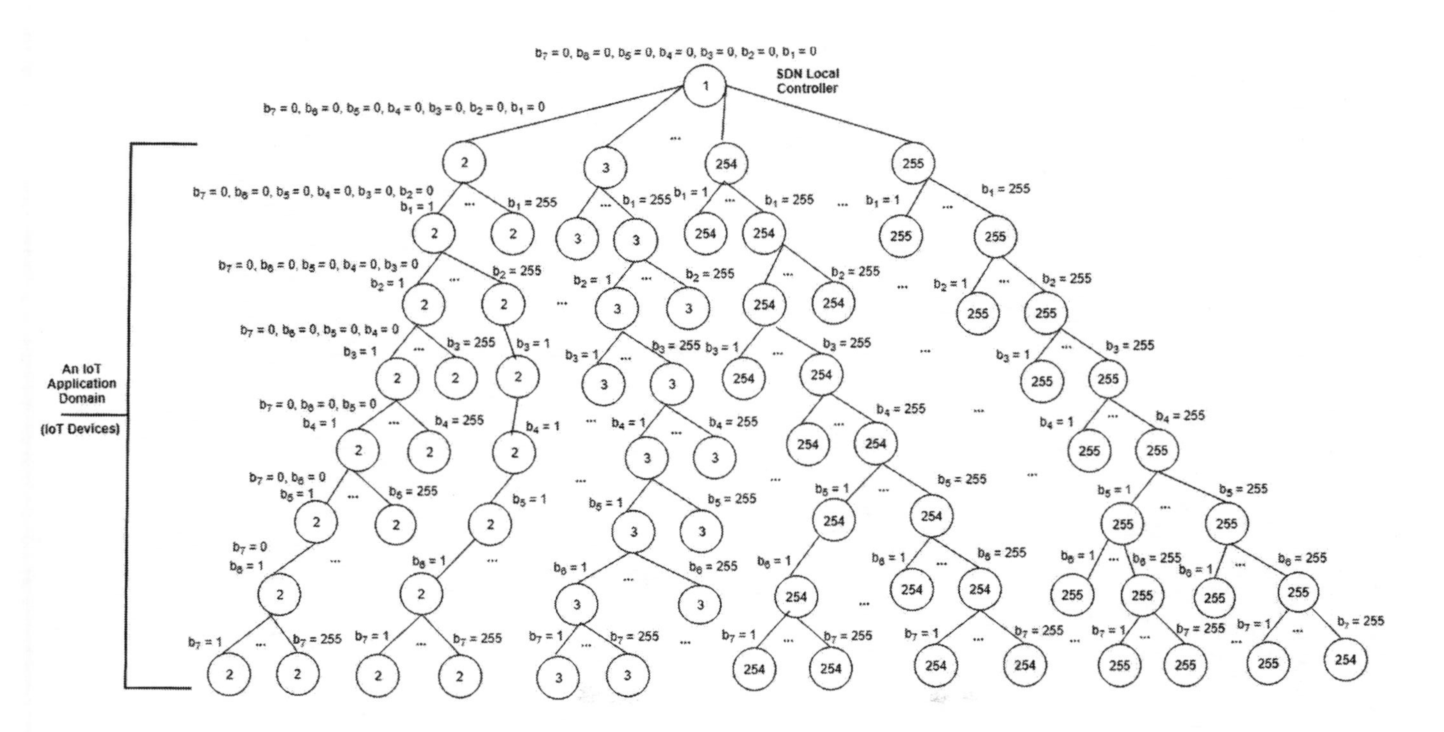

Figure 12.12 Address allocation tree in the SDN-IoT–based smart city: a simplified example.

the last byte ($b0$) of an IP address is presented within the circle and the remaining bytes ($b_7, b_6, b_5, b_4, b_3, b_2, b_1$) outside the circle. In the event that a proxy (i.e., the IoT device) does not have available IP address for nodes that have just joined the infrastructure, then this particular proxy will need to request for new IP address(es) from their parent device. Similarly, in the unlikely event that the parent device does not have any available IP address for allocation, then a similar request will be made to the parent of this particular parent device. This allows the network to scale easily. Thus, in our proposed addressing scheme, address can be uniquely allocated from $b_{15}.b_{14}.b_{13}.b_{12}.b_{11}.b_{10}.b_9.b_8$.0.0.0.0.0.0.0.1 to $b_{15}.b_{14}.b_{13}.b_{12}.b_{11}.b_{10}.b_9.b_8$.255.255.255.255.255.255.255.254 in the network.

We also remark that in our proposed addressing scheme, the *allocation status* is maintained by the individual device. Such a status records the last assigned address (i.e., *count* value), to avoid proxy devices from generating the same IP address. This allows us to avoid the need to introduce complex DAD mechanism during the process of address resolution. Further, new device obtains an IP address from its neighbor; therefore, the proposed scheme has minimal addressing overhead and latency.

12.4.2 Performance Evaluation

Table 12.1 compares the proposed address allocation scheme between the traditional DHCP and DAD schemes. Here, n be the total number of IoT devices, l the average number of links between devices, d the network diameter, and t be the average one-hop latency. We consider the following parameters to analyze the performance of our proposed addressing scheme along with DHCP and DAD schemes:

Uniqueness: The most important metrics in address allocation scheme is to guarantee the uniqueness of the allocated addresses of each device. This unique address is needed to identify the device uniquely and also for unicast communication, and routing in a smart city. DAD does not guarantee the uniqueness of the allocated address, whereas the proposed scheme and DHCP provide unique address allocation to each IoT device.

Addressing Latency: This parameter is the time difference between points when a new device sends the request for an address and when it receives the address from the network. In DHCP, the new device needs to discover the DHCP server where an address request message is flooded

TABLE 12.1 Comparison of Address Allocation Approaches in Smart Cities

Scheme	**IP Family**	**Uniqueness**	**Addressing Latency**	**Addressing Overhead**	**Scalability**	**Complexity**
DHCP	IPv4, IPv6	Yes	$O(2*t*d)$	$O(n^2)$	Low	Low
DAD	IPv4, IPv6	No	$O(2*t*d)$	$O(n^2)$	Low	Medium
Proposed	IPv6	Yes	$O(2*t)$	$O(2 * l/n)$	High	Low

in the whole network. The DHCP server sends the address to the new device in response. Therefore, the addressing latency of DHCP is $O(2 * t * d)$. In DAD, the new device floods an address request message in the whole network and sets a timer based on the diameter of the network for receiving the address reply message. The new device configures itself when the timer expires. Thus, the addressing latency of DAD is $O(2 * t * d)$, whereas the new device acquires an address from a neighbor in our proposed addressing scheme. Therefore, the addressing latency of the proposed scheme is $O(2 * t)$.

Addressing Overhead: Addressing overhead of an addressing protocol refers to the average number of messages required for an address allocation to a new device. In DHCP, the new device floods a message throughout the smart city to discover the DHCP server. Therefore, the addressing overhead of DHCP is $O(n^2)$. In DAD, the new device randomly picks a temporary address and floods a message in the whole smart city network. Therefore, the addressing overhead of DAD considered to be $O(n^2)$. In our proposed scheme, the new device obtains an address from one of its neighbors; thus, the addressing overhead is $O(2 * l/n)$.

Scalability: The scalability of an addressing scheme is considered to be high if the scheme does not degrade much its performance with respect to addressing latency and overhead even when the size of the network is large. The addressing overhead and the addressing latency of DHCP and DAD schemes are $O(n^2)$ and $O(2 * t * d)$, respectively. Therefore, these schemes are considered to be low scalability, whereas the proposed addressing scheme is considered to be highly scalable as it has $O(2 * l/n)$ and $O(2 * t)$ as the addressing overhead and latency, respectively.

Complexity: The addressing scheme should use the network resources (e.g., energy and memory of IoT devices, network bandwidth) as minimal as possible at the time of address allocation. The complexity of DAD scheme is considered to be medium as it generates address from a random number and assigns to a new device, whereas the proposed addressing scheme has low complexity as it does not need to maintain the address blocks and complex functions to generate addresses. In the proposed scheme, the existing devices (already configured with addresses) in the network act as proxies and are capable of generating addresses for new devices. This reduces the complexity and memory requirement of the proposed scheme even further.

12.5 CONCLUSION

In this chapter, we proposed an SDN-IoT–based smart city framework, and a distributed IPV6-based address allocation scheme. In the latter, each device in the city acts as a proxy and is capable of assigning IP addresses to new devices dynamically. We explained how the proposed approach achieves bandwidth and energy savings in IoT devices, as well as having low addressing overhead and latency since new devices obtain their addresses from their neighbors.

Bibliography

[1] R. Jalali, K. El-khatib, and C. McGregor, "Smart City Architecture for Community Level Services Through the Internet of Things," in *18th International Conference on Intelligence in Next Generation Networks*, 2015.

[2] E. S. Madhan, U. Ghosh, D. K. Tosh, K. Mandal, E. Murali, and S. Ghosh, "An Improved Communications in Cyber Physical System Architecture, Protocols and Applications," in *IEEE SECON STP-CPS*, Boston, MA, USA, 2019.

[3] V. Moustaka, A. Vakali, and L. G. Anthopoulos, "A systematic review for smart city data analytics," *ACM Computing Surveys*, vol. 51, no. 5, Article no. 103, 2018.

[4] U. Ghosh, P. Chatterjee, S. S. Shetty, C. Kamhoua, and L. Njilla, "Towards Secure Software-Defined Networking Integrated Cyber-Physical Systems: Attacks and Countermeasures," in *Cybersecurity and Privacy in Cyber Physical Systems*, 1st ed.; CRC Press Taylor & Francis: Boca Raton, FL, USA, 2019.

[5] R. Droms, "Dynamic host configuration protocol," RFC 2131, March 1997.

[6] K. Weniger, "Passive duplicate address detection in mobile ad hoc networks," in *Proceedings of IEEE WCNC*, Florence, Italy, February 2003.

[7] S. Ni, Y. Tseng, Y. Chen, and J. Sheu, "The broadcast storm problem in a mobile ad hoc network," in *Proceedings of the ACM/IEEE MOBICOM*, pp. 151–162, 1999.

[8] Y. Yang, L. Wu, G. Yin, L. Li, and H. Zhao, "A Survey on Security and Privacy Issues in Internet-of-Things," in *IEEE Internet of Things Journal*, vol. 4, no. 5, pp. 1250–1258, October 2017.

[9] A. Humayed, J. Lin, F. Li, and B. Luo, "Cyber-Physical Systems Security—A Survey," in *IEEE Internet of Things Journal*, vol. 4, no. 6, pp. 1802–1831, December 2017.

[10] Y. Lu and L. D. Xu, "Internet of Things (IoT) Cybersecurity Research: A Review of Current Research Topics," in *IEEE Internet of Things Journal*, vol. 6, no. 2, pp. 2103–2115, April 2019.

[11] A. Zanella, N. Bui, A. Castellani, L. Vangelista, and M. Zorzi, "Internet of Things for Smart Cities," in *IEEE Internet of Things Journal*, vol. 1, no. 1, pp. 22–32, Feburary 2014.

[12] A. A. Malik, D. K. Tosh, and U. Ghosh, "Non-Intrusive Deployment of Blockchain in Establishing Cyber-Infrastructure for Smart City," in *2019 16th Annual IEEE International Conference on Sensing, Communication, and Networking (SECON)*, Boston, MA, USA, 2019, pp. 1–6.

[13] P. Singh, A. Nayyar, A. Kaur, and U. Ghosh,"Blockchain and Fog Based Architecture for Internet of Everything in Smart Cities," in *Future Internet* vol. 12, p. 61, 2020.

[14] Y. Dong, S. Guo, J. Liu, and Y. Yang, "Energy-Efficient Fair Cooperation Fog Computing in Mobile Edge Networks for Smart City," in *IEEE Internet of Things Journal*, vol. 6, no. 5, pp. 7543–7554, October 2019.

[15] X. Wang, C. Wang, J. Zhang, M. Zhou, and C. Jiang, "Improved rule installation for real-time query service in software-defined internet of vehicles," *IEEE Transactions on Intelligent Transportation Systems*, vol. PP, no. 99, pp. 1–11, 2016.

[16] P. K. Sahoo and Y. Yunhasnawa, "Ferrying vehicular data in cloud through software defined networking," in *IEEE 12th International Conference on Wireless and Mobile Computing, Networking and Communications*, 2016.

[17] E. Bozkaya and B. Canberk, "Qoe-based flow management in software defined vehicular networks," in *2015 IEEE Globecom Workshops*, 2015.

[18] U. Ghosh, P. Chatterjee, and S. Shetty, "A Security Framework for SDN-enabled Smart Power Grids," in *IEEE ICDCS CCNCPS 2017*, Atlanta, USA, 2017.

[19] M. Mavani and K. Asawa, "Privacy Preserving IPv6 Address Auto-Configuration for Internet of Things." In Hu Y.C., Tiwari S., Mishra K., Trivedi M. (eds) *Intelligent Communication and Computational Technologies.* Lecture Notes in Networks and Systems, vol. 19. Springer, Singapore, 2018.

[20] M. Mavani and K. Asawa, "Privacy enabled disjoint and dynamic address auto-configuration protocol for 6Lowpan," in *Ad Hoc Networks*, vol. 79, pp. 72–86, 2018.

[21] M. Mavani and K. Asawa, "Resilient against Spoofing in 6LoWPAN Networks by Temporary-Private IPv6 Addresses," in *Peer-to-Peer Networks Applications*, vol. 13, pp. 333–347, 2020.

[22] L. Brilli, T. Pecorella, L. Pierucci, and R. Fantacci, "A Novel 6LoWPAN-ND Extension to Enhance Privacy in IEEE 802.15.4 Networks," in *2016 IEEE Global Communications Conference (GLOBECOM)*, Washington, DC, USA, pp. 1-6, 2016.

[23] X. Wang, H. Cheng, and Y. Yao, "Addressing-Based Routing Optimization for 6LoWPAN WSN in Vehicular Scenario," in *IEEE Sensors Journal*, vol. 16, no. 10, pp. 3939–3947, 2016.

[24] X. Wang, D. Le, H. Cheng, and Y. Yao, "Location-based address configuration for 6LoWPAN wireless sensor networks," in *Wireless Networks*, vol. 21, no. 6, pp. 2019–2033, 2015.

[25] X. Wang, H. Chen, and D. Le,"A Novel IPv6 Address Configuration for a 6LoWPAN-Based -WBAN," in *Journal of Network and Computer Applications*, vol. 61, pp. 33–45, 2016.

[26] K. G. Khair, M. Kantarci, and H. T. B. Mouftah, "Cellular IP address provisioning in a heterogeneous wireless network," in *International Journal of Communication Systems*, vol. 27, no. 10, pp. 2007–2021, 2014.

[27] Y. Li and X. Wang, "A Novel and Efficient Address Configuration for MANET," in *International Journal of Communication Systems*, vol. 32, no. 13, pp. e4059, 2019.

[28] X. Wang and X. Wang, "Vehicular content-centric networking framework," in *IEEE Systems Journal*, vol. 13, no. 1, pp. 519–529, 2019.

[29] I. El-Shekeil, A. Pal, and K. Kant, "IP Address Consolidation and Reconfiguration in Enterprise Networks," in *25th International Conference on Computer Communication and Networks (ICCCN)*, Waikoloa, HI, USA, pp. 1–9, 2016.

[30] U. Ghosh and R. Datta, "ADIP: An Improved Authenticated Dynamic IP Configuration Scheme for Mobile Ad Hoc Networks," in *International Journal of Ultra Wideband Communications and Systems (IJUWBCS)*, vol. 1, no. 2, pp. 102–117, 2009.

[31] U. Ghosh and R. Datta, "A Secure Dynamic IP Configuration Scheme for Mobile Ad Hoc Networks," in *Ad Hoc Networks*, vol. 9, no. 7, pp. 1327–1342, 2011.

[32] U. Ghosh and R. Datta, "IDSDDIP: A Secure Distributed Dynamic IP Configuration Scheme for Mobile Ad Hoc Networks," in *International Journal of Network Management*, vol. 23, no. 6, pp. 424–446, 2013.

[33] U. Ghosh and R. Datta,"An ID Based Secure Distributed Dynamic IP Configuration Scheme for Mobile Ad Hoc Networks," In Bononi L., Datta A.K., Devismes S., Misra A. (eds) in *Distributed Computing and Networking*. ICDCN 2012. Lecture Notes in Computer Science, vol 7129. Springer, Berlin, Heidelberg.

[34] U. Ghosh and R. Datta,"Mmip: A New Dynamic ip Configuration Scheme with mac Address Mapping for Mobile Ad Hoc Networks," in *Proceedings of Fifteenth National Conference on Communications 2009*, IIT Guwahati, India, 2009.

[35] U. Ghosh, P. Chatterjee, R. Datta, A. S. K. Pathan, and D. B. Rawat,"Secure Addressing Protocols for Mobile Ad Hoc Networks," in *Security Analytics for the Internet of Everything*, 1st ed.; CRC Press Taylor & Francis: Boca Raton, FL, 2020.

[36] N. Akhtar, M. A. Khan, M, A. Ullah, and M. Y. Javed, "Congestion Avoidance for Smart Devices by Caching Information in MANETS and IoT," in *IEEE Access*, vol. 7, pp. 71459–71471, 2019.

[37] U. Ghosh, P. Chatterjee, D. Tosh, S. Shetty, K. Xiong, and C. Kamhoua, "An SDN Based Framework for Guaranteeing Security and Performance in Information-Centric Cloud Networks," in *2017 IEEE 10th International Conference on Cloud Computing (CLOUD)*, Honolulu, CA, USA, pp. 749–752, 2017.

[38] U. Ghosh, X. Dong, R. Tan, Z. Kalbarczyk, D. K. Y. Yau, and R. K. Iyer, "A Simulation Study on Smart Grid Resilience under Software-Defined Networking Controller Failures," in *2nd ACM International Workshop on Cyber-Physical System Security (CPSS '16)*, pp. 52–58.

[39] W. Alnumay, U. Ghosh, and P. Chatterjee, "A Trust-Based Predictive Model for Mobile Ad Hoc Network in Internet of Things," in *Sensors*, vol. 19, p. 1467, 2019.

[40] U. Ghosh and R. Datta, "A Secure Addressing Scheme for Large-Scale Managed MANETs," *IEEE Transactions on Network and Service Management*, vol.12, no.3, pp. 483–495, September 2015.

Index